Fundamentals of Number Theory

WILLIAM J. LEVEQUE

DOVER PUBLICATIONS, INC.
New York

Copyright

Bibliographical Note

This Dover edition, first published in 1996, is an unabridged, unaltered republication of the work first published by the Addison-Wesley Publishing Company, Reading, Massachusetts, 1977.

The portrait of Adolph Hurwitz on page 225 is reproduced by permission of the Library of Eidgenössische Technische Hochschule, Zurich, Switzerland.

All the other portraits in this book are from the David Eugene Smith Collection, Columbia University Libraries and are reprinted by permission of Columbia University.

Library of Congress Cataloging-in-Publication Data

LeVeque, William Judson.
Fundamentals of number theory / William J. LeVeque.
p. cm.
Originally published: Reading, Mass. ; Addison-Wesley, c1977.
Includes bibliographical references and index.
ISBN-13: 978-0-486-68906-7 (pbk.)
ISBN-10: 0-486-68906-9 (pbk.)
1. Number theory. I. Title.
QA241.L573 1996
512′.74—dc20 95-3713
CIP

Manufactured in the United States by LSC Communications
68906913 2020
www.doverpublications.com

Preface

I started to write the present book as a second edition of my *Topics in Number Theory,* Volume 1, but I soon became aware that the changes were too extensive to allow it to be considered the same book, especially because of the back-references from Volume 2 which would become inappropriate. The two books are still quite similar, and I have carried over some passages verbatim; nevertheless, the differences are both substantial and significant.

Perhaps the most striking change is the introduction of the language of abstract algebra. Many students will already have met the concepts of group, ring, field, and domain, *but this is not assumed.* All terms used are defined and examples are given, and the book is intended to be self-contained in this respect. I have taught the material to mixed classes, containing students who had had algebra and students who had not, and did not find that students of the first kind had any great advantage.

Second, there is a new emphasis on the history of number theory. In part, this reflects my own growing interest in the history of mathematics, but it also recognizes what I take to be more concern among students with the human side of mathematics. A closely related change is the inclusion of many notes and bibliographical references, selected with an eye to their readability as well as their relevance.

The reader acquainted with my earlier book will find a number of new topics: factorization and primality of large integers, p-adic numbers, algebraic number fields, Brun's theorem on twin primes, and the transcendence of e, to mention some. The number of problems has also been substantially increased. The importance of problem-solving as an integral part of a first course in number theory can hardly be overemphasized, and I hope that the additional problems will prove to be useful. Hints are supplied with the more difficult ones.

As was true of its predecessor, this book is directed primarily toward undergraduate majors and beginning graduate students in mathematics. No post-calculus prerequisite is supposed, although here and there comments are included for the benefit of students with additional background.

All the portraits except that of Hurwitz were selected from the superb David Eugene Smith collection in the Columbia University Library; Hurwitz's came from the Eidgenössische Technische Hochschule in Zürich. The Factor Table on pp. 266–267 is a reproduction of a small portion of Burkhardt's *Table des Diviseurs* of 1814–17, as it was reprinted in G. S. Carr's *Formulas and Theorems in Pure Mathematics,* Second Edition, Chelsea Publishing Company, New York 1886/1970. Permission for all these reproductions is gratefully acknowledged.

My first draft benefited from the criticism of a number of friends. In particular, H. W. Gould supplied a detailed critique, and J. Brillhart provided authoritative advice on computational matters, as well as general comments. I am indebted to R. J. LeVeque for preparing the computer plots on pp. 268–269.

April 1977
Claremont, California

W.J. L.

Contents

Chapter 5 Quadratic Residues

Chapter 6 Number-Theoretic Functions and the Distribution of Primes

Chapter 7 Sums of Squares

Chapter 8 Quadratic Equations and Quadratic Fields

Chapter 9 Diophantine Approximation

1

Introduction

1.1 WHAT IS NUMBER THEORY?

This could serve as a first attempt at a definition: it is the study of the set of integers 0, ± 1, $\pm 2, \ldots$, or some of its subsets or extensions, proceeding on the assumption that integers are interesting objects in and of themselves, and disregarding their utilitarian role in measuring. This definition might seem to include elementary arithmetic, and in fact it does, except that the concern now is to be with more advanced and more subtle aspects of the subject. A quick review of elementary properties of the integers is incorporated with some other material, which may or may not be new to the reader, in Sections 1.2 and 1.3.

To get some idea of what the subject comprises, let us go back to the seventeenth century, when the modern epoch opened with the work of Pierre de Fermat [*fair-mah*]. One of Fermat's most beautiful theorems is that every positive integer can be represented as the sum of the squares of four integers, for example,

$$\begin{aligned}
1 &= 1^2 + 0^2 + 0^2 + 0^2, \\
2 &= 1^2 + 1^2 + 0^2 + 0^2, \\
4 &= 1^2 + 1^2 + 1^2 + 1^2 = 2^2 + 0^2 + 0^2 + 0^2, \\
7 &= 2^2 + 1^2 + 1^2 + 1^2, \\
188951 &= 371^2 + 226^2 + 15^2 + 3^2.
\end{aligned}$$

He announced this theorem in 1636, but the first published proof of it was given by Joseph-Louis Lagrange in 1770. It could serve as the ideal example of a theorem in number theory: it is elegant and immediately comprehensible; it reveals a subtle and unexpected relationship among the integers; it is the best theorem of its kind (7 cannot be represented with fewer than four squares); and it says something about an infinite class of integers. The last is an important qualification, as it distinguishes between theorems and numerical facts. It is a fact, and perhaps even an interesting one, that 1729 is the smallest positive integer having two distinct representations as the sum of two cubes ($10^3 + 9^3$ and $12^3 + 1^3$), but this would hardly be called a theorem since it can be verified by examining the finite set $1, 2, 3, \ldots, 1729$. On the other hand, the assertion that there are only finitely

Pierre de Fermat (1601–65)

Fermat was a lawyer by profession, well versed in ancient languages and steeped in classical culture. There were no scientific journals then, and he was not inclined to write out proofs. Instead, he communicated his results by letter, especially to Father M. Mersenne, who maintained an enormous correspondence throughout Europe. Fermat anticipated Descartes in analytic geometry and Newton and Leibniz in differential calculus, but his work was not well known because he failed to publish his books on these subjects. His fame rests chiefly on his work in number theory, where he was without peer. The groundwork that had been laid for him by the Greeks and others is discussed in the final section of this chapter.

many integers having two or more such representations is deceptive; it seems to say something about a finite set, but in fact it cannot be proved by examining any specific finite set, nor can it be disproved in this way. Thus it would be a significant theorem if it were true. (It is not; that, too, is a significant theorem.)

An even more famous assertion credited to Fermat is what is sometimes called his Last Theorem, which says that if n is an integer larger than 2, then the equation $x^n + y^n = z^n$ has no solution in positive integers x, y, z. Fermat claimed to have proved this, but as was his habit he did not reveal the proof. This seems to be the only recorded instance in which he claimed a result that has never been verified (although he did announce an erroneous conjecture, discussed below). Lacking a proof, mathematicians today tend to call it the Fermat Problem, rather than Theorem; it is the oldest, and possibly the most famous, unsolved problem in mathematics. A single counterexample would suffice to destroy it, of course, but finding such a quadruple x, y, z, n, if there is one, might well be beyond the capacity of present or future computers, since the equation is now known to have no solution for $n < 100{,}000$, and in any case to have only solutions with one of x, y, or z larger than n^{2n}. (The known universe would accommodate only about 10^{123} proton-sized objects, close-packed.)

One of the basic concepts in number theory is that of a **prime number**. An integer p is prime if $p \neq \pm 1$ and the equation $p = ab$ has no solution in integers a and b except those for which $a = \pm 1$ or $a = \pm p$. Briefly, then, a prime is an integer $\neq \pm 1$ which has no nontrivial divisors. Euclid already knew that the sequence of positive primes 2, 3, 5, 7, . . . does not terminate (his proof is given in

Section 1.3), but the pattern of occurrence is very irregular. Both Fermat and Mersenne looked for some regularity, and both guessed wrong. Fermat conjectured that all the numbers $f_n = 2^{2^n} + 1$ are prime, this being true for $n = 0, 1, 2, 3, 4$. It turned out that he had stopped just too soon, for Leonhard Euler [*oiler*] showed in 1739 that f_5 has the divisor 641. In fact, no further prime value of f_n has been found, for $5 \leq n \leq 16$. This story would hardly be worth recounting, since false conjectures about the primes are very common, if it were not for the fact that the Fermat primes occurred again, almost 200 years later, in an entirely different context: Carl Friedrich Gauss settled one of the ancient Greek problems by proving that a regular polygon of m sides can be constructed with ruler and compass if and only if m factors as $m = 2^k f_{n_1} \cdots f_{n_r}$, where $k, n_1, \ldots, n_r$ are some nonnegative integers and the f_{n_i} are distinct Fermat primes. Thus it would be interesting to know whether there are further primes of this sort.

Although Mersenne's chief contribution to mathematics was in the dissemination of, rather than in the creation of, new results, he did himself engage in the study of primes among the numbers of the form $M_n = 2^n - 1$. If $n = rs$, then M_n is divisible by M_r and M_s, so M_n can be prime only for prime values of n. In 1644 Mersenne asserted that of the 55 numbers M_p with prime index $p \leq 257$, those which are prime correspond to $p = 2, 3, 5, 7, 13, 17, 19, 31, 67, 127$, and 257. He thereby made five mistakes, by including 67 and 257 and by excluding 61, 89, and 107. What was remarkable was not that he made some mistakes but that, without even a desk calculator, he could do anything at all with numbers having up to 78 digits. We are again dealing with numerical facts and not theorems, but it must be obvious that some theorems are lurking in the background, of the form

Carl Friedrich Gauss (1777–1855)

Gauss is considered by many to be the greatest mathematician who ever lived. He conjectured the prime number theorem when he was 15, characterized the constructible polygons at 18, proved at age 22 that a polynomial of degree n has n zeros, and published his masterpiece *Disquisitiones Arithmeticae* when he was 24. This book changed number theory from a collection of isolated problems to a coherent branch of mathematics. After 1801 he turned to other fields—geometry, analysis, astronomy, and physics, chiefly—except for two articles on biquadratic reciprocity. He spent his entire mature life at the University of Göttingen. His collected works fill 12 volumes.

"N is prime if . . . " or "N is not prime if . . .", and that additional theorems of this sort would be useful. There has always been a strong interplay between facts and theorems in number theory; calculations provide the data from which to infer patterns, find counterexamples or guess at theorems, and they also bring into focus the need for constructive theorems which yield usable algorithms (i.e., systematic procedures for computations).

The Fermat and Mersenne numbers are so sparse that even if they were all prime, one would know very little about the distribution of primes in general. A much more fruitful study, again empirical, was initiated by Gauss in 1792 with the help of a table of the primes less than 102,000, published by Johann Lambert a few years before. If, as is customary, $\pi(x)$ denotes the number of positive primes not exceeding x, then what Gauss did was to consider how $\pi(x)$ grows with x. He began by counting the primes in successive intervals of fixed length, obtaining a table of the following sort, in which $\Delta(x) = \{\pi(x) - \pi(x - 1000)\}/1000$:

x	$\pi(x)$	$\Delta(x)$
1000	168	.168
2000	303	.135
3000	430	.127
4000	550	.120
5000	669	.119
6000	783	.114
7000	900	.117
8000	1007	.107
9000	1117	.110
10000	1229	.112

The "frequency" $\Delta(x)$ of primes in successive intervals seemed to be slowly decreasing, on the average, so Gauss took the reciprocal of $\Delta(x)$ and compared it with various elementary functions. For the natural logarithm of x, this gives the following chart:

x	1000	2000	3000	4000	5000	6000	7000	8000	9000	10,000
$\Delta(x)$	.168	.135	.127	.120	.119	.114	.117	.107	.110	.112
$1/\log x$	.145	.132	.125	.121	.117	.115	.113	.111	.110	.109

The strikingly good match strongly supports the guess that $\Delta(x)$ is approximately $1/\log x$. Since $\Delta(x)$ is the slope of a chord on the graph of $y = \pi(x)$, the hypothetical approximate equality $\Delta(x) \approx 1/\log x$ should be integrated to obtain $\pi(x)$ itself, and thus Gauss conjectured that

$$\pi(x) \approx \int_2^x \frac{dt}{\log t}.$$

The integral occurring here, which happens not to be an elementary function, is commonly denoted by li(x); its values are readily calculated, and more recent calculations for $\pi(x)$ yield the following comparison (where li(x) is given to the nearest integer):

x	$\pi(x)$	$\mathrm{li}(x)$	$\mathrm{li}(x) - \pi(x)$	$\pi(x)/\mathrm{li}(x)$
10^3	168	178	10	.94382
10^4	1,229	1,246	17	.98636
10^5	9,592	9,630	38	.99605
10^6	78,498	78,628	230	.99835
10^7	664,579	664,918	339	.99949
10^8	5,761,455	5,762,209	754	.99987
10^9	50,847,534	50,849,235	1701	.99997
10^{10}	455,052,512	455,055,614	3102	.99999

What Gauss intended, then, in conjecturing that li(x) is a good approximation to $\pi(x)$ for large x, was presumably not that $\mathrm{li}(x) - \pi(x) \to 0$, nor even that

$$\mathrm{li}(x) - \pi(x)$$

remains bounded, but that the *relative* error becomes small:

$$(\mathrm{li}(x) - \pi(x))/\pi(x) \to 0,$$

or

$$\lim_{x\to\infty} \frac{\pi(x)}{\mathrm{li}(x)} = 1. \tag{1}$$

He made this conjecture in 1793, when he was about 15 years old, but it was not proved until more than 100 years later, by J. Hadamard and C. de la Vallée Poussin (independently, in 1896). The proof is too difficult to include in this book, but it will be shown later that if the limit in (1) exists, it must have the value 1. It is not difficult to show that (1) implies and is implied by the relation

$$\lim_{x\to\infty} \frac{\pi(x)}{x/\log x} = 1, \tag{1'}$$

as is indicated in Problem 2 at the end of this section. Because of its central position in the theory of primes, relation (1), or more traditionally (1′), is simply known as the prime number theorem.

One reason for including data in the last table above for such large values of x, for which Gauss had not computed $\pi(x)$, is to emphasize the point that no amount of numerical evidence will substitute for a proof. It appears from the table that li(x) always overestimates $\pi(x)$, in the sense that up to $x = 10^{10}$ at least, $\mathrm{li}(x) -$

$\pi(x)$ is positive and increasing. But this does not continue, for Littlewood* [1914] has shown that $\mathrm{li}(x) - \pi(x)$ changes sign infinitely many times. No one knows when the first change of sign occurs, but Skewes [1955] proved that it happens for some x such that

$$x < 10^{10^{10^{10^3}}}$$

Quite possibly, no specific value of x will ever be known for which $\mathrm{li}(x) < \pi(x)$.

There are many questions concerning primes which have resisted all assaults for two centuries and more, for example, whether there are infinitely many *twin primes* such as 17 and 19, or 4967 and 4969, which differ by 2, and whether every even integer larger than 4 is the sum of two odd primes. Here, for variety, is a less well-known question that was raised much more recently, and which also seems to be very difficult. Form the doubly-infinite array

$$\begin{array}{l}
2\quad 3\quad 5\quad 7\quad 11\quad 13\quad 17\quad 19\quad 23\quad 29\quad 31\quad 37\quad 41\quad 43\quad \cdots \\
\quad 1\quad 2\quad 2\quad 4\quad 2\quad 4\quad 2\quad 4\quad 6\quad 2\quad 6\quad 4\quad 2\quad \cdots \\
\qquad 1\quad 0\quad 2\quad 2\quad 2\quad 2\quad 2\quad 2\quad 4\quad 4\quad 2\quad 2\quad \cdots \\
\qquad\quad 1\quad 2\quad 0\quad 0\quad 0\quad 0\quad 0\quad 2\quad 0\quad 2\quad 0\quad \cdots \\
\qquad\qquad 1\quad 2\quad 0\quad 0\quad 0\quad 0\quad 2\quad 2\quad 2\quad 2\quad \cdots \\
\qquad\qquad\quad 1\quad 2\quad 0\quad 0\quad 0\quad 2\quad 0\quad 0\quad 0\quad \cdots \\
\qquad\qquad\qquad 1\quad 2\quad 0\quad 0\quad 2\quad 2\quad 0\quad 0\quad \cdots \\
\qquad\qquad\qquad\quad 1\quad 2\quad 0\quad 2\quad 0\quad 2\quad 0\quad \cdots \\
\qquad\qquad\qquad\qquad 1\quad 2\quad 2\quad 2\quad 2\quad 2\quad \cdots \\
\qquad\qquad\qquad\qquad\quad 1\quad 0\quad 0\quad 0\quad 0\quad \cdots \\
\qquad\qquad\qquad\qquad\qquad 1\quad 0\quad 0\quad 0\quad \cdots \\
\qquad\qquad\qquad\qquad\qquad \ddots
\end{array}$$

in which the first row contains the primes and each entry below is the absolute value of the difference of the two numbers above it. Is it the case that every row after the first begins with 1? The portion of the array exhibited above shows that this is true up to $p = 43$, and it has been verified up to $p = 792{,}721$.

There are branches of number theory in which the integers enter much less explicitly than in the theory of primes. This is the case, for example, in questions concerning the nature of numbers such as π and e. The question of whether either of these is rational is simply the question of whether either is a solution of (that is, is defined implicitly by) a linear equation $ax + b = 0$ with integral coefficients a and b. The Swiss mathematician Lambert, mentioned earlier, proved in 1761 that π is not rational, and we give in Section 1.3 the much simpler proof that e is irrational. More generally, one could ask whether e or π is algebraic, that is, whether either of them satisfies a polynomial equation $a_0x^n + a_1x^{n-1} + \cdots + a_n = 0$

* A bracketed date indicates that the complete reference is to be found in the Bibliography at the end of the book. Many facts stated in the text with neither proof nor reference are discussed further in the Notes and References at the ends of chapters.

with integral coefficients $a_0 \neq 0, a_1, \ldots, a_n$. Again the answer turns out to be "no" in both cases, but the proofs are rather more difficult (see Section 9.7). Turning the matter around, one can study the numbers that *are* algebraic, and it turns out that a very elaborate theory can be constructed which is interesting in its own right and which also provides some powerful tools for the study of the integers. Too many definitions intervene to allow for examples at this point, but the interested reader may look ahead to Chapter 8.

Johann H. Lambert (1728–1777)

Lambert's family was poor, so he had to leave school at age 12, and had no further schooling. Nevertheless, he made important contributions in philosophy (epistemology and metaphysics), astronomy (existence of nebulae), physics (photometry, hygrometry and pyrometry) and cartography. In mathematics, his major work outside number theory was in geometry, where his books on the parallel postulate and on perspective foreshadowed nineteenth-century developments in noneuclidean geometry and descriptive geometry. He held a scientific position, as a colleague of Euler in the Prussian Academy of Sciences in Berlin, only during the last twelve years of his life. Earlier, he had mainly been a children's tutor in his native Switzerland.

PROBLEMS

1. Show that if $n = rs$, then M_r divides M_n. Why could Fermat restrict attention to the numbers f_n rather than considering the more general class of numbers $2^k + 1$, in hunting for primes?
2. Apply l'Hôpital's rule to show that

$$\lim_{x \to \infty} \frac{\mathrm{li}(x)}{x/\log x} = 1,$$

and deduce that each of (1) and (1′) implies the other.

3. Making use of a logarithm table, compare $\mathrm{li}(x)$ and $x/\log x$ as approximations to $\pi(x)$ for $x = 10^n$, $3 \leq n \leq 10$. What do you conclude from this?
4. For each positive integer n, let $\tau(n)$ be the number of positive integers which divide n. (For example, $\tau(6) = 4$, since the positive divisors of 6 are 1, 2, 3, and 6.) Construct a table of values of $\tau(n)$ for $1 \leq n \leq 50$, formulate some conjectures, and test them

for other suitable values of n. In particular, can you characterize the n for which $\tau(n)$ is odd? Can you prove it? Can you find a connection between $\tau(m)$, $\tau(n)$, and $\tau(mn)$? [*Note:* The Greek alphabet is to be found at the end of the book.]

1.2 ALGEBRAIC PROPERTIES OF THE SET OF INTEGERS

In this section we organize familiar arithmetic facts with the aid of three central concepts from abstract algebra, the notions of group, ring, and field. For brevity, we henceforth use $\mathbf{Z}$ as the name of the set of all integers, $0, \pm 1, \pm 2, \ldots$, and $\mathbf{Z}^+$ for the set of positive integers, $1, 2, 3, \ldots$. As usual, the symbol "$\in$" means "belongs to, is an element of."

Let S be a set of finitely or infinitely many elements, and suppose that any two elements a and b can be combined by an operation "$\circ$" (think of addition or multiplication) to give a unique result, $a \circ b$. Then S is called a **group** (under the operation $\circ$) if the following four conditions are satisfied:

(G)

1. If $a \in S$ and $b \in S$, then $a \circ b \in S$ (in words, S is **closed** under the operation).
2. If $a, b, c \in S$, then $a \circ (b \circ c) = (a \circ b) \circ c$ (the **associative law** holds).
3. There is a unique element $e \in S$ (the **identity element**) such that for all $a \in S$, $a \circ e = e \circ a = a$.
4. Each $a \in S$ has an **inverse** $a^{-1} \in S$ such that $a \circ a^{-1} = a^{-1} \circ a = e$.

For example, $\mathbf{Z}$ is a group under addition (that is, it is an **additive** group), with $e = 0$ and $a^{-1} = -a$. In fact, it is an **abelian** (= **commutative**) group under addition, because it satisfies the further hypothesis:

(A) For all $a, b \in S$, $a \circ b = b \circ a$.

It is also an **infinite** group, because it has infinitely many elements. $\mathbf{Z}^+$ is not an additive group. (Why?) Other familiar examples of infinite additive abelian groups are

$\mathbf{Q}$: the set of rational numbers a/b, with $a, b \in \mathbf{Z}$ and $b \neq 0$;
$\mathbf{R}$: the set of real numbers;
$\mathbf{C}$: the set of complex numbers.

We shall have no occasion to deal with noncommutative groups, but finite groups will play an important role. Here is an example of such a group: the elements are 0, α, and β, and sums are defined in the following table.

$+$	0	α	β
0	0	α	β
α	α	β	0
β	β	0	α

It is not necessary to tell what α, β, and addition "really" are, merely how they behave, and then to verify the required properties. Closure is obvious. That 0 satisfies property G3 is shown by the fact that the first two columns in the above table are identical; the existence of an inverse of each element results from the fact that 0 occurs in every row, and commutativity is implied by the symmetry of the table with respect to the main diagonal. Associativity is the only property requiring any effort to verify, and it involves the examination of eight equations, of which $\alpha + (\alpha + \beta) = (\alpha + \alpha) + \beta$ is typical. (When one or more of a, b, c is 0, the equation $a + (b + c) = (a + b) + c$ follows from G3, the defining property of 0.) The number of equations can be further reduced, using commutativity.

Z does not form a group under multiplication: properties G1, G2, G3 hold, with $e = 1$, but only 1 and -1 have multiplicative inverses. On the other hand, the sets **Q***, **R***, and **C***, obtained by deleting 0 from **Q**, **R**, and **C**, are all multiplicative groups. Two examples of finite multiplicative groups are the sets $\{1, -1\}$ and $\{1, -1, i, -i\}$, where $i^2 = -1$.

If a group is finite, the number of elements is called its **order**.

By a **Ring** we mean a set S having the following properties:

(R)

1. S is an additive abelian group; call the identity element 0.
2. A second commutative operation (call it multiplication) is also defined on S, with respect to which properties G1, G2, and G3 hold (call the multiplicative identity 1).
3. $1 \neq 0$.
4. The distributive law holds: for all $a, b, c \in S$, $a(b + c) = ab + ac$.

(The capital letter R is used to signal the fact that our Rings have two properties not usually required of rings in abstract algebra, namely the commutativity of multiplication and the existence of the multiplicative identity.) **Z**, **Q**, **R**, and **C** are all Rings, and so is the set $S = \{0, 1\}$, if 0 and 1 have all their usual properties except that $1 + 1$ is defined to be 0. **Z** has many other special properties not shared by Rings in general, of course, of which three are fundamental:

(C) The **cancellation law** holds: if $ab = ac$ and $a \neq 0$, then $b = c$ (equivalently, if $ab = 0$, then $a = 0$ or $b = 0$; a Ring having either of these properties is called an integral domain, or simply a **domain**.)

(O) **Z** is **ordered**: the subset $\mathbf{Z}^+$ of positive integers (namely 1, $1 + 1$, $1 + 1 + 1, \ldots$) is closed under addition and multiplication, and for every $a \neq 0$ exactly one of a and $-a$ belongs to $\mathbf{Z}^+$. (The one in $\mathbf{Z}^+$ is called $|a|$; by defining the relation $a > b$ to mean $a - b \in \mathbf{Z}^+$, all the usual properties of inequalities can be derived.)

(W) $\mathbf{Z}^+$ **is well-ordered**: every nonempty set of positive integers contains a smallest element.

These three properties are basic in the following sense: Every ordered domain whose positive elements are well-ordered is abstractly the same as (technically, is isomorphic to) the domain of integers. To understand what this means, consider two Rings S and S' with elements $0, 1, a, b, \ldots$ and $0, 1, a', b', \ldots$. Suppose it is possible to find a map* ϕ such that to every element $a \in S$ there corresponds a unique image $\phi a \in S'$, and for which

(H)
every element of S' is the image of some element of S;
ϕ preserves addition: if $a + b = c$, then $\phi a + \phi b = \phi c$;
ϕ preserves multiplication: if $ab = c$, then $\phi a \cdot \phi b = \phi c$.

Then ϕ is called a **homomorphism** of S onto S', and if the correspondence is 1–1 (meaning that also for each $a' \in S'$ there is a unique $a \in S$ such that $\phi a = a'$), then ϕ is called an **isomorphism**, and S and S' are then said to be isomorphic. Roughly speaking, isomorphic Rings are the same Ring, but the elements are named differently. (The name of a, in S', is ϕa.) In this sense *there is only one ordered domain whose positive elements are well-ordered, and it is* $\mathbf{Z}$. For a proof, starting from a shorter list of axioms called the Peano postulates, see McCoy [1975], p. 69ff. It serves our purposes, however, to regard the assertions (R), (C), (O), and (W) as the axioms which define $\mathbf{Z}$, and we shall do so. Thus they do not require proof.

For later purposes, we note that a homomorphic image of a Ring S (that is, a set S' for which there is a homomorphism of S onto S') is itself automatically a Ring if $\phi 1 \neq \phi 0$; for all other properties of a Ring are defined in terms of the operations $+$ and $\cdot$, and these are preserved. For example,

$$\begin{aligned}
&\text{if } a + (b + c) = (a + b) + c, \text{ then } \phi\{a + (b + c)\} = \phi\{(a + b) + c\},\\
&\text{so } \phi a + \phi(b + c) = \phi(a + b) + \phi c,\\
&\text{so } \phi a + (\phi b + \phi c) = (\phi a + \phi b) + \phi c,
\end{aligned}$$

so the associative property is preserved.

$\mathbf{Q}$, $\mathbf{R}$, and $\mathbf{C}$ are all domains, but none is well-ordered, and $\mathbf{C}$ is not even ordered. They all, however, have a common feature not shared by $\mathbf{Z}$: nonzero elements have multiplicative inverses, so quotients $a/b = ab^{-1}$ are always defined in these sets, for $b \neq 0$. Any domain with this last property, in which all four rational arithmetic operations can be carried out, is called a **field**. The two-element Ring $\{0, 1\}$ mentioned earlier is a finite field, and we shall meet others later.

---

* A map from a set A *into* a set B is simply a function defined on A, taking values in B. The map is said to be *onto* if every $b \in B$ is the image (function-value) of some $a \in A$. That f maps A into B is sometimes expressed by writing "$f\colon A \to B$". The exact nature of the map can sometimes be described by indicating what the image of a typical element of A is, by means of the symbol "$\mapsto$". For example, the map $f\colon \mathbf{Z} \to \mathbf{Z}$ defined by $f\colon a \mapsto a^2$ carries every integer into its square. f is actually of a map of $\mathbf{Z}$ into the non-negative integers, but it is not a map onto this set since 3 is not an image.

PROBLEMS

1. The group axioms (G) were deliberately made redundant, as an aid to understanding them. Show, for example, that if $G1$ and $G2$ hold, and if $a \circ e = a$ and there exists a^{-1} such that $a \circ a^{-1} = e$, for all $a \in G$, then the equalities $e \circ a = a$ and $a^{-1} \circ a = e$, as well as the uniqueness of e, can be deduced. [*Hint:* Put $c = a^{-1} \circ a$ and show that $c^2 = c$.]

†2. A **subgroup** of a group G is a subset of G which is itself a group under the operation of G.

   a) Calling the operation multiplication, show that the set of powers $e = 1 = a^0$, $a, a^{-1}, a^2, a^{-2}, \ldots$ of an arbitrary element $a \in G$ forms a subgroup, called the subgroup **generated by** a.

   b) If G is finite, show that it suffices to consider positive exponents to obtain the entire subgroup generated by a. [*Hint:* Show that in this case, $a^n = 1$ for some n.]

   c) The **order of an element** $a \in G$ is the smallest positive exponent h, if there is one, such that $a^h = 1$; if there is no such h, then a is said to be of infinite order. Show that if G is finite, the order h of a is the same as the order (say k) of the subgroup generated by a. [*Hint:* Show that the powers $a, a^2, \ldots, a^h$ are all distinct, and that the rows in the array

$$\begin{array}{cccc} a & a^2 & \cdots & a^h \\ a^{h+1} & a^{h+2} & \cdots & a^{2h} \\ a^{2h+1} & a^{2h+2} & \cdots & a^{3h} \\ & \cdots & & \\ & \cdots & & \end{array}$$

   are identical.]

   d) Rephrase (a), (b), and (c) for the case in which the operation is addition.

†3. Show that the order of any subgroup H of a finite group G is a divisor of the order of G. [*Hint:* Let the elements of H be $a_1, \ldots, a_h$. If $H \neq G$, there is a $b_1 \in G$ with $b_1 \notin H$, and then $b_1 a_1, \ldots, b_1 a_h$ are all distinct, and none belongs to H. If some element of G has not yet occurred, continue.]

4. Is the set of even integers an additive group? a Ring? How about the set of positive integers?

5. Prove the equivalence, in any Ring, of the two characterizing properties of a domain, given in (C).

6. Show that the continuous real-valued functions on $[0, 1]$ constitute a Ring but not a domain.

7. Prove the usual rules regarding inequalities, in $\mathbf{Z}$:

   if $a < b$ and $b < c$, then $a < c$;
   if $a < b$, then $a + c < b + c$ for all c;
   if $a < b$ and $c > 0$, then $ac < bc$;
   $a^2 \geq 0$ for all a;
   for every a and b, exactly one of the relations $a < b$, $a = b$, $a > b$ is true.

---

† This symbol means that the definitions and results in this problem will be used later in the text.

8. Formulate a definition saying when two groups are isomorphic, which is sufficiently broad to allow for an isomorphism between two groups with different operations. (Note that when an algebraic system has only one operation, as is true in a group, it matters very little what the operation is called; but the two operations in a Ring or field enter nonsymmetrically in the distributive law, and they are almost invariably called addition and multiplication.)

9. Show that any two groups of order 3 are isomorphic.

10. Here are three groups of order 4. Decide which are isomorphic to each other.
    a) $G_1 = \{\pm 1, \pm i\}$; the operation is ordinary multiplication.
    b) $G_2 = \{1, 2, 3, 4\}$; the operation is addition followed by reduction to the smallest positive remainder upon division by 4 (for example, $2 + 3 = 1$, $1 + 2 + 3 + 4 = 2$.)
    c) $G_3 = \{1, 3, 5, 7\}$; the operation is multiplication followed by reduction to the smallest positive remainder upon division by 8, so that $3^2 = 5^2 = 1$.

11. (Assumes matrix algebra.) Consider the set of 2×2 matrices, with the usual addition and multiplication. Is this set an additive group? a Ring? a domain? a field? Does your answer depend on what set the entries are drawn from? If so, consider separately matrices with entries from (a) $\mathbf{Q}$; (b) $\mathbf{Z}$; (c) the even integers, $2\mathbf{Z}$; (d) the field with two elements.

†12. a) Designate by $\mathbf{Z}[x]$ the set of polynomials with coefficients in $\mathbf{Z}$ (briefly, the polynomials over $\mathbf{Z}$), so that $f(x) \in \mathbf{Z}[x]$ if and only if

$$f(x) = a_0 + a_1 x + \cdots + a_n x^n, \qquad n \geq 0, \quad \text{all } a_i \in \mathbf{Z}.$$

If $a_n \neq 0$, then $f(x)$ is of exact **degree** n, and we write $\partial f = n$. If all $a_i = 0$, $f(x)$ is called the **zero polynomial** and it is not given a degree. Sometimes it is useful to write $f(x) = \sum_0^\infty a_k x^k$, with the understanding that only finitely many a_k are different from 0. Addition and multiplication are defined as usual:

$$\sum_{k=0}^{\infty} a_k x^k + \sum_{k=1}^{\infty} b_k x^k = \sum_{k=0}^{\infty} (a_k + b_k) x^k,$$

$$\sum_{k=0}^{\infty} a_k x^k \cdot \sum_{k=0}^{\infty} b_k x^k = \sum_{l=0}^{\infty} \left(\sum_{m=0}^{l} a_m b_{l-m} \right) x^l.$$

Show that $\mathbf{Z}[x]$ is a Ring, and then that it is a domain.

b) Define $\mathbf{Q}[x]$ in just the same way as $\mathbf{Z}[x]$, except that the coefficients of the polynomials are now allowed to be rational numbers rather than integers. Show that $\mathbf{Q}[x]$ is also a domain. Which elements of $\mathbf{Q}[x]$ have multiplicative inverses in this Ring? (These are called the invertible elements, or **units**, in the domain. Every nonzero element in a field is a unit.)

c) Let R be an arbitrary Ring, and define $R[x]$ in an analogous way, as the collection of polynomials over R. Show that $R[x]$ is a domain if and only if R is. Also, show that for no R is $R[x]$ a field. (Continued in Section 1.3, Problem 9.)

1.3 TYPES OF PROOFS, AND SOME EXAMPLES

Proofs loom large in elementary number theory for two reasons. First, many of the theorems (including those contained in some of the problems in this book) are so simple to state and so easy to understand that one is deceived into thinking that they must also be easy to prove, and this is not always the case. Also, there is much more variety in the kinds of reasoning invoked than is the case in more elementary mathematics, the latter, more or less by definition, being restricted to material that is both useful and relatively easy to master. It is not merely because number theory is not useful in commerce or engineering that it is not customarily learned before calculus, as it logically could be. The variety of proof techniques sometimes seems so large that students regard number theory as a "bag of tricks," but of course this is a matter of familiarity. What is a trick the first time one meets it is a device the second time and a method the third time.

In any case, since proofs will be at the heart of the matter, we pause briefly to discuss them. One technique, not especially connected with number theory but also not very common in more elementary mathematics, is proof by contradiction. This depends on the principle that a (properly formulated) assertion is either true or false, so to prove that it is not false is to prove that it is true. For a first example, consider the ancient theorem that *there are infinitely many positive primes*. Suppose the opposite, that there are only finitely many, and let them be $p_1, p_2, \ldots, p_k$. Form the integer $N = p_1 p_2 \cdots p_k + 1$. Clearly, N is not divisible by any of $p_1, p_2, \ldots, p_k$ (that is, for each i, $N/p_i \notin \mathbf{Z}$). But since $N > 1$, N is divisible by some prime, as we shall prove in the next paragraph. So there is a positive prime different from all of $p_1, p_2, \ldots, p_k$, contradicting the supposition that this was the complete list. The proof will therefore be complete as soon as we know that *every integer $n > 1$ is divisible by some positive prime.*

We shall also prove this by contradiction, but with a slightly different twist. First, we need more precise definitions. The nonzero integers are customarily partitioned into three classes, namely the **units** ± 1 (the integers with reciprocals in $\mathbf{Z}$), the **prime** numbers, and the **composite** numbers, the prime numbers being those integers n for which the conditions

$$n = ab; \quad a, b \in \mathbf{Z}; \quad a, b \text{ not units} \tag{2}$$

cannot be satisfied simultaneously, and the composite numbers being the remaining nonzero integers. Also, if c and d are integers, we say that d **divides** c, or is a **divisor** of c, and we write $d \mid c$, if there is an integer f such that $c = df$. (If there is no such integer f, we write $d \nmid c$.)

Now to the proof. Suppose that $n > 1$ but that n has no positive prime divisor. Then n is not prime, since $n \mid n$, so (2) has a solution a, b. Then, also, $n = |a| \cdot |b|$, and $|b| > 1$ so $|a| < n$. If we put $n_1 = |a|$, then n_1, like n, is an integer larger than 1 and it has no prime divisor; for divisibility is transitive: if $p \mid n_1$ and $n_1 \mid n$, then $p \mid n$. Also, $n_1 < n$. Now repeating the same argument with n_1 in place of n gives an integer n_2, $1 < n_2 < n_1$, with no positive prime divisor, and so on. But this yields a nonempty set of positive integers, $n_1, n_2, \ldots$, having no smallest element, in contradiction to the well-ordering principle (W). This contradiction shows that the assertion cannot be false, and hence must be true. ■*

Thus in a proof by contradiction the theorem in question is always assumed to be false, but the resulting contradiction may either be internal (assuming that there were only finitely many primes gave us a way to describe a new one) or external, such as the conflict with (W) in the second proof above.

A simple but useful fact, on which a surprisingly large body of mathematics is based, is that *there is no integer between* 0 *and* 1. Here is a proof similar to the last one. If the assertion is false, then there is an $a \in \mathbf{Z}$ with $0 < a < 1$. Multiplying through by the positive integer a gives $0 < a^2 < a$, and similarly $0 < a^3 < a^2$, and so on, again contradicting (W).

This result can be used, for example, to prove that *the real number* $e = 2.71828\cdots$ *is irrational.* From the series expansion

$$e = 1 + \frac{1}{1!} + \frac{1}{2!} + \frac{1}{3!} + \cdots,$$

we obtain, for each positive integer n,

$$n!\,e = \left\{\frac{n!}{1} + \frac{n!}{1!} + \frac{n!}{2!} + \cdots + \frac{n!}{n!}\right\} + \frac{n!}{(n+1)!} + \cdots$$

and the quantity in braces is an integer, say q_n. Suppose e is rational, perhaps $e = a/b$ with $a, b \in \mathbf{Z}$. Then the quantity

$$r_n = n!\,a - q_n b = b\left(\frac{1}{n+1} + \frac{1}{(n+1)(n+2)} + \frac{1}{(n+1)(n+2)(n+3)} + \cdots\right)$$

is a positive integer. But, keeping only two factors in the denominators, we have

$$\begin{aligned} r_n &< \frac{b}{n+1} + b\left\{\frac{1}{(n+1)(n+2)} + \frac{1}{(n+2)(n+3)} + \cdots\right\} \\ &= \frac{b}{n+1} + b\left\{\left(\frac{1}{n+1} - \frac{1}{n+2}\right) + \left(\frac{1}{n+2} - \frac{1}{n+3}\right) + \cdots\right\} \\ &= \frac{b}{n+1} + \frac{b}{n+1} = \frac{2b}{n+1}, \end{aligned}$$

so $0 < r_n < 1$ for $n > 2b - 1$. This is impossible by the preceding theorem, so e is not rational. ■

* This symbol signals the end of a proof.

Fermat was the first to recognize the power of the technique of obtaining a contradiction to the well-ordering principle. He called it the method of **infinite descent**, and used it very skillfully to prove that various equations, such as $x^4 + y^4 = z^4$, have no solutions $x, y, z \in \mathbf{Z}^+$. The technique can be formulated more positively as a theorem, the **principle of induction**: *If a set S of integers contains n_0, and if S contains $n + 1$ whenever it contains n, then S contains all integers greater than or equal to n_0.* The proof of this principle should have a familiar form by now. Suppose that $b \in \mathbf{Z}$, $b > n_0$ and $b \notin S$. Then $b - 1 \notin S$ (otherwise $(b - 1) + 1$ would belong to S), so $b - 1 \neq n_0$ (since $n_0 \in S$), so $b - 1 > n_0$, so $b - 1$ has all the properties attributed to b, and the argument can be repeated indefinitely, contrary to (W). ■

The induction principle as just stated can be recast as a rule for theorem-proving. Let $P(n)$ be a proposition involving an integer variable n, and suppose we wish to prove the theorem "for every integer $n \geq n_0$, $P(n)$", where n_0 is a specific integer. This can be accomplished by proving the following two statements instead: "$P(n_0)$" and "for every integer $k \geq n_0$, $P(k)$ implies $P(k + 1)$". That this rule follows from (and, in fact, is equivalent to) the principle of induction is seen simply by taking S to be the set of integers n for which $P(n)$. The rule itself is frequently called the induction principle; it was first used explicitly by B. Pascal (1623–1662). It must be emphasized that the three statements in quotation marks are different from one another. The last of them is usually proved by assuming that $n \geq n_0$ is an integer for which $P(n)$, and deducing $P(n + 1)$, but this is quite different from assuming that *for all* $n \geq n_0$, $P(n)$, which is the theorem. The assumption "$n \geq n_0$ is an integer for which $P(n)$" is called the **induction hypothesis**.

As it is a topic having some independent interest, and is one in which proofs by induction are immediately encountered, consider the so-called **Fibonacci sequence**

$$1, 1, 2, 3, 5, 8, 13, 21, \ldots,$$

where every element after the second is the sum of the two preceding elements. Thus if F_n denotes the nth element, then the sequence is said to be defined **recursively** by the equations

$$F_1 = 1, \quad F_2 = 1, \quad F_{n+1} = F_n + F_{n-1} \qquad \text{for } n \geq 2.$$

The following theorem is very easy to prove by induction: *For every $n \geq 1$, $F_n < 2^n$.* For the inequality is true for $n = 1$ and $n = 2$, and it is clear from the definition that each F_k is larger than the preceding one. (The extracareful reader may wish to prove *this* by induction!) Hence if $k \geq 2$ and $F_k < 2^k$, then

$$F_{k+1} = F_k + F_{k-1} < F_k + F_k < 2 \cdot 2^k = 2^{k+1},$$

so that $F_k < 2^k$ implies $F_{k+1} < 2^{k+1}$ for all $k \geq 2$. ■

Although the stronger theorem,

(3) $$\text{for all } n \geq 1,\ F_n < (7/4)^n,$$

is also true, it cannot be proved in exactly the same way, for the argument above gives only $F_{k+1} < 2F_k < 2(7/4)^k$, and unfortunately $2(7/4)^k > (7/4)^{k+1}$. Here the induction hypothesis is too weak to yield the desired conclusion. In this instance the difficulty can be circumvented by choosing for $P(n)$ not the assertion "$F_n < (7/4)^n$", but the equivalent assertion "$F_{n-1} < (7/4)^{n-1}$ and $F_n < (7/4)^n$"; for the induction hypothesis then gives

(4) $$F_{k+1} < (7/4)^{k-1} + (7/4)^k = (7/4)^{k-1}(1 + 7/4) < (7/4)^{k+1},$$

which yields $P(k + 1)$. Usually this artificial procedure is avoided by a further reformulation of the induction principle: *If $P(n_0)$ and if, for all $k \geq n_0$,*

$$\{P(n_0)\ \&\ P(n_0 + 1)\ \&\cdots\&\ P(k)\} \textit{ implies } P(k + 1),$$

then for all $n \geq n_0$, $P(n)$. This is what results from the second version by taking $P(n)$ to be $\{P(n_0)\ \&\ P(n_0 + 1)\ \&\cdots\&\ P(n)\}$.

Using this third version, the proof of (3) would go as follows. The inequality $F_n < (7/4)^n$ holds for $n = 1$ and $n = 2$. Suppose it holds for $n = 2, \ldots, k$, where $k \geq 2$. Then (4) holds, and hence (3) is true.

Just as the proof of the irrationality of e depended on the simple fact that no positive integer is smaller than 1, several nontrivial theorems will be proved later with the help of the following, which is known as **Dirichlet's principle**: *If a set having n elements is partitioned into m subsets* (*that is, each element belongs to exactly one subset*), *and if $1 \leq m < n$, then some subset contains more than one of the elements.* Obvious as it may seem, this is a theorem and it requires proof.

If it were false, there would be a smallest integer n for which it fails. Necessarily $n \geq 2$, since otherwise the hypothesis cannot be true that $1 \leq m < n$. (Proof?) Let S be a set with n elements which violates the assertion, so that there are m subsets $S_1, S_2, \ldots, S_m$ which partition S and none of which contains two or more elements of S. Our assumption is that no set with fewer than n elements has this property. Obviously $m > 1$. If a is any element of S, then a belongs to exactly one of the subsets, say S_1, and in fact S_1 can have only this one element. Deleting a from S leaves a set S' of $n - 1$ elements, and $S_2, \ldots, S_m$ partition S', and it is still true that none of $S_2, \ldots, S_m$ have as many as two elements. This contradicts the minimality of n, so the theorem is not false. ∎

Dirichlet's principle is also known as the pigeon-hole or box principle, since it is sometimes phrased less formally to say that if $n + 1$ things are put into n boxes, then some box contains at least two of the things. One reason for its importance is that it is a rather general existence principle—it says that something exists, without providing a method for constructing or finding it, and what is more important, without *requiring* such a method. Naturally, a constructive proof is also desirable, for any theorem, but sometimes none is available.

PROBLEMS

1. Verify the following elementary properties of divisibility, where a, b, and c are integers:
   i) $a \mid 0$, $a \mid a$, and $\pm 1 \mid a$;
   ii) if $a \mid b$ and $b \mid c$, then $a \mid c$;
   iii) if $a \mid b$ and $a \mid c$, then $a \mid (bx + cy)$ for all $x, y \in \mathbf{Z}$.
2. Prove that no two successive Fibonacci numbers F_n and F_{n+1} have a common divisor $a > 1$.
3. a) Show that for $n > k + 1 > 1$,

$$F_n = F_{k+1}F_{n-k} + F_kF_{n-k-1}.$$

   [*Hint:* Eliminate F_{n-1} from the defining equation $F_n = F_{n-1} + F_{n-2}$, then eliminate F_{n-2}, etc.]
   b) Show that $F_n \mid F_{rn}$ for all $n, r \geq 1$.
4. Let α be any real number larger than $\beta = \frac{1}{2}(1 + \sqrt{5}) = 1.61\ldots$. Prove by induction or otherwise that for $n \geq 1$, $F_n < \alpha^n$. [Note that β is a solution of the equation $x^2 = x + 1$.]
5. Prove by induction or otherwise that

$$F_n = \frac{1}{\sqrt{5}}\left\{\left(\frac{1 + \sqrt{5}}{2}\right)^n - \left(\frac{1 - \sqrt{5}}{2}\right)^n\right\}.$$

   [*Hint:* $(1 - \sqrt{5})/2$ is another solution of $x^2 = x + 1$.]
6. If n is an integer, show that there is no integer a such that $n < a < n + 1$.
7. Prove that for $n \geq 1$,

   †a) $\sum_{m=1}^{n} m = \frac{1}{2}n(n + 1)$,

   †b) $\sum_{m=1}^{n} (2m - 1) = n^2$,

   c) $\sum_{m=1}^{n} m^2 = \frac{1}{6}n(n + 1)(2n + 1)$,

   d) $\sum_{m=1}^{n} m^3 = \frac{1}{4}n^2(n + 1)^2$.
8. Suppose that n pairs of gloves of different sizes are mixed together in a drawer. How many individual gloves must you take out if you are to be sure of having at least one complete pair?
9. In the text it is shown that every integer $n > 1$ has a prime divisor. A related assertion is that if d is the smallest divisor of n which is larger than 1, then d is prime. Discuss the relationship between these two assertions, and their relative advantages, and prove the second.

10. (Continued from Section 1.2, Problem 12.) As before, let $\mathbf{Q}[x]$ be the domain of polynomials over $\mathbf{Q}$, in one variable. Formulate definitions similar to those in the text for prime and composite polynomials and for divisibility. Extend the results of Problems 1 and 9 above to the present context. (Continued in Section 1.4, Problem 8.)

11. a) Let n be a positive integer and let $\alpha_1, \alpha_2, \ldots, \alpha_{n+1}$ be any real numbers in the interval $[0, 1)$ (that is, the interval $0 \leq \alpha < 1$). Show that indices i and j exist such that $i \neq j$ and $|\alpha_i - \alpha_j| < 1/n$. [*Hint:* Use the informal version of Dirichlet's principle. What are the set and the subsets in the formal version, in this case?]

    b) Show that if α is real and $q_0, \ldots, q_n$ are distinct positive integers, then there exist an integer r and indices i and j with $0 \leq i < j \leq n$ such that $|\alpha(q_i - q_j) - r| < 1/n$.

    c) Show that if α is real and n is a positive integer, then there are integers q and r with $0 < q \leq n$ and $|q\alpha - r| < 1/n$.

1.4 REPRESENTATION SYSTEMS FOR THE INTEGERS

We conclude this review of "elementary arithmetic from an advanced standpoint" with a discussion of several methods of representing the integers. So far we have an intrinsic description of the entire set $\mathbf{Z}$, as a domain with certain properties, and we have the descriptions $1, 1 + 1, 1 + 1 + 1, \ldots$ of the positive integers, which are useful for a definition but little more, at least out beyond 20 or so. A much more efficient system is the one everyone learns in elementary school, the decimal representation; it is so useful that every educated person has devoted five or more years to becoming familiar with it and learning to use the associated algorithms (carrying, borrowing, long division, etc.) for performing the four rational operations. Let us develop a slight generalization of it, by thinking about just how it works.

The decimal representation of an integer less than 1000, say, is determined by partitioning the interval $[0, 1000)$ (containing the integers $0, 1, \ldots, 999$) into 10 equal parts $[0, 100), [100, 200), \ldots, [900, 1000)$, finding which subinterval the integer belongs to, partitioning that subinterval into 10 equal parts, and continuing until the integer is uniquely identified. If, for example, $n \in [8 \cdot 100, 9 \cdot 100)$, and then $n - 8 \cdot 100 \in [5 \cdot 10, 6 \cdot 10)$, and then in turn $n - 8 \cdot 100 - 5 \cdot 10 \in [7, 8)$, then the decimal representation of n is 857. Thus the 1000 integers from 0 to 999 are partitioned into 10 classes, each of which is partitioned into 10 subclasses, each of which is again partitioned into 10 subclasses, each containing a unique integer. Put this way, the generalization is obvious. Let $m_1, m_2, \ldots, m_n$ be integers larger than 1 (in the decimal case, each $m_i = 10$), and define

$$M_1 = m_1, \quad M_2 = m_1 m_2, \quad \ldots, \quad M_n = m_1 m_2 \cdots m_n.$$

Partition the interval $[0, M_n)$ into the m_n subintervals

$$[0, M_{n-1}), \ldots, [(m_n - 1)M_{n-1}, M_n)$$

of length M_{n-1}, partition each of these into m_{n-1} subintervals of length M_{n-2}, etc. This eventually yields $m_n m_{n-1} \cdots m_1 = M_n$ intervals, each of length 1 and hence containing a unique integer, and each integer $k \in [0, M_n)$ is then uniquely described in the form

$$k = a_{n-1}M_{n-1} + a_{n-2}M_{n-2} + \cdots + a_1 M_1 + a_0, \quad \text{where } 0 \le a_i < m_{i+1} \text{ for } 0 \le i < n.$$

With an infinite sequence of integers $m_1, m_2, \ldots$ larger than 1, instead of the finite set $m_1, m_2, \ldots, m_n$, we apparently obtain a unique representation of every integer $k \ge 0$ in the form

$$k = a_0 + a_1 m_1 + a_2 m_1 m_2 + a_3 m_1 m_2 m_3 + \cdots, \tag{5}$$

where $0 \le a_i < m_{i+1}$ for each $i \ge 0$, and of course $a_i = 0$ for all i for which $m_1 m_2 \cdots m_i > k$. When $m_i = 10$ for all i, this reduces to the decimal system, and when $m_i = r$, a fixed integer larger than 1, for all i, it is called the expansion to the **base** or **radix** r. Base 2 and base 8 representations are commonly employed in computing machines. The general representation (5) is sometimes useful, both in theoretical considerations (as we shall see in Chapter 3) and in parallel computational work.

The derivation given above for (5) was informal—not incorrect, but involving one or two hidden inductions. A formal proof is most easily founded on the following result, which is fundamental to much of this book. **Here, and elsewhere unless the context indicates otherwise, lower-case Latin letters will denote integers.**

Theorem 1.1 (The division theorem) *If a is positive and b is any integer, there is exactly one pair of integers q and r such that the conditions*

$$b = aq + r, \qquad 0 \le r < a, \tag{6}$$

hold.

Remark. If r satisfies the inequality in (6), then q is called the **quotient** and r the **remainder** when b is divided by a.

Proof. First we show that (6) has at least one solution. Consider the set D of integers of the form $b - ua$, where u runs over all integers, positive and nonpositive. For the particular choice

$$u = \begin{cases} -1 & \text{if } b \ge 0, \\ b & \text{if } b < 0, \end{cases}$$

the number $b - ua$ is nonnegative, so that D contains nonnegative elements. The subset consisting of the nonnegative elements of D has a smallest element. Take r to be this number, and q the value of u which corresponds to it. Then

$$r = b - qa \ge 0, \qquad r - a = b - (q + 1)a < 0,$$

so that (6) is satisfied.

To show the uniqueness, assume that also

$$b = q'a + r', \qquad 0 \le r' < a.$$

Then if $q' < q$,

$$b - q'a = r' \ge b - (q - 1)a = r + a \ge a,$$

while if $q' > q$,

$$b - q'a = r' \le b - (q + 1)a = r - a < 0.$$

Hence $q' = q$, $r' = r$. ■

Returning to (5), we see that a_0 is the remainder when k is divided by m_1. Thus $(k - a_0)/m_1$ is an integer, say k_1, and a_1 is the remainder when k_1 is divided by m_2. Similarly,

$$k_2 = \frac{k_1 - a_1}{m_2} = \frac{k - a_0 - a_1 m_1}{m_1 m_2}$$

is an integer, and a_2 is the remainder when k_2 is divided by m_3. And so on—meaning a proof by induction.

Here is the same thing, in simpler form. Given k and $m_1 > 1, m_2 > 1, \ldots,$ apply Theorem 1.1 repeatedly to obtain

$$\begin{aligned} k &= m_1 q_1 + a_0, & 0 &\le a_0 < m_1 \\ q_1 &= m_2 q_2 + a_1, & 0 &\le a_1 < m_2 \\ &\vdots \end{aligned}$$

The q_i form a strictly decreasing sequence as long as they are positive; so by the well-ordering principle, an index $n > 0$ exists such that $q_1 > 0, \ldots, q_n > 0$ but $q_{n+1} = 0$, so that

$$\begin{aligned} q_{n-1} &= m_n q_n + a_{n-1}, & 0 &\le a_{n-1} < m_n,\ q_n > 0, \\ q_n &= m_{n+1} \cdot 0 + a_n, & 0 &\le a_n = q_n < m_{n+1}. \end{aligned}$$

Eliminating q_1 between the first two of this sequence of equations, and then q_2, etc., gives

$$\begin{aligned} k &= a_0 + m_1(m_2 q_2 + a_1) \\ &= a_0 + a_1 m_1 + m_1 m_2(m_3 q_3 + a_2) \\ &= \cdots \\ &= a_0 + a_1 m_1 + a_1 m_1 m_2 + \cdots + a_n m_1 m_2 \cdots m_n, \qquad 0 \le a_i < m_{i+1}, \end{aligned}$$

as in (5). (There is a simple induction in the last step, of course.)

However important radix or similar representations may be for computations, they are not central to a serious study of the integers because they depend on an extraneous element, the choosing of the sequence $\{m_i\}$. Studying the decimal representations of integers is doing just that; it is not the same as studying the integers themselves.

Here, on the other hand, is a theorem which gives an *intrinsic* property of the integers, as well as a representation system for them.

Theorem 1.2 (The unique factorization theorem) *Every integer a either is 0, or is a unit ± 1, or has a representation in the form*

$$a = up_1p_2 \cdots p_r, \tag{7}$$

where u is a unit and $p_1, p_2, \ldots, p_r$ are (one or more) positive primes, not necessarily distinct. The representation (7) is unique, except for the order in which the primes occur.

We shall derive this fundamental theorem from the following result.

Theorem 1.3 *If p is prime and $p \mid ab$, then $p \mid a$ or $p \mid b$.*

Proof. The theorem is obviously true if $a = 0$ or $b = 0$. Otherwise, it suffices to consider the case in which p, a, and b are positive, since $p \mid c$ if and only if $|p|$ divides $|c|$. Suppose that the theorem is false, and let p be the smallest positive prime for which there are positive integers a and b such that $p \mid ab$, $p \nmid a$, $p \nmid b$. By Theorem 1.1, since $p \nmid a$ and $p \nmid b$,

$$\begin{aligned} a &= pq_1 + a', & 0 < a' < p, \\ b &= pq_2 + b', & 0 < b' < p, \end{aligned} \tag{8}$$

for suitable q_1, q_2. Since $p \mid ab$, the quantity

$$a'b' = ab + p(pq_1q_2 - bq_1 - aq_2)$$

is divisible by p, say $a'b' = pc$. By the inequalities in (8), $0 < c < p$. Thus if p' is a positive prime such that $p' \mid c$, then $p' < p$, and since $p' \mid a'b'$, either $p' \mid a'$ or $p' \mid b'$. Hence every prime divisor of c can be divided out of both sides of the equation $a'b' = pc$, leaving $a''b'' = p$, where $a'' \mid a'$ and $b'' \mid b'$. By the definition of a prime number, one of a'' or b'' is 1 and the other is p; suppose $b'' = p$. Then $p \mid b'$, contradicting (8). ∎

Corollary *If $p, p_1, \ldots, p_k$ are positive primes and $p \mid p_1 \cdots p_k$, then $p = p_i$ for some i.*

For either $p \mid p_1$, in which case $p = p_1$, or $p \mid p_2 \cdots p_k$. Thus there is no minimal k for which the corollary is false. ∎

Proof of Theorem 1.2. Suppose that a is not 0 or ± 1; then, as before, we may restrict attention to the case $a > 1$. We saw in the preceding section that then a has a positive prime divisor p_1, so

$$a = p_1a_1, \qquad 1 \le a_1 < a.$$

If $a_1 > 1$, then it in turn has a positive prime divisor, etc. By the well-ordering principle, this process must terminate, so there is a representation of a as a finite product of primes, $a = p_1p_2 \cdots p_r$.

Now suppose that a is the smallest positive integer for which there are two representations,

$$a = p_1 p_2 \cdots p_r = p'_1 p'_2 \cdots p'_s, \tag{9}$$

where of course either r or s could be 1. Then $p'_1 \mid p_1 \cdots p_r$, so by the corollary above, $p'_1 = p_i$ for some i, and

$$\frac{a}{p_i} = \frac{p_1 \cdots p_r}{p_i} = p'_2 \cdots p'_s.$$

This contradicts the minimality of a unless these two representations of a/p_i differ at most in the order in which the primes occur, and in that case the same was already true in (9). ■

As we shall see, the unique decomposability of integers into primes plays an absolutely fundamental role in many, if not most, branches of number theory.

For many purposes it is useful to shorten the prime factorization of an integer to what we shall simply call the **prime-power decomposition** of $a > 1$,

$$a = p_1^{e_1} p_2^{e_2} \cdots p_l^{e_l}, \tag{10}$$

where the p_i are now *distinct* primes and the e_i are positive integers. The factor p^e corresponding to a particular prime p in this decomposition is called the **p-component** of a. Thus $p^e \mid a$, while $p^{e+1} \nmid a$; this relationship is described in symbols by writing $p^e \parallel a$, and we use this notation even when $e = 0$, that is, when $p \nmid a$.

Returning to Theorem 1.1, we see that it too provides a representation system of sorts. For each fixed a, it partitions $\mathbf{Z}$ into the various arithmetic progressions with difference a. For $a = 2$ this gives two classes, the set of "even" numbers, $\{2k\}$, and the "odd" numbers, $\{2k + 1\}$. For $a > 2$ the classes are not given names, but there are, for example, six arithmetic progressions with difference 6, namely the set of integers of the form $6k$, those of the form $6k + 1, \ldots$, and those of the form $6k + 5$. Much more will be said about progressions in Chapter 3 and elsewhere; we content ourselves for the moment with the statement of a beautiful theorem, due to P. L. Dirichlet, concerning them.

Consider the various progressions with fixed difference b: $\{bk\}, \{bk + 1\}, \ldots, \{bk + (b - 1)\}$. If r and b have a common divisor $d > 1$, then clearly the progression $\{bk + r\}$ consists entirely of integers divisible by d, so no prime numbers occur in the progression if d is not prime, and at most the two primes $\pm d$ if d is prime. (For example, in the case $b = 9$, the progression $\{9k\}$ contains no primes, $\{9k + 3\}$ contains the prime 3, and $\{9k + 6\}$ contains the prime -3.) If, on the other hand, b and r have no common factor larger than 1, it is possible that the progression $\{bk + r\}$ might contain infinitely many primes. *Dirichlet's theorem on primes in a progression* asserts that it does always contain infinitely many in that case. The proof of the complete theorem is beyond the scope of this book, but several special cases are given as problems, as the tools needed become available.

PROBLEMS

1. The long-division algorithm learned in elementary school yields, for example, the computation

$$\begin{array}{r} 243 \\ 11\overline{)2679} \\ \underline{22} \\ 47 \\ \underline{44} \\ 39 \\ \underline{33} \\ 6 \end{array}$$

and the conclusion that $2679 = 11 \cdot 243 + 6$. Discuss the interplay between long division and Theorem 1.1.

2. a) Show that any unknown integral weight less than 2^{n+1} can be weighed on a pan balance using the standard weights $1, 2, 2^2, \ldots, 2^n$, by putting the unknown weight on one pan and a suitable combination of standard weights on the other.
   b) Prove that no other combination of $n + 1$ standard weights will do this. (But note, for example, that either of two sets of four standard weights will weigh all integral unknowns from 1 to 14.)

3. Let $1, 1, 2, 3, 5, \ldots, F_n, \ldots$ be the Fibonacci sequence.
   a) Show that for $n > 0$,

$$F_n > \sum_k F_{n-2k-1},$$

   where the sum extends over all $k \geq 0$ for which $n - 2k - 1 > 1$.
   b) Show that every positive integer can be represented uniquely in the form $F_{n_1} + F_{n_2} + \cdots + F_{n_m}$, where $m \geq 1$, $n_{j-1} \geq n_j + 2$ for $j = 2, 3, \ldots, m$, and $n_m > 1$.

4. a) Show that if $p = 6k + r$ is a prime different from 2 and 3, then $r = 1$ or 5 (assuming, of course, that $0 \leq r \leq 5$).
   b) Show that the progression $\{6k + 1\}$ is closed under multiplication.
   c) Show that the progression $\{6k + 5\}$ contains a prime. Then show that it contains infinitely many primes.
   d) Find a second arithmetic progression $\{bk + r\}$, with $b \neq 2, 3, 6$, for which this same argument works.

5. The squares, of course, are the numbers $1, 4, 9, \ldots$. The **square-free numbers** are the integers $1, 2, 3, 5, 6, \ldots$ which are not divisible by the square of any prime (so that 1 is both square and square-free). Show that every positive integer is uniquely representable as the product of a square and a square-free number. Show that there are infinitely many square-free numbers.

6. Suppose that p is prime, $p \mid b$ and $p \mid r$, where $0 \leq r < b$. Give the exact conditions under which $\{bk + r\}$ contains no primes, one prime, or two primes.

7. A simple hand-held calculator has only the four rational operations, and it presents $a \div b$ as the largest 8-digit decimal number which is less than or equal to a/b. For positive integers a and b less than 10^8, tell how to use the calculator to find q and r in Theorem 1.1.

†8. (Continued from Section 1.3, Problem 10.) a) Let D be a domain, so that $D[x]$ is also. Here is an analogue of Theorem 1.1: *If $f = f(x)$ and $g = g(x)$ are elements of $D[x]$, where $g \neq 0$ and the leading coefficient of g (i.e., the coefficient of the highest power of x) is a unit in D, then there is a unique pair $q, r \in D[x]$ such that $f = gq + r$, where $r = 0$ or $\partial r < \partial g$.* The following is a sketch of a proof; fill in all details.

Proof by induction on ∂f. If $f = 0$ or $\partial f < \partial g$, take $q = 0$. Suppose the result is true for fixed g of degree m and for all f with $\partial f < k$, for some $k \geq m$. Suppose $f(x) = ax^k + \cdots + a_k$ and $g(x) = bx^m + \cdots + b_m$, where $ab \neq 0$. Then $h = f - ab^{-1}x^{k-m}g \in D[x]$ and either $h = 0$ or $\partial h < k$; the induction hypothesis can be applied to h. Uniqueness is easy.

When D is a field F, all nonzero coefficients are units, and this division theorem can always be applied.

b) Any polynomial with leading coefficient 1 is said to be **monic**. Show that each monic $f \in F[x]$ can be written as a product of monic prime polynomials. (Uniqueness is not yet asserted.)

9. Prove that a positive odd integer $N > 1$ has a unique representation in the form $N = x^2 - y^2$ if and only if N is prime. Here x and y range over the nonnegative integers.

10. If a has the prime-power decomposition (10), describe the set of all positive divisors of a. Show that $\tau(a)$, as defined in Section 1.1, Problem 4, depends only on l and $e_1, \ldots, e_l$ and not at all on the values of $p_1, \ldots, p_l$. Reconsider the questions asked earlier about the τ-function in Problem 4 of Section 1.1.

Let $\sigma(a)$ be the sum of the positive divisors of $a > 0$, so that $\sigma(1) = 1$, $\sigma(2) = 3$, $\sigma(6) = 6$. Show that for a as in (10),

$$\sigma(a) = \prod_{i=1}^{l} (1 + p_i + \cdots + p_i^{e_i}),$$

where the symbol $\prod_{i=1}^{l}$ is the multiplicative analogue of $\sum_{i=1}^{l}$.

11. a) Show that when $a = 100$ and $b > 0$, the r of Theorem 1.1 is the final 2-digit number at the end of the decimal expansion of b.

b) Show that the last two digits of $(50k + l)^2$ are the same as those of l^2, for all $k, l \in \mathbf{Z}$.

c) Show that every positive integer has a unique representation in the form $50k + l$ with $-24 < l \leq 25$. Conclude that all final 2-digit numbers of the decimal expansions of squares are to be found among those of $0^2, 1^2, 2^2, \ldots, 25^2$.

d) By direct computation, show that squares must end in 00, 25, e1, e4, e9, or o6, where e is an even digit and o is an odd digit.

12. It will be shown later that a positive integer n of the form $4k + 1$ is prime if and only if it has a unique representation in the form $n = x^2 + y^2, 0 < x < y, (x, y) = 1$. Explain why the following computation shows that 137 is prime:

$$\begin{array}{rrl} 137 - & 1 = & 136 \\ - & 3 & 133 \\ - & 5 & 128 \\ - & 7 & 121 = 11^2 \\ - & 9 & 112 \\ - & 11 & 101 \\ - & 13 & 88 \\ - & 15 & 73. \end{array}$$

(Recall Problem 7(b) of Section 1.3.) If 137 were replaced by an interestingly large number—say of 6 or 8 digits—explain how Problem 11(d) would be of use.

13. Consider the progression $\{6k + 1\}$ as an algebraic entity, forgetting all the other integers. It is not a domain but it is multiplicatively closed, has a unit, and satisfies the cancellation law. The concepts of divisor and prime can be defined within it; do this. Does the unique factorization theorem hold in this set?
14. Reconsider Problem 1(c) of Section 1.2, now that the division theorem is available.

1.5 THE EARLY HISTORY OF NUMBER THEORY

Up to now we have been considering scraps and snippets, much of it perhaps well-known to the reader, most of it necessary or at least useful in later chapters, and all of it chosen to illustrate typical methods and results in number theory. It is almost time to begin a more methodical development of the subject. But before doing so, it might be enlightening to supplement the foregoing description of the mathematical foundations of the subject with a brief account of its historical foundations. We therefore pause a moment to give a quick sketch of what had happened before Fermat in the early seventeenth century.

The Mesopotamian civilization (ca. 3000–200 B.C.) was the earliest from which records remain showing mathematical activity. Fortunately, there was no paper then and writing, in cuneiform, was done on clay tablets, of which tens of thousands have been preserved. There are calendars dating from the beginning of the period, and tablets from about 2100 B.C. demonstrate the Sumerians' understanding of topographical measurement, simple and compound interest, the solution of specific quadratic equations (with square roots determined from tables), and the use of negative numbers. The first convincing demonstration that learned men of this era had knowledge of a topic that would now be regarded as number theory was discovered only in 1945, when O. Neugebauer and A. Sachs analyzed the tablet known as Plimpton 322 (from the Plimpton Library of Columbia University). From the language in which it is inscribed it can be dated rather accurately as

1900–1600 B.C., near the beginning of the first Babylonian dynasty, and at least 1000 years before the Pythagorean school. But it contains a table of 15 solutions in integers of the equation $x^2 + y^2 = z^2$, ranging in complexity from (3, 4, 5) and then (65, 72, 97) on up to (12709, 13500, 18541). Moreover, the order in which the solutions are listed is such as to make one angle of the right triangle with sides (x, y, z) decrease nearly steadily from 45° to 31°. Evidently, then, the early Babylonians not only knew the so-called Pythagorean theorem and may have had the concept of trigonometric functions, but they had a rule for solving the Pythagorean equation in integers. As if all this were not sufficiently impressive, these "primitive" people did all this with no algebraic symbolism at all and, as far as is known, without the concept of proof or formal derivation!

Egyptian mathematics, some of which is preserved on papyri, seems not to have been as sophisticated as that of the Mesopotamians in any direction, and none with genuine number-theoretic content is known. Records from the pre-Christian era in India and China are extremely fragmentary, and about all that is clear is that whatever was done then had little or no impact on later developments elsewhere.

Mathematics as we know it today—deduction, proof, theorem, abstraction—started with the Greeks. Deduction may have been used by the geometer Thales of Miletus (ca. 624–548 B.C.), and almost certainly was used by members of the Pythagorean school. Pythagoras (ca. 580–500 B.C.) traveled to Babylonia, Egypt, and possibly India; he was a mystic and philosopher who gave the integers numerological and philosophical importance. He and his school were probably responsible for the notion of "figurate" numbers (the triangular numbers 1, 3, 6, 10, . . . ; the squares; etc.), "perfect" numbers (28 is perfect because it is equal to the sum $1 + 2 + 4 + 7 + 14$ of its proper divisors), "amicable" numbers (220 and 284, for example, because each is the sum of the proper divisors of the other), and so on. Whether they actually proved theorems involving these concepts is not known.

Records are incomplete for the 200 years following the death of Pythagoras. In about 300 B.C. the first institution similar to a university, called the Museum, was established in Alexandria, and one of its first faculty members was Euclid. Although Euclid was a fine mathematician himself, much of what he included in his *Elements* synthesized earlier work. Books VII, VIII, and IX of the *Elements* are devoted to the theory of numbers. Theorem 1.3 above (if a prime divides a product, it divides a factor) is Proposition 30 of Book VII, and the unique factorization theorem is nearly equivalent to Proposition 14 of Book IX; the theorem that there are infinitely many primes is Proposition 20 of Book IX. Propositions 1 and 2 of Book VIII give a method for finding the greatest common divisor of two integers; we shall examine this "Euclidean algorithm" in the next chapter.

Of the three mathematical giants who created the Golden Age of Greek mathematics (300–200 B.C.)—Euclid, Archimedes, and Apollonius—only the first seems to have concerned himself to any great extent with number theory. Mechanics and geometry were much more in the air, and more than three centuries passed before

Diophantus of Alexandria broke new ground with his remarkable *Arithmetica*. In this treatise of about thirteen Books, of which only six have survived, the author began the study of indeterminate equations: equations in two or more unknowns of which the solutions are required to be in $\mathbf{Q}^+$, or (nowadays) in $\mathbf{Z}$. (Today these are called **Diophantine equations**, slightly inaccurately since it is not the form of the equation but the nature of the solutions which is restricted.) Also included were a few theorems of other kinds, such as the proposition that if each of two integers is a sum of two squares, then their product is also a sum of two squares, and in fact (generally) in two ways. Diophantus's systematic use of a letter as an abbreviated name for the unknown quantity, extended in the sixteenth century by the introduction of literal coefficients, exponents, and pictograms (+, −, etc.) for the operations, finally yielded the algebraic notation familiar today.

On the basis of indirect evidence it seems likely that the Chinese knew a substantial amount of mathematics long before it was discovered elsewhere; this would include Pascal's triangle and some simple magic squares. But, perhaps because it was mostly rendered sterile by lack of communication with the outside world, the Chinese contribution is now acknowledged only in the name "Chinese remainder theorem" which is commonly attached to Theorem 3.14, dating from the first few centuries of our era. In India, Brahmagupta (ca. 628) found the general integral solution of the linear Diophantine equation $ax + by = c$, as described in Theorem 2.9; Diophantus had treated only equations of higher degree, since a linear equation is trivial when rational solutions are allowed, and he always contented himself with single solutions. Somewhat later, Bhaskara (1114–ca. 1185) solved certain cases of the equation $x^2 - dy^2 = 1$, which we shall treat in Chapter 8. (Much earlier, instances of this equation had been solved by Archimedes or one of his contemporaries, and also by Diophantus.)

With the decline of Greek influence and the subsequent demise of the Roman Empire (which had produced nothing of mathematical interest), the hub of civilization moved in the eighth century to Baghdad. The resulting confluence of Babylonian, Egyptian, Greek, and Hindu knowledge yielded less than might have been hoped for, perhaps. From the present point of view the chief contributions of Arabian mathematicians lay in the adoption of the Hindu numeral system, positional notation and algebra of irrational numbers, and in their preservation of the Greek classics, although to be sure they also made numerous original contributions in algebra (the word itself comes from the Arabic al-jabr), trigonometry, and astronomy. There was also some work on Diophantine equations, including an unsuccessful attempt to show that the equation $x^3 + y^3 = z^3$ has no solution in positive integers.

After several hundred years had passed, Europe began to stir; the capture of the Spanish Muslim city of Toledo by the Christians in 1085 opened up a major center of Arabic culture to Western scholars, and by 1202 we have the first significant European book on mathematics, the *Liber Abaci* by Leonardo of Pisa, also called Fibonacci (son of Bonaccio). This book on algebra and arithmetic introduced the

Fibonacci sequence already defined in Section 1.3, and was one of the principal means by which Hindu-Arabic numeration and computational algorithms, to which Leonardo had grown accustomed in his travels as a merchant in the Orient, were promulgated in Europe. (Incidentally, "algorithm" is another Arabic word: Al-Khwarizmi wrote an early book on algebra.) In 1225 Fibonacci published the *Liber Quadratorum*, containing problems involving indeterminate equations, mostly borrowed from an Arabian book and thus ultimately from Diophantus. He was an original and creative mathematician, but the times were not yet ripe for theoretical mathematics. What with the Black Death, the Crusades, and the Inquisition, it is hardly surprising that, even though the ancient universities were established at Paris, Oxford, and Cambridge during Fibonacci's lifetime, rather little of any interest happened in mathematics during the next 250 years.

Movable type was invented in Europe in 1447, and within 50 years over 30,000 editions of various works had been published, including mathematical classics in Greek, Latin, and Arabic versions. Modern number theory might have started at about that time, for Johannes Müller (also called Regiomontanus, 1436–1476), a German astronomer and translator-publisher, discovered a Greek manuscript of the first six Books of Diophantus in the Vatican library. He wrote to a friend that he found the contents astonishing, and that he intended to translate it as soon as he found the missing seven Books. Unfortunately he died suddenly, possibly by poison, before making the translation. One hundred years passed before translations were published, a fragment in Bombelli's *Algebra* in 1572 and the whole of the surviving six Books, in Latin, by W. Holzmann (Xylander) in 1575. François Viète was familiar with the latter, and in his *Zetetika* of 1575 he published the solution of a problem mentioned but not solved by Diophantus, by finding the complete rational solution of the equation $x^3 + y^3 = a^3 - b^3$ with $a > b > 0$ and positive x, y. The method was that of Diophantus. It was also Viète who introduced into mathematics the use of letters for general coefficients and parameters, again taking the lead from Diophantus who, as mentioned, had used letters for unknowns.

Holzmann's translation of the *Arithmetica* was imperfect, and in 1621 C. G. Bachet de Méziriac published a substantially superior one, also in Latin, with some annotations and accompanied by the original Greek text. It was this new work that drew Fermat to number theory. He studied it deeply, and wrote many of his own results in the margin of his copy; the latter, complete with comments, was published by his eldest son five years after his death.

NOTES AND REFERENCES

Section 1.1

The assertion that every positive integer is a sum of four squares is sometimes called Bachet's Theorem. But Bachet merely noted, in his translation of Diophantus's *Arith-*

metica, that Diophantus seemed to assume the truth of this assertion and remarked that he, Bachet, would welcome a proof. Fermat definitely claimed that he had proved it.

That Fermat's equation $x^n + y^n = z^n$ has no solution if $n < 100{,}000$ was proved by Wagstaff [1976]. The lower bound n^{2n} for x, y, and z is due to Inkeri [1946]. For additional literature on the Fermat Problem, see LeVeque [1974], vol. 2. Regarding long computations and the size of the universe, see *Science* **192** (1976), 989–990.

The study of the first 55 Mersenne numbers M_p was completed by Uhler [1948]. The largest p for which M_p is at present known to be prime is 19937. For further literature on Fermat and Mersenne numbers, see LeVeque [1974], vol. 1.

The values given for $\pi(10^n)$ for $n \leq 9$ were computed in the nineteenth century by E. Meissel. The value found for $n = 9$ was erroneous, according to Lehmer [1959], who also supplied the value of $\pi(10^{10})$. These computations of $\pi(x)$ do not require knowledge of the individual primes up to x. On the other hand Gauss, who was a prodigious mental calculator, wrote to a friend that "I have (since I lacked the patience to go through the whole series systematically) often used a spare quarter of an hour to investigate a thousand numbers here and there; at last I gave it up altogether, without ever finishing the first million. . . . The thousand numbers between 101,000 and 102,000 bristles with errors in Lambert's table . . . ."

Gauss did not communicate his conjecture about $\pi(x)$ until 1849. In the meantime A. Legendre, in his famous book *Essai sur la théorie des nombres* of 1798, made conjectures closely related in form to (1′), so it was the latter which traditionally bore the name "prime number theorem," rather than Gauss's form (1).

A computer plot of the graph of the step function $y = \pi(x)$ and its approximations $y = \mathrm{li}(x)$ and $y = x/\log x$ is given in Graph 1 at the back of this book. The distribution of twin primes is illustrated in Graph 2.

The conjecture concerning the successive absolute differences of the primes was made by N. L. Gilbreath in 1958 (unpublished). The mentioned computation was carried out by Killgrove and Ralston [1959].

Section 1.3

Our first example of a proof by contradiction is, in fact, one of the earliest known instances of its type.

Section 1.4

The proof given here of Theorem 1.3 is due to Korselt [1940].

Section 1.5

Devotees of number theory are extremely fortunate in having a nearly complete guide to the literature in their subject (from antiquity until almost 1920 at least) in L. E. Dickson's [1919] monumental *History of the Theory of Numbers*, in three volumes. The period 1940–1972 is covered in LeVeque [1974]. For the period 1920–1940 one must go to specialized monographs, or to the abstracting journals, *Jahrbuch über die Fortschritte der Mathematik* and *Zentralblatt für Mathematik und ihre Grenzgebiete*. For books and articles since 1972, see *Mathematical Reviews*.

There are charming (if occasionally slightly romanticized) biographies of several of the mathematicians mentioned in this book in Bell [1937]. Many more are to be found in standard reference works, e.g., the *Encyclopaedia Britannica* and the *Dictionary of Scientific Biography*. Such books as Bell [1945] and Boyer [1968] include information on number theory in their broader accounts of the history of mathematics.

The assertion in the text that Euclid's *Elements* contains the unique factorization theorem needs a slight qualification. It is shown that a given product of primes does not have a second (different) factorization, but it is not shown that every integer larger than 1 is such a product.

2

Unique Factorization and the GCD

2.1 THE GREATEST COMMON DIVISOR

The following basic result is an immediate consequence of the unique factorization theorem proved in Chapter 1.

Theorem 2.1 *Given any two integers a and b, not both zero, there is a unique integer d, called their* **greatest common divisor**, *with these properties:*

i) $d \mid a$ *and* $d \mid b$;

ii) *if* $d_1 \mid a$ *and* $d_1 \mid b$, *then* $d_1 \mid d$;

iii) $d > 0$.

Proof. If there is such an integer d, it is obviously unique. For if both d and d' have properties (i) and (ii), then each must divide the other, so that both d/d' and $(d/d')^{-1}$ are integers, so $d/d' = \pm 1$, so $d = d'$ by (iii). Thus it suffices, in any proof of the theorem, to find an integer having all three properties.

If $a = 0$, then $d = |b|$; and if $b = 0$, then $d = |a|$. Suppose then that $ab \neq 0$. Let $p_1, p_2, \ldots, p_s$ be the set of all primes which divide a or b (or both, of course). Then for suitable nonnegative exponents $\alpha_1, \ldots, \alpha_s, \beta_1, \ldots, \beta_s$,

$$a = \pm p_1^{\alpha_1} \cdots p_s^{\alpha_s}, \qquad b = \pm p_1^{\beta_1} \cdots p_s^{\beta_s}.$$

Since $p^m \mid p^n$ if and only if $m \leq n$, the number

$$d = p_1^{\min(\alpha_1,\beta_1)} \cdots p_s^{\min(\alpha_s,\beta_s)} \tag{1}$$

clearly has the properties (i), (ii), and (iii). (As usual, $\min(\cdots)$ means the smallest of the numbers listed in the parentheses.) ■

Remark. Here and sometimes elsewhere we use the common convention that "prime" means "positive prime." It is not invariably intended; the meaning will always be obvious from the context if nothing more explicit is said. Similarly, we might for example refer to $\tau(n)$ (Section 1.1, Problem 4) simply as the number of divisors of n, rather than as the number of positive divisors.

The greatest common divisor, or **GCD**, of two integers plays such an important role in number theory that it is dignified as the no-name function, simply (a, b). (No confusion arises, as one might fear, with other uses of this symbol, since, for example, the left side of the equation $(8, 12) = 4$ could hardly refer to a point in the xy-plane.) When $(a, b) = 1$, a and b are said to be **relatively prime**. This has nothing whatever to do with whether a and b are prime; 6 and 35 are relatively prime, but neither is prime.

A possible alternative to the definition of GCD given in the theorem would be that it is the numerically largest of the common divisors of a and b. But "largest" involves inequalities, which are not available in nonordered domains, and we prefer a definition which brings out instead the multiplicative maximality of the GCD, as in (ii), and thus can be extended to other domains than **Z**. (The inequality (iii) could be omitted without serious loss; see Theorem 2.4, for example.)

The proof given above for Theorem 2.1 is somewhat unsatisfactory as a constructive method for actually evaluating a GCD, since it requires knowing the prime decompositions of a and b, whereas Euclid already knew how to find the first without knowing the second. The key to his method is the simple observation that $(a, b) = (a, b - a)$, which is true since every divisor of one side is a divisor of the other. By repeatedly subtracting the smaller entry from the larger this gives, for example,

$$\begin{aligned}(741, 715) &= (715, 26)\\ &\qquad [= (26, 715 - 26) = (26, 715 - 2\cdot 26) = \cdots = (26, 715 - 27\cdot 26)]\\ &= (26, 13) = (13, 0) = 13.\end{aligned}$$

Here is a tidier version of the same calculation:

$$\begin{aligned}\underline{741} &= \underline{715}\cdot 1 + \underline{26}\\ 715 &= 26\cdot 27 + \underline{13},\\ 26 &= 13\cdot 2 + \underline{0},\end{aligned}$$

in which the 27 bracketed steps in the first calculation have been compressed into one. Taking successive pairs of underlined integers from the second calculation gives the essence of the first, and the second is simply the repeated application of the division theorem. In general, this process is called the **Euclidean algorithm**; for $b \neq 0$, it is described by a set of equations such as this:

$$\begin{aligned}a &= bq_1 + r_1, & 0 &< r_1 < |b|,\\ b &= r_1q_2 + r_2, & 0 &< r_2 < r_1,\\ &\vdots\\ r_{k-3} &= r_{k-2}q_{k-1} + r_{k-1}, & 0 &< r_{k-1} < r_{k-2},\\ r_{k-2} &= r_{k-1}q_k + r_k, & 0 &< r_k < r_{k-1}\\ r_{k-1} &= r_kq_{k+1}.\end{aligned} \tag{2}$$

The remainder could be 0 at any stage, of course; if $r_1 = 0$ then $(a, b) = |b|$, while if $r_1 \neq 0$ then k, by definition, is the index of the last nonvanishing remainder, r_k, and we assert that $(a, b) = r_k$. Indeed, since the GCD exists,

$$\begin{aligned}(a, b) &= (a - bq_1, b)\\ &= (b, r_1) = (b - r_1q_2, r_1)\\ &= (r_1, r_2) = \cdots\\ &= (r_{k-1}, r_k) = (r_{k-1} - r_kq_{k+1}, r_k)\\ &= (r_k, 0) = r_k.\end{aligned}$$

(The algorithm always terminates, since the r_i form a strictly decreasing sequence of positive integers.)

It is a simple matter to reverse this whole argument and make Theorem 2.1 depend on the Euclidean algorithm rather than on the unique factorization theorem. For reading equations (2) from the bottom up, we see that $r_k \mid r_{k-1}$, and hence $r_k \mid r_{k-2}, \ldots$, and hence $r_k \mid a$ and $r_k \mid b$, so that (i) holds with $d = r_k$. Similarly, (ii) holds with $d = r_k$, as is immediately seen by reading equations (2) from the top down. Thus d exists, and Theorem 2.1 follows as before.

There is still another way of proving Theorem 2.1, involving an idea of considerable importance in domains other than $\mathbf{Z}$. It depends on the observation that if $d \mid a$ and $d \mid b$, then $d \mid (ax + by)$ for all $x, y \in \mathbf{Z}$, or rather, on the more exact statement that the largest common factor of all the numbers $ax + by$, for x and y in $\mathbf{Z}$, is precisely (a, b).

Theorem 2.2 *If a and b are integers, not both zero, then the smallest positive integer of the form $ax + by$, with $x, y \in \mathbf{Z}$, has properties* (i)–(iii), *and thus is* (a, b). *The set of all integers $ax + by$ is exactly the set of all integral multiples of* (a, b).

Proof. Let d be the smallest positive integer of the form $ax + by$, say $d = ax_0 + by_0$. (x_0 and y_0 might not be unique, of course.) Obviously d has properties (ii) and (iii). Suppose (i) is false; by symmetry, we may suppose $d \nmid a$. Then $a = dq + r$, with $0 < r < d$, and then

$$r = a - dq = a - (ax_0 + by_0)q = a(1 - qx_0) + b(-qy_0),$$

contradicting the minimality of d. Thus (i) is also true. Hence (a, b) exists and is the number d. Obviously, every number $ax + by$ is a multiple of d. Furthermore, every multiple of d is of the form $ax + by$, since $md = a(mx_0) + b(my_0)$, for all m. ∎

In proving Theorem 2.2 we have proved Theorem 2.1 also (at least the hard part of it—uniqueness is separate). This proof is elegant, but it is also nonconstructive: no method is provided for finding coefficients corresponding to the smallest positive element of the set $\{ax + by; x, y \in \mathbf{Z}\}$. But this again can be obtained from the Euclidean algorithm, knowing that $d = r_k$. Thus in the numerical example used earlier,

$$\begin{aligned} 13 &= 715 - 26 \cdot 27 = 715 - (741 - 715 \cdot 1)27 \\ &= -27 \cdot 741 + 28 \cdot 715, \end{aligned}$$

so in this case a possible choice is $x_0 = -27$, $y_0 = 28$.

Corollary 1 *The GCD of two integers (not both zero) can be expressed as a linear combination of them, with coefficients in* $\mathbf{Z}$.

Corollary 2 *If* $(a, b) = 1$ *and* $a \mid bc$, *then* $a \mid c$.

For by Corollary 1, $ax + by = 1$ for some $x, y \in \mathbf{Z}$, and so $acx + bcy = c$; since $a \mid ac$ and $a \mid bc$, also $a \mid c$. ■

Corollary 2 generalizes the lemma (Theorem 1.3) we used in proving the unique factorization theorem, to the effect that if p is prime, $p \mid bc$, and $p \nmid b$, then $p \mid c$; in Corollary 2 we have replaced p by a nonzero integer a, and the condition $p \nmid b$ by $(a, b) = 1$. Thus the unique factorization theorem, Theorem 1.2, can be deduced from Corollary 2, just as it was from Theorem 1.3, and we have gone full circle: we began this chapter by deducing Theorem 2.1 from Theorem 1.2, and now we have made the opposite deduction.

A small warning: The Euclidean algorithm is superficially similar to that developed in Chapter 1 to find the generalized radix representation

$$k = a_0 + a_1 m_1 + a_2 m_1 m_2 + \cdots.$$

The reader should be clear in his mind as to the difference between them.

PROBLEMS

1. Evaluate $(4655, 12075)$, and express the result as a linear combination of 4655 and 12075 with coefficients in $\mathbf{Z}$.

†2. Show that if $(a, b) = 1$, then $(a - b, a + b) = 1$ or 2. Exactly when is the value 2?

†3. Show that if $a + b \neq 0$, $(a, b) = 1$ and p is an odd prime, then

$$\left(a + b, \frac{a^p + b^p}{a + b}\right) = 1 \text{ or } p.$$

[*Hint:* Write a^p as $(a + b - b)^p$ and use the binomial theorem.]

†4. Show that if $b \mid a$ and $c \mid a$ and $(b, c) = 1$, then $bc \mid a$.

5. Show that if $(b, c) = 1$, then $(a, bc) = (a, b)(a, c)$, and that $(bx + cy, bc) = (b, y)(c, x)$ for all integers x and y.

6. The GCD of $n \geq 2$ integers $a_1, \ldots, a_n$, not all 0, is defined as that positive common divisor which is divisible by every common divisor; it is denoted by $(a_1, \ldots, a_n)$. Order the a's so that $a_1 \neq 0$, and define

$$D_2 = (a_1, a_2), \quad D_3 = (D_2, a_3), \quad \ldots, \quad D_n = (D_{n-1}, a_n).$$

Show that $(a_1, \ldots, a_n) = D_n$.

7. Show that the assertion "$(a_1, \ldots, a_n) = 1$" is not the same as "$(a_i, a_j) = 1$ whenever $1 \le i < j \le n$." [Vocabulary: In the first case we say that the a_i are relatively prime, in the second that they are **relatively prime in pairs.**]
8. Show that the sum of two reduced fractions a/b and c/d is not an integer unless $b = d$.
9. Decide whether each of the following statements is true or false, and give a proof or counterexample:
   a) $(a, b) = (a, c)$ implies $(a^2, b^2) = (a^2, c^2)$.
   b) $(a, b) = (a, c)$ implies $(a, b) = (a, b, c)$.
   c) $\{p \mid (a^2 + b^2)$ and $p \mid (b^2 + c^2)\}$ implies $p \mid (a^2 + c^2)$.
10. Let a, m, n be positive integers with $a \ge 2$ and $n \ge m$. With the help of the identity
$$a^n - 1 = (a^m - 1)(a^{n-m} + \cdots + a^{n-km}) + a^{n-km} - 1,$$
or otherwise, show that $(a^n - 1, a^m - 1) = a^{(n,m)} - 1$.
11. Show that the fraction
$$\frac{a_1 + a_2}{b_1 + b_2}$$
is in reduced form if $a_1 b_2 - a_2 b_1 = \pm 1$.
12. In the Euclidean algorithm (2), show that each nonzero remainder r_m $(m \ge 2)$ is less than $\frac{1}{2} r_{m-2}$. (Consider separately the cases in which r_{m-1} is less than, equal to, or greater than $\frac{1}{2} r_{m-2}$.) Deduce that the number of steps in the algorithm is less than
$$\frac{2 \log b}{\log 2} = (2.88\ldots) \log b,$$
where b is the larger of the two numbers whose GCD is being found. (Here and elsewhere, "log" means the natural logarithm.) This theorem, with a somewhat smaller constant factor, is due to G. Lamé (1795–1870). It is probably the first theorem ever proved about what is now known as "computational complexity."
13. a) Prove the following modification of Theorem 1.1: If $a \ne 0$, there are unique integers q and r such that
$$b = aq + r, \qquad -\tfrac{1}{2}|a| < r \le \tfrac{1}{2}|a|.$$
The Euclidean algorithm that ensues is called the "least remainder algorithm." Use it to evaluate (4655, 12075).
    b) Show that the number of steps in the least remainder algorithm is at most $(\log b)/\log 2$.
14. If you know programming, write programs for the Euclidean and least remainder algorithms.

2.2 UNIQUE FACTORIZATION IN OTHER DOMAINS

In the course of this book we shall need to know that several other integral domains besides **Z** have unique factorization. Each case could be handled individually, but by proving a general theorem at this stage we can deduce all needed cases as simple corollaries, and at the same time bring more clearly into view some essential features of the proofs already given for $\mathbf{Z}$.

A domain D is said to be **Euclidean** if to each nonzero element $a \in D$ there corresponds a nonnegative integer $s(a) \in \mathbf{Z}$, such that for every two nonzero elements $a, b \in D$,

i) $s(ab) \geq s(a)$, with strict inequality unless b is a unit (an invertible element) in D;

ii) there are $q, r \in D$ such that $a = bq + r$, where either $r = 0$ or $s(r) < s(b)$.

Note that $\mathbf{Z}$ itself is a Euclidean domain, since $s(a) = |a|$ has the required properties.

For a second example of a Euclidean domain, consider the set of what are called the **Gaussian integers**, consisting of the complex numbers $\alpha = a + bi$, where $i^2 = -1$ and $a, b \in \mathbf{Z}$. Since the set $\mathbf{C}$ of *all* complex numbers is a field (and so in particular is a domain), we can verify that the set of Gaussian integers, which we shall denote by $\mathbf{Z}[i]$, is a domain simply by noting that it is closed under addition and multiplication and contains 0 and ± 1. For then all the other required properties are inherited from those of $\mathbf{C}$. The units of $\mathbf{Z}[i]$ are the invertible elements, and for $\alpha \neq 0$,

$$\frac{1}{\alpha} = \frac{1}{a + bi} = \frac{a - bi}{a^2 + b^2} = \frac{a}{a^2 + b^2} - \frac{b}{a^2 + b^2}\, i.$$

This is an element of $\mathbf{Z}[i]$ if and only if $(a^2 + b^2) \mid a$ and $(a^2 + b^2) \mid b$. Now $a^2 + b^2 > \max(|a|, |b|)$ if $a \neq 0$ and $b \neq 0$, or if $|a| > 1$, or if $|b| > 1$, so it must be that either $a = \pm 1$ and $b = 0$, or that $a = 0$ and $b = \pm 1$. Thus the units are ± 1 and $\pm i$.

It is now easy to verify that the quantity

$$s(\alpha) = |\alpha|^2 = a^2 + b^2 \qquad \text{for } \alpha = a + bi \in \mathbf{Q}[i]$$

has properties (i) and (ii). Property (i) is left to the reader. To verify (ii), recall that the equation $\alpha = \beta\kappa + \rho$ is supposed to reveal κ as some sort of approximation to the exact quotient α/β, although the latter may itself not lie in $\mathbf{Z}[i]$. This provides the key. Form the quotient of $\alpha = a + bi$ and $\beta = c + di$, in $\mathbf{C}$,

$$\begin{aligned}\frac{\alpha}{\beta} = \frac{a + bi}{c + di} &= \frac{(a + bi)(c - di)}{c^2 + d^2} = \frac{ac + bd}{c^2 + d^2} + \frac{bc - ad}{c^2 + d^2}\, i,\\ &= A + Bi, \text{ say,}\end{aligned}$$

where A and B are rational numbers, not necessarily integers. Let x and y be elements of $\mathbf{Z}$ such that $|A - x| \leq \frac{1}{2}$ and $|B - y| \leq \frac{1}{2}$. Then

$$\begin{aligned}\left|\frac{\alpha}{\beta} - (x + iy)\right|^2 &= |(A - x) + (B - y)i|^2\\ &= (A - x)^2 + (B - y)^2 \leq \tfrac{1}{4} + \tfrac{1}{4} < 1.\end{aligned}$$

Hence if we put

$$\kappa = x + iy, \qquad \rho = \alpha - \beta\kappa,$$

then $\kappa, \rho \in Z[i]$, and if $\rho \neq 0$, then

$$s(\rho) = s(\alpha - \beta\kappa) = s(\beta)s\left(\frac{\alpha}{\beta} - \kappa\right) < s(\beta).$$

This shows that (ii) holds, so that $\mathbf{Z}[i]$ is a Euclidean domain. (Note that κ and ρ may not be unique. But that was not required.)

Extending what was done in $\mathbf{Z}$, we say that a nonzero element a of an arbitrary domain D **divides** a second element b if there exists $c \in D$ such that $b = ac$. The nonzero elements of D can then be partitioned into three classes: the **units**, which divide 1, the **prime** elements, which have no nontrivial divisors (the trivial divisors of a are the units u and the elements au, called the **associates** of a) and the **composite** elements with nontrivial divisors. In a field, every nonzero element is a unit and there are no primes. The situation is quite different in a Euclidean domain.

Theorem 2.3 *If D is a Euclidean domain, then every nonzero element of D either is a unit or can be represented as a finite product of primes.*

Proof. If the theorem were false for some domain D, there would be a nonunit $a \in D$ with minimal value of $s(a)$, having no finite decomposition as product of (one or more) primes. Then a itself is not prime, so by definition it has a nontrivial factorization $a = bc$, in which neither b nor c is a unit. Then by (i), $s(b) < s(a)$ and $s(c) < s(a)$, and not both b and c could have prime factorizations or else a would have. This contradicts the minimality of $s(a)$. ■

Theorem 2.4 *Let D be a Euclidean domain. Then if $a, b \in D$ are not both 0, the set of all elements $ax + by$, with $x, y \in D$, coincides with the set of all multiples of a fixed element $d \in D$, and d is unique except for a unit factor. Moreover, $d \mid a$ and $d \mid b$; and for $d_1 \in D$, if $d_1 \mid a$ and $d_1 \mid b$, then $d_1 \mid d$.*

Proof. Let S be the set of all linear combinations $ax + by$, and let d be any nonzero element of S with minimal value of $s(d)$. If $c \in S$, then by condition (ii) above there are $q, r \in D$ such that $c = dq + r$, and either $r = 0$ or $s(r) < s(d)$. But since $c, d \in S$, also $r = c - dq \in S$, so $r = 0$; hence $d \mid c$. Thus S is the set of all multiples of d, with coefficients in D, which we write as $S = \{md;\ m \in D\}$. If also $S = \{md';\ m \in D\}$, then $d' = m_1d$ and $d = m_2d'$, whence $d' = m_1m_2d'$, $m_1m_2 = 1$, and m_1, m_2 are units.

Since $a \in S$ and $b \in S$, $d \mid a$ and $d \mid b$. Finally, writing $d = ax_0 + by_0$, it is clear that if $d_1 \mid a$ and $d_1 \mid b$, then $d_1 \mid d$. ■

As before, d is called a GCD of a and b, and denoted by (a, b); it is now defined only to within a unit factor.

Theorem 2.5 *Suppose $a, b, c \in D$, where D is Euclidean. If $(a, b) = 1$ and $a \mid bc$, then $a \mid c$.*

The proof is identical with that of Corollary 2 of Theorem 2.2.

Corollary *If $p, p_1, \ldots, p_r$ are primes in D and $p \mid p_1 \cdots p_r$, then $p = up_i$ for some i, where u is a unit of D.*

Theorem 2.6 *If D is a Euclidean domain and $a \in D$ is neither 0 nor a unit, then the representation of a as a product of primes of D is unique except for the order in which the primes occur and the presence of unit factors.*

Proof. By Theorem 2.3, every nonunit $a \neq 0$ has at least one prime factorization. If the present theorem were false, there would be such an a having two distinct factorizations, and for which $s(a)$ is minimal. Suppose

$$a = p_1 p_2 \cdots p_r = p_1' p_2' \cdots p_s'. \tag{3}$$

By the corollary to Theorem 2.5, $p_1' = up_i$ for some i, so

$$\frac{a}{p_i} = \frac{p_1 \cdots p_r}{p_i} = up_2' \cdots p_s'. \tag{4}$$

This contradicts the minimality of $s(a)$ unless the two factorizations in (4) are the same, in the sense of the theorem, and in that case the same was already true in (3). ∎

Thus, speaking briefly, Theorem 2.6 asserts that every Euclidean domain is a **unique factorization domain**. The converse is not true. Weaker conditions than being Euclidean are known, which imply that a domain has unique factorization, but their study would carry us too far afield.

As an immediate consequence of what has been proved in this section, we have the following result.

Theorem 2.7 *The Gaussian integers form a unique factorization domain.*

Another instance that we shall need is this.

Theorem 2.8 *Let F be a field, and let $F[x]$ be the set of all polynomials in one variable over F (i.e., having coefficients in F). Then $F[x]$ is a domain, when addition and multiplication are defined in the usual way, and in fact $F[x]$ is a unique factorization domain.*

Proof. By the "usual" definitions of addition and multiplication we mean those given in Section 1.2, Problem 12. Verifying that $F[x]$ is a domain is straightforward, and it is left to the reader. The nonzero elements of F itself belong to $F[x]$ and are its units. (That they are units is clear. That there are no others is seen as follows: if the degree f of a nonzero element $f \in F[x]$ is defined in the usual way, then

$$\partial(fg) = \partial f + \partial g; \tag{5}$$

so $f(x)g(x) = 1$ implies $\partial f + \partial g = 0$, which implies $\partial f = \partial g = 0$ since degrees are nonnegative.) The prime elements of $F[x]$ are usually referred to as the **irreducible** polynomials over F; x itself is an example of one.

The ring $F[x]$ is a Euclidean domain, for which the associated integer-valued function is $s(f(x)) = \partial f$. Property (i) of the s-function follows immediately from (5). A proof of property (ii) is sketched in Section 1.4, Problem 8, and the details can easily be filled in. ■

More generally, it can be shown that if D is a unique factorization domain, then so is $D[x]$, but we shall not need this. If R is a Ring but not a domain, so that there are nonzero elements $a, b \in R$ such that $ab = 0$, then factorization in $R[x]$ is definitely not unique—for example, $(x - a)(x - b) = x^2 - (a + b)x = x(x - a - b)$. We shall meet this situation in the next chapter.

PROBLEMS

1. Verify that in the Gaussian integers $s(\alpha) = |\alpha|^2$ has property (i) of a Euclidean domain, as asserted in the text.
2. Is a field a Euclidean domain?
3. The so-called **sieve of Eratosthenes** is an algorithm for singling out the primes from among the set of integers k with $|k| \le n$, for arbitrary $n > 0$. It depends on the fact that if $|m| > 1$, and if m has no divisor d with $1 < d \le \sqrt{|m|}$, then m must be prime. (Prove this fact.) First, the smallest integer larger than 1—namely 2—must be prime, and now we know all the primes p with $|p| \le 2$. Suppose we know all the primes p with $|p| \le n$. Then the primes in the set of m with $n < |m| \le n^2$ are the integers left in this set after eliminating all the multiples of those known primes. For example, knowing that 2, 3, and 5 are all the primes in the interval $2 \le k \le 5$, we can eliminate all the multiples of 2, of 3, and of 5 from among 6, 7, . . . , 25, and find 7, 11, 13, 17, 19, 23 as the next block of primes. Extend all of this to $\mathbf{Z}[i]$ and so find all primes γ in this domain for which $|\gamma|^2 \le 9$.
4. Show that the definition given for a Euclidean domain is redundant: the weak inequality in (i), together with (ii), imply that $s(ab) > s(a)$ if b is not a unit of D. [*Hint:* Start from $a = (ab)q + r$.]
5. According to Theorem 2.4, if D is a Euclidean domain, and a and b are relatively prime elements of D, then there are $m, n \in D$ such that $ma + nb = 1$.
   a) Show that 2 and x are relatively prime elements of $\mathbf{Z}[x]$.
   b) Conclude that $\mathbf{Z}[x]$ is not a Euclidean domain. (Nevertheless, according to the remark at the end of the text, above, $\mathbf{Z}[x]$ is a unique factorization domain.)
6. Show that $(3 + \sqrt{10})^n$ is a unit in $\mathbf{Z}[\sqrt{10}]$ for every $n \in \mathbf{Z}$.

   The remaining problems concern the domain $\mathbf{Z}$.
7. Show that if the reduced fraction $a/b \in \mathbf{Q}$ is a root of the equation
$$c_0 + c_1x + \cdots + c_nx^n = 0, \quad c_k \in \mathbf{Z} \text{ for } 0 \le k \le n, \qquad c_n \ne 0,$$

then $a \mid c_0$ and $b \mid c_n$. In particular, show that if m is an integer, then $\sqrt[n]{m}$ is rational if and only if it is an integer. More generally, a zero of a monic polynomial is irrational or is an integer.

8. Show that the following identity is formally correct (ignoring problems connected with convergence and rearrangement of terms):

$$\sum_{k=0}^{\infty} \frac{1}{2^{2k}} \cdot \sum_{k=0}^{\infty} \frac{1}{3^{2k}} \cdot \sum_{k=0}^{\infty} \frac{1}{5^{2k}} \cdots = \sum_{n=1}^{\infty} \frac{1}{n^2}.$$

The denominators on the left are the even powers of the primes. If you know the requisite analysis, give a complete proof.

9. Show that for $n > 1$, the partial sum

$$1 + \frac{1}{2} + \frac{1}{3} + \cdots + \frac{1}{n}$$

of the harmonic series is not an integer. [*Hint:* Think about the powers of 2.]

10. Given positive integers a and b such that $a \mid b^2$, $b^2 \mid a^3$, $a^3 \mid b^4$, $b^4 \mid a^5, \ldots$, prove that $a = b$.

11. Show that every large integer has a large prime-power factor. That is, if $P(n)$ designates the largest number p^a which divides n, then $\lim_{n\to\infty} P(n) = \infty$.

12. If $1 < a_1 < \cdots < a_k \leq x$ and no a_i divides the product of all the rest, then $k \leq \pi(x)$. [*Hint:* Find a prime corresponding to each a_i.]

13. Suppose $1 \leq a_1 < \cdots < a_{n+1} \leq 2n$. Show that $a_i \mid a_j$ for some i and j with $i \neq j$. [*Hint:* Consider the maximal odd divisors of the a_i.]

2.3 THE LINEAR DIOPHANTINE EQUATION

By exploiting the connection we have established between the GCD of two integers and linear combinations of them, it is easy to analyze the linear Diophantine equation in two unknowns x and y,

$$ax + by = c, \qquad a, b, c, x, y \in \mathbf{Z}. \tag{6}$$

First suppose that $(a, b) = 1$. Then we know that for suitable $x, y \in \mathbf{Z}$, $ax + by = 1$, so cx, cy give a solution of (6). If x, y and x_0, y_0 are any two distinct solutions of (6), then $a(x - x_0) + b(y - y_0) = c - c = 0$. Since $(a, b) = 1$, the equation

$$\frac{a}{b} = -\frac{y - y_0}{x - x_0}$$

shows that for some $t \in \mathbf{Z}$,

$$y - y_0 = -at, \qquad x - x_0 = bt. \tag{7}$$

Contrariwise, if x_0, y_0 satisfy (6) and t is an integer (including 0), then the x, y determined by (7) also satisfy (6). Hence (7) provides a general solution of (6), in this case. As a passing remark to readers who have studied linear systems of algebraic or differential equations, we remark that a general solution of (6) has thus been given as the sum of a particular solution of the inhomogeneous equation and a general solution of the homogeneous equation, $ax + by = 0$.

Now suppose that in (6), $(a, b) = d$. If $d \nmid c$, there are obviously no solutions. If $d \mid c$, then (6) has exactly the same set of solutions as the simplified equation

$$\frac{a}{d}x + \frac{b}{d}y = \frac{c}{d},$$

and since $(a/d, b/d) = 1$, we already know these. Hence we have the following theorem.

Theorem 2.9 *A necessary and sufficient condition that the equation* (6) *have a solution x, y is that $d \mid c$, where $d = (a, b)$. If there is one solution—say x_0, y_0—there are infinitely many; they are exactly the numbers*

$$x = x_0 + \frac{b}{d}t, \quad y = y_0 - \frac{a}{d}t, \qquad t \in \mathbf{Z}. \tag{8}$$

This leaves the problem of actually finding a solution x_0, y_0 of the reduced equation in which $(a, b) = 1$. Various devices are available, of which the simplest using only the tools developed so far is to exploit the Euclidean algorithm once again. Omitting the last equation from (2), the others can be rewritten, in reverse order, in the form

$$\begin{aligned}
(a, b) = r_k &= r_{k-2} - r_{k-1}q_k,\\
r_{k-1} &= r_{k-3} - r_{k-2}q_{k-1},\\
&\vdots\\
r_3 &= r_1 - r_2q_3,\\
r_2 &= b - r_1q_2,\\
r_1 &= a - bq_1.
\end{aligned}$$

The first equation gives (a, b) as a linear combination of r_{k-2} and r_{k-1}; eliminating r_{k-1} with the help of the second equation gives (a, b) as a linear combination of r_{k-2} and r_{k-3}, etc. Eventually, (a, b) is a linear combination of a and b, and the coefficients lead to a particular solution x_0, y_0 of (6).

Example To decide whether the equation $69x + 39y = 15$ is solvable, we first find (69, 39):

$$\begin{aligned}
69 &= 39 \cdot 1 + 30,\\
39 &= 30 \cdot 1 + 9,\\
30 &= 9 \cdot 3 + 3,\\
9 &= 3 \cdot 3.
\end{aligned}$$

Thus $(69, 39) = 3$, and since $3 \mid 15$, the equation is solvable, and is equivalent to the reduced equation $23x + 13y = 5$. From the Euclidean algorithm,

$$\begin{aligned} 3 &= 30 - 9 \cdot 3 = 30 - (39 - 30 \cdot 1)3 \\ &= 4 \cdot 30 - 3 \cdot 39 = 4(69 - 39 \cdot 1) - 3 \cdot 39 \\ &= 4 \cdot 69 - 7 \cdot 39, \end{aligned}$$

so that $4, -7$ is a solution of $69x + 39y = 3$, or of $23x + 13y = 1$. Hence $x_0 = 4 \cdot 5$, $y_0 = -7 \cdot 5$ is a solution of $69x + 39y = 15$, so a general solution of this last equation is

$$x = 20 + 13t, \qquad y = -35 - 23t.$$

The parametric equations in Theorem 2.9 have a simple geometric interpretation. If x and y are thought of as real rather than integer variables, the equation $ax + by = c$ is that of a line in the xy-plane, of slope $-a/b$. Consecutive integer values of t in (8) give the equally spaced points on this line obtained by starting from an integral point and repeatedly moving b/d units horizontally and a/d units vertically. The distance between successive points is $(a^2 + b^2)^{1/2}/d$.

Here is a simple application of this geometry. If $a, b, c > 0$, the line $ax + by = c$ cuts through the first quadrant and has intercepts at $(c/a, 0)$ and $(0, c/b)$. The distance between intercepts is

$$\sqrt{\left(\frac{c}{a}\right)^2 + \left(\frac{c}{b}\right)^2} = \frac{c}{ab}\sqrt{a^2 + b^2},$$

so if $d \mid c$, there is surely an integral point x, y on the line and in the first quadrant if

$$\frac{c}{ab}\sqrt{a^2 + b^2} > \frac{1}{d}\sqrt{a^2 + b^2},$$

or $c > ab/d$. This shows, for example, that if $a > 0$, $b > 0$, and $(a, b) = 1$, then every sufficiently large integer c is a linear combination, with positive integral coefficients, of a and b.

The linear equation in more than two variables arises only infrequently, so we treat it somewhat cavalierly. The equation

$$a_1x_1 + \cdots + a_nx_n = c \qquad (a_1 \cdots a_n \neq 0) \tag{9}$$

clearly is solvable only if $D_n \mid c$, where $D_m = \mathrm{GCD}(a_1, \ldots, a_m)$ for $2 \leq m \leq n$. We induct on n. Suppose that this necessary condition is also sufficient, for every equation in fewer than n variables; this is so for $n = 3$, by Theorem 2.9. We know that the set of linear combinations $a_{n-1}x_{n-1} + a_nx_n$ is identical with the set of multiples of (a_{n-1}, a_n), and from a particular solution of $a_{n-1}x_{n-1} + a_nx_n =$

(a_{n-1}, a_n) we can obtain a general solution of $a_{n-1}x_{n-1} + a_n x_n = (a_{n-1}, a_n)u$, where u is a parameter. Then (9) reduces to

$$a_1 x_1 + \cdots + a_{n-2}x_{n-2} + (a_{n-1}, a_n)u = c,$$

and c is divisible by $(a_1, \ldots, a_{n-2}, (a_{n-1}, a_n)) = D_n$, so this indeterminate equation in $n - 1$ unknowns is solvable, by the induction hypothesis. Hence the condition $D_n \mid c$ is necessary and sufficient for solvability, and the proof indicates how to go about finding a general solution.

PROBLEMS

1. Find a general solution of the equation

$$2072x + 1813y = 2849.$$

2. Let m and n be positive integers, with $m \le n$, and let $x_0, x_1, \ldots, x_k$ be all the distinct numbers among the two sequences

$$\frac{0}{m}, \frac{1}{m}, \ldots, \frac{m}{m} \quad \text{and} \quad \frac{0}{n}, \frac{1}{n}, \ldots, \frac{n}{n},$$

arranged so that $x_0 < x_1 < \cdots < x_k$. Describe k as a function of m and n. What is the shortest distance between successive x's? How are the other distances related to this shortest one?

3. Find all solutions of $19x + 20y = 1909$ with $x > 0$, $y > 0$.
4. Suppose $(a, b) = 1$, $a > 0$, $b > 0$, and let x_0, y_0 be any integral solution of the equation $ax + by = c$. Find a necessary and sufficient condition, possibly depending on a, b, c, x_0, y_0, that the equation have a solution with $x > 0$, $y > 0$.
5. Show that if $a, b, c \in \mathbf{Z}^+$ and $(a, b) = 1$, then the number n of nonnegative solutions of $ax + by = c$ satisfies the inequality

$$\frac{c}{ab} - 1 < n \le \frac{c}{ab} + 1.$$

6. a) Let $N = (a - 1)(b - 1)$, where $a, b \in \mathbf{Z}^+$ and $(a, b) = 1$. Show that every integer $c \ge N$ is representable in the form $c = ax + by$ with $x, y \ge 0$, while $c = N - 1$ is not so representable.
   b) Show that exactly half the integers $0, 1, \ldots, N - 1$ are so representable.

[*Remark:* The problem of finding the analogue of $(a - 1)(b - 1)$ for expressions in more than two variables is still unsolved.]

7. Suppose that $(a, b, c) \mid d$ and that x_0, u_0, y_0, z_0 satisfy the equations

$$ax_0 + (b, c)u_0 = d,$$
$$by_0 + cz_0 = (b, c).$$

Show that a general solution of $ax + by + cz = d$ is given by

$$x = x_0 + \frac{(b, c)}{(a, b, c)} t,$$

$$y = y_0 u_0 - \frac{a y_0}{(a, b, c)} t + \frac{c}{(b, c)} s,$$

$$z = z_0 u_0 - \frac{a z_0}{(a, b, c)} t - \frac{b}{(b, c)} s,$$

for $s, t \in \mathbf{Z}$.

8. (Requires familiarity with $n \times n$ determinants.) From the $(n - 1) \times n$ array of integers

$$\begin{matrix} a_{11} & \cdots & a_{1n} \\ & \cdots & \\ a_{n-1,1} & \cdots & a_{n-1,n} \end{matrix}$$

form the determinants $d_1, \ldots, d_n$ of the arrays resulting by deleting successively the 1st, ..., nth column. Show that if $(d_1, \ldots, d_n) = 1$, then there are integers $a_{n1}, \ldots, a_{nn}$ such that $\det(a_{ij})_{i,j=1}^n = 1$.

9. Here is a technique for solving a linear Diophantine equation. Solve for the unknown, say x, whose coefficient a is numerically smallest, obtaining that unknown as a linear polynomial over $\mathbf{Q}$ in the remaining variables. By splitting each coefficient into an integer plus a proper fraction, obtain x as a linear polynomial over $\mathbf{Z}$ plus a "fractional part," a linear polynomial over $\mathbf{Q}$ with coefficients numerically smaller than 1. Since $x \in \mathbf{Z}$, the fractional part must be an integer, say t, which leads to a new linear equation in which the coefficient of t is a, and a is now maximal rather than minimal. Iterate, and eventually eliminate intermediate variables. Apply this method to find general solutions of

   a) $811x - 547y = 39$; b) $53x - 103y - 150z = 13$.

10. Under what circumstances do two arithmetic progressions $\{b_1 k + r_1\}$ and $\{b_2 l + r_2\}$ intersect?

2.4 THE LEAST COMMON MULTIPLE

In complete analogy with the definition given in Theorem 2.1 for the GCD, we define the **least common multiple** or **LCM** of two integers a and b, not both 0, as that positive common multiple of a and b such that every common multiple of a and b is a multiple of it. We designate it by $[a, b]$. In the notation used in the proof of Theorem 2.1, it is clear that for $ab \neq 0$,

$$[a, b] = p_1^{\max(\alpha_1, \beta_1)} \cdots p_s^{\max(\alpha_s, \beta_s)}.$$

No algorithm corresponding to that for finding the GCD is necessary for the LCM, because the latter can be evaluated immediately when the GCD is known, using the relation

$$[a, b] = \frac{|ab|}{(a, b)}.$$

This identity follows from the fact that

$$|ab| = p_1^{\alpha_1+\beta_1} \cdots p_s^{\alpha_s+\beta_s},$$

together with the observation that for any two real numbers x and y,

$$\min(x, y) + \max(x, y) = x + y.$$

PROBLEMS

1. Evaluate [198061, 231896].
2. Show that if g and m are positive integers, there are integers a and b such that $(a, b) = g$ and $[a, b] = m$ if and only if $g \mid m$.
3. a) Show that for any three real numbers x, y, z,

$$\min(x, \max(y, z)) = \max(\min(x, y), \min(x, z)).$$

[*Hint:* Since y and z enter symmetrically, there is no loss in generality in supposing that $y \geq z$.]

b) Show that $(a, [b, c]) = [(a, b), (a, c)]$.

c) It is not entirely accidental that if GCD (as an operation on two integers) is replaced by multiplication and LCM by addition, the relation in (b) becomes the ordinary distributive law, $a(b + c) = ab + ac$. On the other hand, show that the "dual" relation

$$[a, (b, c)] = ([a, b], [a, c])$$

also holds, although the analogue $a + bc = (a + b)(a + c)$ is generally false.

4. a) Let $x_1, \ldots, x_m$ be real numbers. Show that

$$\max(x_1, \ldots, x_m) = \sum_{1 \leq i \leq m} x_i - \sum_{1 \leq i < j \leq m} \min(x_i, x_j) + \sum_{1 \leq i < j < k \leq m} \min(x_i, x_j, x_k) - \cdots + (-1)^{m-1} \min(x_1, \ldots, x_m),$$

where, on the right side, the summands are successively the minima of the various 1-element, 2-element, ..., m-element subsets of $\{x_1, \ldots, x_m\}$. [*Hint:* There is no loss in generality in supposing that $x_1 \geq x_2 \geq \cdots \geq x_m$, and, if equality should occur, in always choosing the x with larger subscript as the minimum. Then $\max(x_1, \ldots, x_m) = x_1$, and x_1 occurs once in the first sum on the right, but is not the value of any later min symbol. If $m \geq 3$, x_3 is a term-value once in the first sum, twice in the second, once in the third, and never again, and $1 - 2 + 1 = 0$.]

b) Show that for nonzero integers $a_1, \ldots, a_m$,

$$[a_1, \ldots, a_m] = \frac{\prod_{1 \leq i \leq m} a_i \cdot \prod_{1 \leq i < j < k \leq m} (a_i, a_j, a_k) \cdots}{\prod_{1 \leq i < j \leq m} (a_i, a_j) \cdot \prod_{1 \leq i < j < k < l \leq m} (a_i, a_j, a_k, a_l) \cdots},$$

in which the numerator on the right is the product of GCD's of all subsets of $a_1, \ldots, a_m$ with odd numbers of elements, and the denominator is the product of GCD's of subsets with even numbers of elements.

NOTES AND REFERENCES

Section 2.1

Lamé proved that the number of steps in the Euclidean algorithm is less than $5 \log_{10} b = (2.17\ldots) \log b$. For a proof, see Uspensky and Heaslet [1939]. (Lamé was primarily a mathematical physicist. His only other known contributions to number theory were the first proof of Fermat's conjecture for the exponent 7 and a fallacious "proof" for general n. See Section 8.3.)

Section 2.2

Complex numbers had occurred long before Gauss, of course. Some systematic use of them was made in Bombelli's *Algebra* of 1572, mentioned earlier, and the relation $e^{ix} = \cos x + i \sin x$ was known to Euler and others in the eighteenth century. But Gauss was the first to study and use the arithmetic properties of those complex numbers whose real and imaginary parts are integers, in his work (1825, 1831) on quartic residues. (Cf. Chapter 4.) No doubt his use of them helped to make complex numbers respectable in the eyes of other mathematicians.

For more information on unique factorization domains in general, see Herstein [1964] or Jacobson [1975].

Eratosthenes, a contemporary of Archimedes, was a many-sided scholar. He gave a mechanical solution of the problem of duplicating the cube, and he calculated the diameter of the earth with considerable accuracy. Chief librarian of the Museum in Alexandria, he became blind in his old age and committed suicide by starvation.

Section 2.3

Systems of m linear Diophantine equations in n unknowns have also been studied. The definitive theorem is that of G. Frobenius [1878], who proved that a necessary and sufficient condition for solvability is that the rank l, and the GCD of all the l-rowed determinants of the coefficient matrix, are the same as those of the augmented matrix. See Dickson [1919], vol. 2.

The unsolved question mentioned in Problem 6 is also commonly ascribed to Frobenius. There is an extensive literature, some of it covered in LeVeque [1974], vol. 2. In a more recent paper, Erdös and Graham [1972] have obtained upper and lower bounds differing only by a constant factor.

Section 2.4

For more information on the relation between GCD and LCM on the one hand, and multiplication and addition on the other, see, for example, Birkhoff and MacLane [1965], Chap. 11.

3

Congruences and the Ring $\mathbf{Z}_m$

3.1 CONGRUENCE AND RESIDUE CLASSES

The problem of solving the Diophantine equation $ax + by = c$ is just that of finding an $x \in \mathbf{Z}$ such that c and ax leave the same remainder when divided by b, since then $y = (c - ax)/b$ is an integer. Another way of saying this is that c and ax must lie in the same arithmetic progression $\{bk + r\}$. It is hardly surprising that one linear problem should translate into a second (an arithmetic progression being the set of values of a linear function); what is less predictable is the breadth of the spectrum of problems in number theory which in the end reduce to the question of whether certain integers can or do lie in certain arithmetic progressions. Two more examples will illustrate this.

The first example concerns sums of squares. We mentioned in Section 1.1 Fermat's theorem that every positive integer is the sum of four squares. An equally beautiful theorem, also due to Fermat, is that every prime of the form $4k + 1$ is the sum of two squares. (We shall give a proof of this in Chapter 7.) It is easy to see that no prime (in fact no integer, prime or not) of the form $4k + 3$ is of the form $x^2 + y^2$, just by observing that for any two integers k and l,

$$\begin{aligned} (2k)^2 + (2l)^2 &= 4m, \\ (2k)^2 + (2l + 1)^2 &= 4n + 1, \\ (2k + 1)^2 + (2l + 1)^2 &= 4q + 2, \end{aligned} \tag{1}$$

for suitable m, n, q. Thus the complete theorem is that the equation $p = x^2 + y^2$ is solvable if and only if the prime p is 2 or lies in the progression $\{4k + 1\}$.

For a second example consider the Fermat equation $x^n + y^n = z^n$ with $n > 2$. If this equation has a solution and $p \mid n$, then $(x^{n/p})^p + (y^{n/p})^p = (z^{n/p})^p$, so it would suffice to show that no equation $x^p + y^p = z^p$ is solvable, with $xyz \neq 0$, in order to settle the Fermat Problem. The case in which it is further assumed that $p \nmid xyz$—the so-called Case I—is somewhat easier than the general problem; for example, it is known that there is no solution in Case I for $p < 3 \cdot 10^9$. The problem is especially simple when $p = 3$; for then each of x, y, z is of one of the forms $3k \pm 1$, and $(3k \pm 1)^3 = 9(3k^3 \pm 3k^2 + k) \pm 1$, so (allowing independent choices of the $\pm$ signs),

$$(2) \qquad \begin{aligned} x^3 + y^3 - z^3 &= (9r \pm 1) + (9s \pm 1) - (9t \pm 1) \\ &= 9w \pm 1 \text{ or } 9w \pm 3, \end{aligned}$$

which rules out $x^3 + y^3 - z^3 = 0$. This time arithmetic progressions served their purpose and then disappeared from the final result.

Note that in (1) and (2), the quantities $k, l, m, n, q, r, s, t, w$ were of no interest at all—they were only there to aid in describing certain arithmetic progressions. Gauss not only noticed this, he did something about it, in the *Disquisitiones*. For a given integer $m > 1$, he rephrased the assertion "a and b lie in the same arithmetic progression with difference m" as "**a is congruent to b modulo m**," and he wrote "$a \equiv b \pmod{m}$." Thus $a \equiv b \pmod{m}$ means that $m \mid (a - b)$, or that a and b differ only by a multiple of m. What a magnificent invention this was can hardly be appreciated at this point, but it will soon become clear.

For any fixed a, the integers x such that $x \equiv a \pmod{m}$ are those of the form $x = a + mk$, $k \in \mathbf{Z}$. This arithmetic progression will be called a **residue class** (mod m) and, when the **modulus** $m > 1$ is understood, it will be designated by $\bar{a}$. For fixed m, the full set $\mathbf{Z}$ of integers is thus partitioned into m residue classes, $\bar{0}, \bar{1}, \ldots, \overline{m - 1}$: each integer belongs to exactly one of them. (In general, partitioning a set means splitting it into disjoint subsets.) As is the case with every partitioning of any set whatever, "belonging to the same subset" (in this case, same residue class) is an equivalence relation, in the sense of the following theorem.

Theorem 3.1 *Congruence* (mod m) *is reflexive, symmetric, and transitive: for every* $a, b, c \in \mathbf{Z}$,

$a \equiv a \pmod{m}$;
if $a \equiv b \pmod{m}$, *then* $b \equiv a \pmod{m}$;
if $a \equiv b \pmod{m}$ *and* $b \equiv c \pmod{m}$, *then* $a \equiv c \pmod{m}$.

What is more important, because it allows us to do algebra and arithmetic with the residue classes, is the following.

Theorem 3.2 *For fixed* $m > 1$, *if* $a \equiv b \pmod{m}$ *and* $c \equiv d \pmod{m}$, *then* $a + c \equiv b + d \pmod{m}$ *and* $ac \equiv bd \pmod{m}$. *Hence if we define the sum and product of residue classes by the equations* $\bar{a} + \bar{c} = \overline{a + c}$ *and* $\bar{a} \cdot \bar{c} = \overline{ac}$, *then the set* $\mathbf{Z}_m$ *of residue classes* (mod m) *becomes a Ring. The mapping* $\phi a = \bar{a}$ *is a homomorphism of* $\mathbf{Z}$ *onto* $\mathbf{Z}_m$.

Proof. If $m \mid (a - b)$ and $m \mid (c - d)$ then $m \mid ((a + c) - (b + d))$. Also $m \mid (a - b)(c - d)$ and

$$(a - b)(c - d) = ac - bd + b(d - c) + d(b - a),$$

so $m \mid (ac - bd)$. This proves the first sentence.

The meaning of the equation $\bar{a} + \bar{c} = \overline{a + c}$ is this: the sum of the residue classes (mod m) determined by two integers a and c is the residue class containing their sum, $a + c$. There is a possible difficulty lurking here which is brought out by phrasing the definition slightly differently: to add two residue classes (mod m), say C and C', select an element in each, add these integers, and define $C + C'$ to be the residue class C'' containing the latter sum. This gives a unique sum only if (and if) the same C'' results no matter how the elements of C and C' are selected. But that is exactly what is guaranteed by the first sentence of the theorem, which says that if a and b are both in C, and c and d are both in C', then $a + c$ and $b + d$ are in the same residue class. Thus addition of residue classes is uniquely defined, and similarly for multiplication; furthermore, $\mathbf{Z}_m$ is closed under these operations, since sums and products of residue classes (mod m) are again residue classes (mod m).

Most of the other requirements which must be met if $\mathbf{Z}_m$ is to be a Ring are inherited from $\mathbf{Z}$. For example,

$$\begin{array}{llll} a + b = b + a & \text{in } \mathbf{Z} & \text{implies} & \bar{a} + \bar{b} = \bar{b} + \bar{a} \text{ in } \mathbf{Z}_m, \\ a + 0 = a & \text{in } \mathbf{Z} & \text{implies} & \bar{a} + \bar{0} = \bar{a} \quad \text{in } \mathbf{Z}_m, \\ a \cdot 1 = a & \text{in } \mathbf{Z} & \text{implies} & \bar{a} \cdot \bar{1} = \bar{a} \quad \text{in } \mathbf{Z}_m, \end{array}$$

and so on. The only condition that requires some care is that $\bar{1} \neq \bar{0}$, and this is true since $m > 1$ and hence $m \nmid 1$. Thus $\mathbf{Z}_m$ is a Ring.

It is now obvious that the conditions (H) of Section 1.2 are satisfied, so that the map ϕ which sends every $a \in \mathbf{Z}$ into the residue class $\bar{a}$ (mod m) which contains it (more briefly, the map $\phi: a \mapsto \bar{a}$) is a homomorphism of $\mathbf{Z}$ onto $\mathbf{Z}_m$. ■

Students acquainted with the necessary algebra will recognize this theorem as standard fare. For $a \equiv b \pmod{m}$ if and only if $a - b \in m\mathbf{Z}$, where $m\mathbf{Z}$ means the set of all integral multiples of m, and of course $m\mathbf{Z}$ is an ideal. (A nonempty subset I of a Ring R is an **ideal** if I is an additive group and $ka \in I$ whenever $a \in I$ and $k \in R$.) Thus Theorem 3.2 repeats the standard construction of the quotient Ring of a Ring modulo one of its ideals; in the standard notation we could write $\mathbf{Z}/m\mathbf{Z}$ in place of $\mathbf{Z}_m$. No use will be made of this remark, nor of ideals, in this book. For further details, see, for example, Herstein [1964], p. 94ff.

Corollary *If $f(x_1, \ldots, x_n)$ is a polynomial over $\mathbf{Z}$, and $a_j \equiv b_j \pmod{m}$ for $1 \leq j \leq n$, then $f(a_1, \ldots, a_n) \equiv f(b_1, \ldots, b_n) \pmod{m}$.*

One reason for the importance of the residue class rings $\mathbf{Z}_m$ was already exemplified, in disguised form, in the discussion above of the equation $x^3 + y^3 = z^3$. Namely, an arbitrary polynomial Diophantine equation $f(x, y, \ldots, z) = 0$ over $\mathbf{Z}$ has an analogue in each of the rings $\mathbf{Z}_m$, in which the coefficients and variables are interpreted as residue classes (mod m), and a solution of the $\mathbf{Z}$-equation gives a solution of each of the $\mathbf{Z}_m$-equations. Put the other way around; the impossibility of any one (or any collection) of the $\mathbf{Z}_m$-equations implies the impossibility of the original equation, and any necessary conditions that can be found for solvability carry back from $\mathbf{Z}_m$ to $\mathbf{Z}$. For example, in the new language, the argument concerning the cubic Fermat equation could be phrased in this form: the equation $x^3 + y^3 = z^3$ is solvable in $\mathbf{Z}$ only if it is solvable in $\mathbf{Z}_9$. But if $u \not\equiv 0 \pmod 3$, then $u \equiv \pm 1 \pmod 3$, and then it is easily verified that $u^3 \equiv \pm 1 \pmod 9$, and all of the equations $\pm\bar{1} \pm \bar{1} = \pm\bar{1}$ in $\mathbf{Z}_9$ are false, for all choices of signs. Hence if $x^3 + y^3 = z^3$ holds in $\mathbf{Z}$, then $3 \mid xyz$.

(Note that, given $u \equiv \pm 1 \pmod 3$, Theorem 3.2 yields only $u^3 \equiv (\pm 1)^3 \equiv \pm 1 \pmod 3$. The residue class $\bar{1} \pmod 3$ comprises three classes, $\bar{1}$, $\bar{4}$, and $\bar{7}$, (mod 9), and $-\bar{1}$ comprises the classes $-\bar{1}$, $-\bar{4}$, $-\bar{7}$, so the fact that $u \equiv \pm 1 \pmod 3$ implies $u^3 \equiv \pm 1 \pmod 9$ involves something a little deeper than Theorem 3.2. See Section 4.1, Problem 1.)

We see from this last discussion that it is sometimes useful or necessary to consider two or more different moduli in the course of an argument. Frequently (though not above), this happens because the cancellation law, "in $\mathbf{Z}_m$, if $\bar{k} \neq \bar{0}$ and $\bar{k}\bar{a} = \bar{k}\bar{b}$, then $\bar{a} = \bar{b}$," is false unless m is prime, as we see by taking ka to be a nontrivial factorization of m and b to be 0. Here is the correct version.

Theorem 3.3 *Suppose $k \neq 0$. If $(k, m) = d$ and $ka \equiv kb \pmod m$, then $a \equiv b \pmod{m/d}$.*

Proof. If $(k, m) = d$, then $(k/d, m/d) = 1$, and vice versa. If $m \mid (ka - kb)$, then $(m/d) \mid (k/d)(a - b)$, and vice versa. Hence the theorem follows from Corollary 2 of Theorem 2.2. ■

Theorem 3.4 *The Ring $\mathbf{Z}_m$ is not a domain if m is composite. On the other hand, if p is prime, $\mathbf{Z}_p$ is not only a domain, it is a field.*

Proof. Showing that $\mathbf{Z}_p$ is a field requires showing that every nonzero element in $\mathbf{Z}_p$ is invertible—that the equation $\bar{a}x = \bar{1}$ is always uniquely solvable in $\mathbf{Z}_p$ if $\bar{a}$ is not $\bar{0}$, that is, if $p \nmid a$. This equation is equivalent to the equation $ax = 1 + py$ in $\mathbf{Z}$, and if $p \nmid a$, this last equation is solvable and the set of allowable x's exactly fills one residue class (mod p), by Theorem 2.9. ■

PROBLEMS

1. Let $f(x) = a_0x^n + \cdots + a_n$ be a polynomial over $\mathbf{Z}$. Show that if r consecutive values of f (i.e., values for consecutive integers) are all divisible by r, then $r \mid f(m)$ for all $m \in \mathbf{Z}$. Show by an example that this can happen with $r > 1$, even when $(a_0, \ldots, a_n) = 1$.
2. In the days before calculators, a simple rule for (partially) checking arithmetical operations, called "casting out nines," was in common use. It said that if you have formed the sum or product of several integers, do the same thing to the sums of their decimal digits (or to the sums of the digits of those sums, if the latter are multidigit numbers), and you should get the same result as from adding the digits of the alleged answer. For example, the equation $8897 \cdot 2969 = 26{,}445{,}193$ cannot be correct, for

the digit-sums are 32, 26, and 34, of which the digit-sums are 5, 8, and 7, and the digit-sum of $5 \cdot 8$ is not 7. Show how all this hinges on the Corollary to Theorem 3.2 and the fact that $10 \equiv 1 \pmod 9$. The test fails to detect the most common mistake, interchanging two digits.

3. Theorem 3.2 shows that for arbitrary integers $m > 1$ and $n > 0$ the polynomial sequence $1^n, 2^n, 3^n, \ldots$ defines a periodic sequence of residue classes (mod m). (Briefly, $\{k^n\}$ is periodic mod m.) How about the exponential sequence $n, n^2, n^3, \ldots$? Try some small values of m, and a suitable number of values of n, for each m. Gather some empirical data, make some conjectures, try to prove them.
4. Suppose that $d \mid m$. Show that if $a \equiv b \pmod{m/d}$, then
$$a \equiv b \quad \text{or} \quad b + 1 \cdot \frac{m}{d} \quad \text{or} \quad \cdots \quad \text{or} \quad b + (d - 1)\frac{m}{d} \pmod m.$$
5. Rephrase Problem 10 of Section 2.3 in terms of congruences.
6. Rephrase Problem 11 of Section 1.4 in terms of congruences.
7. Show that if r and s are odd, then

   a) $\dfrac{rs - 1}{2} \equiv \dfrac{r - 1}{2} + \dfrac{s - 1}{2} \pmod 2$;

   b) $r^2 \equiv s^2 \equiv 1 \pmod 8$;

   c) $\dfrac{(rs)^2 - 1}{8} \equiv \dfrac{r^2 - 1}{8} + \dfrac{s^2 - 1}{8} \pmod 8$.
8. Show that there are not infinitely many triples $\{n, n + 2, n + 4\}$, all of whose entries are prime.
9. Show that if $n > 1$, then $2^n - 1 \equiv 3 \pmod 4$, and also that $x^m \equiv 3 \pmod 4$ only if $x \equiv m \equiv 1 \pmod 2$, in which case
$$\frac{x^m + 1}{x + 1} = x^{m-1} - x^{m-2} + \cdots - x + 1$$
is an odd integer. Conclude that the equation $2^n - x^m = 1$ has no solution with $x > 1$, $m > 1$, $n > 1$. (Continued in Section 4.1, Problem 18.)
10. Prove that $\{k^k\}$ is periodic (mod 3) and find its period.
11. Show that every function mapping the set of residue classes (mod p) into itself is equal to a (not necessarily unique) polynomial in one variable. More exactly, the same mapping can be obtained with a polynomial. (For this problem, familiarity with either the Lagrange interpolation formula or the Vandermonde determinant would be useful.)

3.2 COMPLETE AND REDUCED RESIDUE SYSTEMS; EULER'S φ-FUNCTION

In the preceding section we denoted the residue classes (mod m) by $\bar{0}, \bar{1}, \ldots, \overline{m - 1}$. Of course $\bar{a} = \bar{b}$ if $a \equiv b \pmod m$, so we were using the fact that $0, 1, \ldots, m - 1$ constitute a system of representatives, exactly one from each class—what is traditionally referred to in number theory as a **complete residue system** (mod m). Thus a complete residue system (mod m) is any set of integers which

a) has m elements, say $a_1, \ldots, a_m$;

b) consists of incongruent integers (mod m): $a_i \equiv a_j$ (mod m) implies $i = j$;

c) represents every residue class (mod m) exactly once: for every $a \in \mathbf{Z}$, there is a unique $a_i \equiv a$ (mod m).

It is easy to see that a set having any two of these properties automatically has the third, and this gives us an easy proof of the following theorem.

Theorem 3.5 *If $a_1, \ldots, a_m$ is a complete residue system* (mod m), *and if $b, k \in \mathbf{Z}$ and $(k, m) = 1$, then $ka_1 + b, \ldots, ka_m + b$ is also a complete residue system* (mod m).

Proof. Since (a) above is obvious, we need only verify (b). If

$$ka_i + b \equiv ka_j + b \pmod{m},$$

then $ka_i \equiv ka_j$ (mod m), so $a_i \equiv a_j$ (mod m) by Theorem 3.3, so $i = j$. ∎

Theorem 3.5 can also be expressed in terms of $\mathbf{Z}_m$ itself; it says that if $(k, m) = 1$, then the linear function $\bar{k}x + \bar{b}$ maps $\mathbf{Z}_m$ 1–1 onto itself, and hence that it simply effects a permutation of the elements of $\mathbf{Z}_m$.

Ideally, every element of every ring $\mathbf{Z}_m$ should have an unambiguous name—for example $(1)_4$, perhaps, for the arithmetic progression $4k + 1$. But the cost, in resulting complication, is too high, and in fact it is customary to move instead in the opposite direction and to omit not only the subscript but even the bar, whenever the context makes clear what is intended. Thus we might refer to 0 and 1 as the neutral elements in $\mathbf{Z}_m$, rather than $\bar{0}$ and $\bar{1}$, and we might say that the polynomial $3x + 4$, rather than $\bar{3}x + \bar{4}$, permutes the elements of $\mathbf{Z}_5$. It follows from Theorem 3.2 and its corollary that, with this convention, $3x + 4$ and $8x - 1$ behave exactly alike, in all respects, (mod 5), and any difficulty that arises is psychological, not mathematical, as long as the modulus is known and fixed.

The ring $\mathbf{Z}_m$ is a finite additive group, of course, and in fact it is a **cyclic** group, in the sense that it is **generated** by a single element, in this case $\bar{1}$:

$$\bar{1} = \bar{1}, \quad \bar{2} = \bar{1} + \bar{1}, \quad \ldots, \quad \overline{m-1} = \underbrace{\bar{1} + \bar{1} + \cdots + \bar{1}}_{m-1}, \quad \bar{0} = \underbrace{\bar{1} + \bar{1} + \cdots + \bar{1}}_{m}.$$

We shall see that the multiplicative group $\mathbf{Z}_p^*$ of nonzero elements of $\mathbf{Z}_p$ is also cyclic, for prime p; for example, in $\mathbf{Z}_5$ it is generated by $\bar{2}$:

$$\bar{2} = \bar{2}, \quad \bar{2}^2 = \bar{4}, \quad \bar{2}^3 = \bar{3}, \quad \bar{2}^4 = \bar{1}.$$

For composite modulus, however, the multiplicative structure of $\mathbf{Z}_m$ is much more complicated; for one thing, $\mathbf{Z}_m^*$ is not even a group. But it contains a very important subset which is a group, namely the set of units or invertible elements of $\mathbf{Z}_m$. (The set of units in any Ring forms a group, as is easily seen.) We designate by U_m the **group of units** of $\mathbf{Z}_m$, and by $\varphi(m)$ the **order** of U_m. These quantities can also be described in $\mathbf{Z}$. For $u \in U_m$ if and only if the equation $ux = 1$ is solvable in $\mathbf{Z}_m$, or, equivalently, the congruence $ux \equiv 1 \pmod{m}$ is solvable, or, equivalently, the equation $ux + my = 1$ is solvable in $\mathbf{Z}$. We already know that this is the case if and only if $(u, m) = 1$, so the units of $\mathbf{Z}_m$ are those residue classes consisting of integers relatively prime to m. (All elements of a fixed residue class have the same GCD with the modulus, of course.) This yields the classical definition of the **Euler φ-function** for $m \geq 2$: $\varphi(m)$ is the number of positive integers $a \leq m$ which are relatively prime to m. The latter description also gives $\varphi(1) = 1$, which we adopt.

A complete system of representatives in $\mathbf{Z}$ of the units of $\mathbf{Z}_m$ is called a **reduced residue system** (mod m). In parallel with (a), (b), and (c) above, a reduced residue system is characterized as a set of integers having any two of the following three properties:

a′) it has $\varphi(m)$ elements, say $a_1, \ldots, a_{\varphi(m)}$, each relatively prime to m;

b′) it consists of incongruent integers (mod m);

c′) for every $a \in \mathbf{Z}$ with $(a, m) = 1$, there is a unique $a_i \equiv a \pmod{m}$.

Theorem 3.6 *If $a_1, \ldots, a_{\varphi(m)}$ is a reduced residue system* (mod m) *and* $(k, m) = 1$, *then $ka_1, \ldots, ka_{\varphi(m)}$ is also a reduced residue system* (mod m).

The proof is essentially the same as that of Theorem 3.5.

As is implied by the name associated with it, the Euler φ-function considerably antedates the algebraic framework in which we first defined it above. It has been one of the central functions of number theory for over two centuries, and we pause to examine it briefly. The φ-function is an example of an **arithmetic** or **number-theoretic function**: a function defined on $\mathbf{Z}$. An arithmetic function f is said to be **multiplicative** if $f(mn) = f(m)f(n)$ whenever $(m, n) = 1$.

Theorem 3.7 *φ is multiplicative.*

Proof. Since $\varphi(1) = 1$, the equation $\varphi(mn) = \varphi(m)\varphi(n)$ holds when either $m = 1$ or $n = 1$, so suppose $m > 1$, $n > 1$. Suppose $(m, n) = 1$, and consider the array

$$\begin{array}{ccccc}
0 & 1 & 2 & \cdots & m-1 \\
m & m+1 & m+2 & \cdots & m+(m-1) \\
 & & \cdots & & \\
 & & \cdots & & \\
(n-1)m & (n-1)m+1 & (n-1)m+2 & \cdots & (n-1)m+(m-1).
\end{array}$$

It consists of mn consecutive integers, so it is a complete residue system (mod mn), and $\varphi(mn)$ of the entries are relatively prime to mn. Now the first row is a complete residue system (mod m), and all the elements in any one column are congruent

Leonhard Euler (1707–1783)

Euler was a fabulous mathematician. He published over 500 papers, and another 350 have appeared posthumously. He worked in practically all branches of mathematics and physics, and wrote outstanding textbooks in most of them: algebra, trigonometry, calculus, mechanics, dynamics, calculus of variations, astronomy, artillery, optics—the list goes on and on. He lost the sight in one eye in his twenties, and became totally blind before he was 60, but his productivity was undiminished. (He hardly needed eyes—he mentally summed 17 terms of a complicated series, with 50-place accuracy!) His original research was of fundamental importance, but of almost equal significance for eighteenth and nineteenth century mathematics was his ability to unify and systematize all the mathematics then known. He was born and educated in Switzerland, but spent his entire adult life at the Academies of St. Petersburg (now Leningrad) and Berlin.

(mod m), so there are $\varphi(m)$ columns consisting entirely of integers prime to m, while the other columns contain no such integers. Take a column of the first type; its elements are $b, m + b, 2m + b, \ldots, (n - 1)m + b$, for some b, and by Theorem 3.5 this is a complete residue system (mod n). So each of these columns contains $\varphi(n)$ integers relatively prime to n, and hence there are $\varphi(m)\varphi(n)$ elements prime to both m and n. Since $(m, n) = 1$, an integer l is relatively prime to mn if and only if it is relatively prime to each of m and n, so there are $\varphi(m)\varphi(n)$ integers relatively prime to mn. Thus $\varphi(mn) = \varphi(m)\varphi(n)$. ■

If f is multiplicative and $n = p_1^{e_1} \cdots p_r^{e_r}$ is the prime-power decomposition of n, then

$$f(n) = f(p_1^{e_1}) \cdots f(p_r^{e_r}).$$

Hence a multiplicative function is completely determined by its values for prime-power arguments.

Theorem 3.8 *For each prime power p^e with $e > 0$, $\varphi(p^e) = p^{e-1}(p - 1) = p^e(1 - 1/p)$. Hence for $n > 1$,*

$$\varphi(n) = n \prod_{p \mid n} \left(1 - \frac{1}{p}\right),$$

where the notation means that the product is extended over all positive primes which divide n.

Proof. Among the integers $1, 2, \ldots, p^e$, those which are not relatively prime to p^e are exactly the multiples of p, namely $p, 2p, 3p, \ldots, p^{e-1} \cdot p$. Obviously their number is p^{e-1}, so $\varphi(p^e) = p^e - p^{e-1}$, as asserted. ∎

For example, $\varphi(10) = 10(1 - \frac{1}{2})(1 - \frac{1}{5}) = 4$, and $U_{10} = \{1, 3, 7, 9\}$. Note that for $m > 1$, $\varphi(m)$ gives the number of reduced fractions a/m with $0 < a/m < 1$.

Returning to U_m, recall that each element a generates a cyclic subgroup of U_m (possibly U_m itself), consisting of the powers of a. This group, which we designate by $\langle a \rangle$, is finite and contains 1; if a^h is the power of a with smallest positive h such that $a^h = 1$, then $a, a^2, \ldots, a^h$ are distinct (for if $a^r = a^s$, then $a^{r-s} = 1$) and every power of a is equal to one of these. By Problem 3 of Section 1.2, $h \mid \varphi(m)$, so $a^{\varphi(m)} = 1$ in U_m. This can again be translated back into $\mathbf{Z}$: if $(a, m) = 1$, then there exists a minimal positive exponent h such that $a^h \equiv 1 \pmod{m}$; in line with the language of Problem 2 of Section 1.2, we call h the **order of a (mod m)**, and designate it by $\operatorname{ord}_m a$. It is of course defined only when $(a, m) = 1$, and is equal to the order of the group $\langle a \rangle$. We have thus proved two famous theorems which long antedate the notion of a group.

Theorem 3.9 (Fermat's "little" theorem) *If p is prime and $p \nmid a$, then $a^{p-1} \equiv 1 \pmod{p}$. Hence for every integer a, $a^p \equiv a \pmod{p}$.*

Theorem 3.10 (Euler's theorem) *If $(a, m) = 1$, then*

$$a^{\varphi(m)} \equiv 1 \pmod{m}.$$

As usual, Fermat did not publish his proof of Theorem 3.9, and Euler gave the first published proof. He later (1760) generalized it to Theorem 3.10, and gave a proof using the ideas, but of course not the language, of the proof we have given. Actually, Leibniz had proved Fermat's theorem before 1683, but also did not publish it; he relied on properties of multinomial coefficients. A very simple proof of Fermat's theorem was given by J. Ivory in 1806, and it extends immediately to Euler's theorem, as follows. If $a_1, \ldots, a_{\varphi(m)}$ is a reduced residue system, then so also is $aa_1, \ldots, aa_{\varphi(m)}$, by Theorem 3.6, so

$$a_1 \cdots a_{\varphi(m)} \equiv (aa_1) \cdots (aa_{\varphi(m)}) \equiv a^{\varphi(m)} a_1 \cdots a_{\varphi(m)} \pmod{m},$$

and the factor $a_1 \cdots a_{\varphi(m)}$ can be cancelled, by Theorem 3.3 with $d = 1$.

Theorem 3.11 *If $(a, m) = 1$, then* $\operatorname{ord}_m a \mid \varphi(m)$.

Proof. As noted above, this follows from Problem 3 of Section 1.2, or it can be proved directly, as follows. By definition, $t = \operatorname{ord}_m a$ is the smallest positive integer such that $a^t \equiv 1 \pmod{m}$. If $\varphi(m) = tq + r$, $0 \leq r < t$, then

$$1 \equiv a^{\varphi(m)} \equiv a^{tq+r} \equiv a^r \pmod{m},$$

so $r = 0$ and $t \mid \varphi(m)$. ∎

A final example, for the present, of the elegant behavior of the φ-function is the following.

Theorem 3.12 *For* $n > 0$, $\sum_{d \mid n} \varphi(d) = n$.

The notation means that the sum is extended over all *positive* divisors of n. For example, for $n = 6$ the possible values of d are 1, 2, 3, and 6, and $\varphi(1) + \varphi(2) + \varphi(3) + \varphi(6) = 1 + 1 + 2 + 2 = 6$.

Proof. This is a special case of the principle that if a finite set is partitioned into subsets, the cardinality (= number of elements) of the set is the sum of the cardinalities of the subsets. Here the set is $S = \{1, 2, \ldots, n\}$, and the subsets are the classes C_d, for $d \mid n$, consisting of the integers $a \in S$ for which $(a, n) = d$. The conditions

$$1 \le a \le n, \qquad (a, n) = d$$

can be rewritten, upon putting $a = bd$, in the equivalent form

$$1 \le b \le \frac{n}{d}, \qquad \left(b, \frac{n}{d}\right) = 1,$$

and the number of such b's is $\varphi(n/d)$. But as d ranges over the divisors of n, so does n/d, in reverse order, so

$$\sum_{d \mid n} \varphi(n/d) = \sum_{d \mid n} \varphi(d). \quad \blacksquare$$

PROBLEMS

1. Show that in every Ring, the set of units forms a multiplicative group.
2. Show that if $a \equiv b \pmod{m}$, then $(a, m) = (b, m)$.
3. Discuss divisibility in $\mathbf{Z}_m$, and the notion of GCD. What happens to the GCD under the homomorphism $\mathbf{Z} \to \mathbf{Z}_m$ of Theorem 3.2?
4. Suppose $(m, n) = 1$. Show that $(mx + ny, mn) = 1$ if and only if $(x, n) = (y, m) = 1$. Use this to give a second proof of Theorem 3.7. [Recall Problem 5 of Section 2.1.]
5. In Problem 4 of Section 1.1, a question was raised about the connection among $\tau(m)$, $\tau(n)$, and $\tau(mn)$. If you did not see any then, reconsider the question in light of the present section. Can you prove your conjecture?
6. Show that if $m > 2$ and $a_1, \ldots, a_m$ is a complete residue system (mod m), then $a_1^2, \ldots, a_m^2$ is not.
7. Show that

$$\sum_{d \mid n} (-1)^{n/d} \varphi(d) = \begin{cases} 0 & \text{for } n \text{ even} \\ -n & \text{for } n \text{ odd.} \end{cases}$$

[*Hint:* For n even, partition the set of divisors of n according to the powers of 2 contained in them, and evaluate the corresponding subsums.]

8. Prove that $\varphi(n) \to \infty$ as $n \to \infty$. [Recall Problem 11 of Section 2.2.]

9. Verifying congruences such as those that occur in Fermat's and Euler's theorems requires the computation of a large power of a (mod m). To find a^r (mod m), one could successively compute $a, a^2, \ldots, a^r$ (mod m), requiring r multiplications and reductions (mod m). Show how this can be reduced to $c \log r$ such operations (where c is some not-very-interesting constant) by utilizing the base-2 representation of r, $r = \varepsilon_0 + \varepsilon_1 \cdot 2 + \cdots + \varepsilon_m 2^m$, each $\varepsilon_i = 0$ or 1.
10. Show that for all a and b, and for p prime,
$$a^p + b^p \equiv (a + b)^p \pmod{p}.$$
Generalize to arbitrarily many terms. Take $a = b = \cdots = 1$ to obtain an additional proof of Fermat's theorem.
11. Prove the following partial converse of Fermat's Theorem: if there is an a for which $a^{m-1} \equiv 1 \pmod{m}$, while none of the congruences $a^{(m-1)/p} \equiv 1 \pmod{m}$ holds, where p runs over the prime divisors of $m - 1$, then m is prime. (The first congruence alone is not sufficient; for example, $4^2 \equiv 1 \pmod{15}$ so $4^{14} \equiv 1 \pmod{15}$.) If you have a 10-digit-display calculator, use this to show that $f_4 = 2^{2^4} + 1$ is prime.
12. Prove that if $(a, b) = d$, then
$$\varphi(ab) = \frac{d\varphi(a)\varphi(b)}{\varphi(d)}.$$
13. Show that if $n > 1$, then the sum of the positive integers less than n and prime to n is $\frac{1}{2}n\varphi(n)$. [*Hint:* If m satisfies the conditions, so does $n - m$.]
14. Show that if $d \mid n$, then $\varphi(d) \mid \varphi(n)$.
15. Show, perhaps with the help of Theorem 3.8, that $\varphi(n) \equiv 2 \pmod{4}$ if and only if $n = p^a$ or $2p^a$, where p is a prime $\equiv 3 \pmod{4}$ and $a > 0$. Conclude that (a) the φ-function does not assume the value 14; (b) if the number of $n \leq N$ such that $4 \nmid \varphi(n)$ is $f(N)$, then $f(N)/N \to 0$ as $N \to \infty$.
16. Let $f(x) \in \mathbf{Z}[x]$, and for $n \geq 1$ let $\psi_f(n) = \psi(n)$ denote the number of values of k, with $1 \leq k \leq n$, such that $(f(k), n) = 1$.

    a) Show that ψ is multiplicative.

    b) Show that
$$\psi(n) = n \prod_{p \mid n} \left(1 - \frac{b_p}{p}\right),$$
    where $b_p = p - \psi(p)$ is the number of multiples of p among $f(1), f(2), \ldots, f(p)$.
17. How many reduced fractions r/s are there with $0 \leq r < s \leq n$?
18. Show that if $ab \equiv 1 \pmod{m}$, then $\operatorname{ord}_m a = \operatorname{ord}_m b$.
19. Show that if $m > 1$ is odd and $\operatorname{ord}_m a = 2t$, then
$$a^t \equiv -1 \pmod{m}.$$
Show that this need not be true if $2 \mid m$.
20. Show that if m is larger than 1 and $a^t \equiv -1 \pmod{m}$, then $\operatorname{ord}_m a = 2u$ and $t = (2k + 1)u$, for some integers u and k.

21. Show that if p is an odd prime and $p \mid (a^{2^r} + 1)$ for some $a > 1$, then $p \equiv 1 \pmod{2^{r+1}}$.
    a) Deduce a special case of Dirichlet's theorem: for each $r \geq 1$, there are infinitely many primes of the form $k \cdot 2^r + 1$.
    b) Find the 5 smallest primes which might divide $f_5 = 2^{2^5} + 1$, and test them. [Slight shortcut: $f_n \nmid f_m$ unless $n = m$.]
22. Show that if p is a prime different from 2 and 5, then it divides infinitely many of the decimal integers 1, 11, 111, 1111, ...
23. Show that for every a such that $(a, 561) = 1$, the congruence $a^{560} \equiv 1 \pmod{561}$ holds. [*Hint:* Use Theorem 3.9.] Show that every m is square-free which has the property that $a^{m-1} \equiv 1 \pmod{m}$ for all a with $(a, m) = 1$.
24. Show that if $N = 2^p - 1$, p prime, then $2^{N-1} \equiv 1 \pmod{N}$.

3.3 LINEAR CONGRUENCES

In the next section we shall enter into a full discussion of polynomial congruences $f(x) \equiv 0 \pmod{m}$. For the present we observe merely that the Corollary to Theorem 3.2 guarantees that the $x \in \mathbf{Z}$ which satisfy such a congruence are not a chaotic set of integers, but consist exactly of (all) the elements of certain residue classes (mod m). Thus the congruence can be considered as an equation over $\mathbf{Z}_m$: the polynomial is defined over $\mathbf{Z}_m$, and the solutions (if any) are elements of $\mathbf{Z}_m$. In particular, by the **number of solutions** of $f(x) \equiv 0 \pmod{m}$, we shall mean the number of solutions of $f(x) = 0$ in $\mathbf{Z}_m$. Naturally, the modulus must be clearly understood.

Let us first study linear congruences in some detail. We have already proved the following theorem, in another guise.

Theorem 3.13 *For $m > 1$, a necessary and sufficient condition that the congruence*

$$ax \equiv b \pmod{m} \tag{3}$$

be solvable is that $d \mid b$, where $d = (a, m)$. If this condition is satisfied, there is a unique solution—say x_0—(mod m/d), and hence there are d solutions (mod m), *namely,*

$$x \equiv x_0 \quad or \quad x_0 + 1 \cdot \frac{m}{d} \quad or \quad \cdots \quad or \quad x_0 + (d-1)\frac{m}{d} \pmod{m}.$$

For the congruence (3) is equivalent to the linear equation $ax + my = b$, and Theorem 3.13 is Theorem 2.9 in different language.

As an example, the congruence

$$15x \equiv 21 \pmod{33}$$

is solvable because $(15, 33) \mid 21$, and in fact it is equivalent to $5x \equiv 7 \pmod{11}$. By inspection, this has the (unique) solution $x \equiv 8 \pmod{11}$, so the solutions (mod 33) are 8, 19, 30. Whether one wants to list solutions (mod 11) or (mod 33) is a matter of choice, or of convenience if some other purpose is to be served.

Solving linear equations over **Z** leads from the integers to the rational numbers; is there an equivalent statement for $\mathbf{Z}_m$? With certain restrictions, yes. The symbol b/a can be used to designate the solution in $\mathbf{Z}_m$ of the equation $ax = b$ *provided that it is uniquely defined*, and this is the case, by Theorem 3.13, if and only if a is a unit in $\mathbf{Z}_m$. There would be nothing incorrect in saying, for example, that in $\mathbf{Z}_{11}$ the equation $5x = 7$ has the solution $x = 7/5$ instead of $x = 8$, but then, of course, one loses the easy interchangeability of the element 8 (literally, $\bar{8}$) of $\mathbf{Z}_{11}$ with the arithmetic progression $11k + 8$ in **Z**. Writing more precisely, $\bar{5}$ has an inverse $\bar{5}^{-1}$ $(= \bar{9})$ in $\mathbf{Z}_{11}$, so $\bar{7} \cdot \bar{5}^{-1}$ is an element of $\mathbf{Z}_{11}$ (namely $\bar{8}$), but there is no sense in which $\bar{7} \cdot \bar{5}^{-1}$ came from $7/5$ under the homomorphism which mapped each $a \in \mathbf{Z}$ into $\bar{a} \in \mathbf{Z}_{11}$, since $7/5 \notin \mathbf{Z}$. (And, in fact, there is no way of extending this homomorphism so as to map **Q** onto $\mathbf{Z}_{11}$ and preserve inverses as well, since 11^{-1} exists in **Q** but $\overline{11} = \bar{0}$ has no inverse in $\mathbf{Z}_{11}$.) The likelihood of wanting to go back from $\mathbf{Z}_m$ to **Z** is sufficiently great that fractional representations in $\mathbf{Z}_m$ are not used much, although we shall use them when it is convenient.

There is an important problem regarding congruences which has no analogue in **Z**, nor in any single ring $\mathbf{Z}_m$ for that matter. It is the question of finding a solution to a system of simultaneous linear congruences with different moduli, in the one-variable case this being the problem of finding all $x \in \mathbf{Z}$ such that

$$\begin{aligned} a_1x &\equiv b_1 \pmod{m_1}, \\ a_2x &\equiv b_2 \pmod{m_2}, \\ &\;\vdots \\ a_rx &\equiv b_r \pmod{m_r}. \end{aligned} \qquad (a_i, b_i \in \mathbf{Z}) \tag{4}$$

For this system to be solvable it is clearly necessary that each congruence individually be solvable. So suppose that for each i, the ith congruence has one or several solutions (mod m_i); then the system (4) is equivalent to one or many systems of the form

$$\begin{aligned} x &\equiv c_1 \pmod{m_1}, \\ x &\equiv c_2 \pmod{m_2}, \\ &\;\vdots \\ x &\equiv c_r \pmod{m_r}, \end{aligned} \tag{5}$$

each such system resulting by choosing one solution of the first congruence in (4), together with one solution of the second, and so on. (Alternatively, one could obtain a unique system similar to (5) but with moduli which are divisors of the original moduli.) So the problem is really that of investigating the intersection of several arithmetic progressions.

When the moduli are relatively prime in pairs the situation is especially simple (and important).

Theorem 3.14 (Chinese remainder theorem) *If* $(m_i, m_j) = 1$ *for* $1 \le i < j \le r$, *then the system* (5) *has as its complete solution a single residue class* $(\bmod\ m_1 \cdots m_r)$.

Proof. Put $m = m_1 \cdots m_r$. With the moduli fixed and in fixed order, the system (5) is completely described by the vector (or r-tuple) of constants, $\{c_1, \ldots, c_r\}$, where actually c_i is a residue class $(\bmod\ m_i)$, that is, $\bar{c}_i \in \mathbf{Z}_{m_i}$. The customary notation for this situation is $\{\bar{c}_1, \ldots, \bar{c}_r\} \in \mathbf{Z}_{m_1} \times \cdots \times \mathbf{Z}_{m_r}$. Clearly, given $x \in \mathbf{Z}$ there is a unique vector $\{\bar{c}_1, \ldots, \bar{c}_r\}$ determined by (5). If $x \in \mathbf{Z}$ satisfies (5), then so does every $x' \equiv x \pmod{m}$, obviously, since also $x' \equiv x \pmod{m_i}$ for $1 \le i \le r$, so we may instead consider x as an element $\bar{x}$ of $\mathbf{Z}_m$. Hence to each $\bar{x} \in \mathbf{Z}_m$ corresponds, by (5), a unique vector in $\mathbf{Z}_{m_1} \times \cdots \times \mathbf{Z}_{m_r}$, and there are m distinct $\bar{x}$'s and $m_1 \cdots m_r = m$ distinct vectors. Finally, each vector determines at most one $\bar{x} \in \mathbf{Z}_m$; for if both x and x' satisfy (5), then $m_i \mid (x - x')$ for $1 \le i \le r$, and the hypothesis of the theorem then implies that $m \mid (x - x')$. Putting all this together, we have established a 1–1 correspondence between the elements of $\mathbf{Z}_m$ and those of $\mathbf{Z}_{m_1} \times \cdots \times \mathbf{Z}_{m_r}$; in particular, to each $\{\bar{c}_1, \ldots, \bar{c}_r\}$ there corresponds a unique $\bar{x}$, which is the theorem. ■

But this is only the beginning: we are in the process of developing not merely a 1–1 correspondence but an isomorphism. To explain this we need the notion of **direct sum**: If $R_1, \ldots, R_r$ are Rings, their direct sum $R = R_1 \oplus \cdots \oplus R_r$ is the set of vectors $R_1 \times \cdots \times R_r$, endowed with the following structure: if $\{a_1, \ldots, a_r\}$ and $\{b_1, \ldots, b_r\} \in R$, then

$$\begin{aligned} &\{a_1, \ldots, a_r\} = \{b_1, \ldots, b_r\} \text{ means } a_1 = b_1, \ldots, a_r = b_r, \\ &\{a_1, \ldots, a_r\} + \{b_1, \ldots, b_r\} = \{a_1 + b_1, \ldots, a_r + b_r\}, \qquad (6) \\ &\{a_1, \ldots, a_r\} \cdot \{b_1, \ldots, b_r\} = \{a_1 b_1, \ldots, a_r b_r\}. \end{aligned}$$

Then R is easily seen to be a Ring, with neutral elements $0 = \{0, 0, \ldots, 0\}$ and $1 = \{1, 1, \ldots, 1\}$.

Theorem 3.15 *If* $m = m_1 \cdots m_r$, *where the* $m_i > 1$ *are relatively prime in pairs, then* $\mathbf{Z}_m$ *is isomorphic to* $\mathbf{Z}_{m_1} \oplus \cdots \oplus \mathbf{Z}_{m_r}$.

Proof. At this point it is worthwhile to use the elaborate notation $(x)_m$ suggested before for the element of $\mathbf{Z}_m$ containing $x \in \mathbf{Z}$. The mapping $x \mapsto (x)_m$ of $\mathbf{Z}$ onto $\mathbf{Z}_m$ is a homomorphism, so for $x, y \in \mathbf{Z}$,

$$x + y \mapsto (x)_m + (y)_m, \qquad xy \mapsto (x)_m (y)_m, \tag{7}$$

and similarly for the residue-class mappings of $\mathbf{Z}$ onto $\mathbf{Z}_{m_i}$:

$$x + y \mapsto (x)_{m_i} + (y)_{m_i}, \qquad xy \mapsto (x)_{m_i} (y)_{m_i}, \qquad 1 \le i \le r. \tag{8}$$

We have already established the fact that the correspondence

$$(x)_m \leftrightarrow \{(x)_{m_1}, \ldots, (x)_{m_r}\} \tag{9}$$

is 1–1, and relations (6), (7), and (8) show that this correspondence preserves sums and products, so it is an isomorphism. ■

We shall use this theorem in the next chapter for a theoretical purpose. For the moment we mention only that it provides a practical way to solve the system (5). For if $\{\bar{c}_1, \ldots, \bar{c}_r\} \in \mathbf{Z}_{m_1} \oplus \cdots \oplus \mathbf{Z}_{m_r}$, then it can be represented in the form

$$\begin{aligned}\{\bar{c}_1, \ldots, \bar{c}_r\} &= \{\bar{c}_1, \bar{0}, \ldots, \bar{0}\} + \{\bar{0}, \bar{c}_2, \bar{0}, \ldots, \bar{0}\} + \cdots + \{\bar{0}, \ldots, \bar{0}, \bar{c}_r\} \\ &= c_1\{\bar{1}, \bar{0}, \ldots, \bar{0}\} + c_2\{\bar{0}, \bar{1}, \bar{0}, \ldots, \bar{0}\} + \cdots + c_r\{\bar{0}, \ldots, \bar{0}, \bar{1}\},\end{aligned}$$

where $c_1, \ldots, c_r$ are representatives in $\mathbf{Z}$ of the corresponding residue classes in $\mathbf{Z}_{m_1}, \ldots, \mathbf{Z}_{m_r}$. So the problem reduces to finding the image in $\mathbf{Z}_m$ of each of the basis vectors $\mathbf{e}_1 = \{\bar{1}, \bar{0}, \ldots, \bar{0}\}, \ldots, \mathbf{e}_r = \{\bar{0}, \ldots, \bar{0}, \bar{1}\}$. For if $(y_i)_m \leftrightarrow \mathbf{e}_i$ under the correspondence (9), for $i = 1, \ldots, m$, then $x = c_1y_1 + \cdots + c_ry_r$ satisfies (5). Now for the vector of constants $\mathbf{e}_i$, (5) requires that x, now called y_i, be a multiple of all the m_j except m_i—and hence that $y_i = z_im/m_i$ for some $z_i \in \mathbf{Z}$—and also that y_i be $\equiv 1 \pmod{m_i}$, or

$$\frac{m}{m_i} z_i \equiv 1 \pmod{m_i}. \tag{10}$$

Thus, for each $i = 1, \ldots, r$, the system (5) for $\mathbf{e}_i$ collapses to the single congruence (10), from which z_i must still be found. The advantage of proceeding in this fashion is that, once the z_i have been determined, we have an explicit representation of the solutions of (5) for all the various constant vectors $\{c_1, \ldots, c_r\}$:

$$\text{if } x \equiv c_1 \frac{m}{m_1} z_1 + \cdots + c_r \frac{m}{m_r} z_r \pmod{m}, \text{ then } x \equiv c_i \pmod{m_i} \text{ for } 1 \le i \le r.$$

Consider, for example, the system

$$x \equiv c_1 \pmod{10}, \quad x \equiv c_2 \pmod{11}, \quad x \equiv c_3 \pmod{13}.$$

The congruences (10) become

$$143z_1 \equiv 1 \pmod{10}, \quad 130z_2 \equiv 1 \pmod{11}, \quad 110z_3 \equiv 1 \pmod{13},$$

and they have solutions $z_1 = 7$, $z_2 = 5$, $z_3 = 11$. Hence

$$\begin{aligned}x &\equiv 143 \cdot 7c_1 + 130 \cdot 5c_2 + 110 \cdot 11c_3 \\ &\equiv 1001c_1 + 650c_2 + 1210c_3 \pmod{1430}.\end{aligned}$$

Now let us turn to the case in which the moduli in (5) are not relatively prime in pairs. The system may then be inconsistent, as the example $x \equiv 1 \pmod 6$, $x \equiv 2 \pmod{10}$ shows, since clearly x would have to be both odd and even. If the ith congruence in (5) holds, then $x = c_i + m_iy$, for some $y \in \mathbf{Z}$; if the jth congruence also holds, then

$$m_iy \equiv c_j - c_i \pmod{m_j}.$$

We know that this last congruence is solvable if and only if $(m_i, m_j) \mid (c_i - c_j)$. And now we have the following elegant result.

Theorem 3.16 *A necessary and sufficient condition that the system* (5) *have a solution is that*

$$(11) \qquad (m_i, m_j) \mid (c_i - c_j) \qquad \text{for } 1 \le i < j \le r.$$

In other words, any finite collection of arithmetic progressions has a nonempty intersection if each pair of them has. If there is a solution, the complete solution in $\mathbf{Z}$ *is a residue class* (mod $[m_1, \ldots, m_r]$).

Proof. The congruence $x \equiv c_i \pmod{m_i}$ holds if and only if $x \equiv c_i$ modulo every prime-power factor of m_i. Concentrate on a single prime p which occurs in at least two of the m_i, and suppose $p^{e_i} \parallel m_i$ for each i. Let p^t be the largest power of p occurring in any m_i. To be definite, suppose $t = e_1$. Then the condition $x \equiv c_1 \pmod{p^t}$ implies every other congruence $x \equiv c_i \pmod{p^{e_i}}$ if $c_1 \equiv c_i \pmod{p^{e_i}}$ for all relevant i, and in that case all these other congruences are redundant and can simply be omitted. Recognizing that p^{e_i} is the p-component of (m_1, m_i), we see that the congruence $c_1 \equiv c_i \pmod{p^{e_i}}$ follows from (11), for all i. So after eliminating the redundant congruences, we have one congruence for each prime p dividing $[m_1, \ldots, m_r]$, and that congruence has modulus p^t, the p-component of $[m_1, \ldots, m_r]$. The theorem now follows from the Chinese remainder theorem. ∎

PROBLEMS

1. Find all x, y, and z such that

   a) $x \equiv 3 \pmod{4}$, $x \equiv 5 \pmod{21}$, $x \equiv 7 \pmod{25}$

   b) $3y \equiv 9 \pmod{12}$, $4y \equiv 5 \pmod{35}$, $6y \equiv 2 \pmod{11}$

   c) $z \equiv 1 \pmod{12}$, $z \equiv 4 \pmod{21}$, $z \equiv 18 \pmod{35}$

2. Write out explicitly the pairing between elements of $\mathbf{Z}_{10}$ and of $\mathbf{Z}_2 \times \mathbf{Z}_5$ which yields the isomorphism in Theorem 3.15.

3. a) Show that if $m_1, \ldots, m_s$ are positive integers, then

$$[(m_1, m_s), \ldots, (m_{s-1}, m_s)] = ([m_1, \ldots, m_{s-1}], m_s).$$

   b) With the help of the relation in part (a) or otherwise, prove sufficiency in Theorem 3.16 by induction on r. [*Hint:* The induction hypothesis shows that the first $r - 1$ congruences are simultaneously solvable.]

4. Give x explicitly in terms of $c_1, \ldots, c_4$ if

$$x \equiv c_1 \pmod{2},$$
$$x \equiv c_2 \pmod{3},$$
$$x \equiv c_3 \pmod{5},$$
$$x \equiv c_4 \pmod{7}.$$

5. Show, with the help of the Chinese remainder theorem or otherwise, that there are arbitrarily long blocks of consecutive integers, none of which is square-free.
6. Prove that for each $k \in \mathbf{Z}^+$ there are infinitely many blocks of k or more successive integers, with none of these integers prime.
7. a) Given $a, b, n \in \mathbf{Z}$ with $(a, b) = 1$, show that there is an x such that $(ax + b, n) = 1$. [*Hint:* For each p dividing n, x can be chosen so that $p \nmid (ax + b)$. Use the Chinese remainder theorem.] Note that this is a very weak consequence of Dirichlet's theorem on primes in a progression.

   b) Show that the arithmetic progression $\{ax + b\}$ contains an infinite subsequence (not necessarily a progression), every two of whose elements are relatively prime.
8. If $a \in U_m$, use Euler's theorem to obtain an explicit solution of $ax \equiv 1 \pmod{m}$ when it is solvable, and hence of $ax \equiv b \pmod{m}$. If you are interested in calculations, compare the length of this algorithm with that of others you know for solving a linear congruence.
9. Show that the mapping from $\mathbf{Z}_m[x]$ to $\mathbf{Z}_m \oplus \cdots \oplus \mathbf{Z}_m$ (m summands) defined by

$$f \mapsto \{\overline{f(0)}, \overline{f(1)}, \ldots, \overline{f(m-1)}\}$$

is a homomorphism. Can there be distinct polynomials having the same image under this mapping? Must there be, for every m? What does all this imply about the relations between the assertions "$f(x) = g(x)$ in $\mathbf{Z}_m[x]$" and "for every $a \in \mathbf{Z}_m$, $f(a) = g(a)$"?

3.4 HIGHER-DEGREE POLYNOMIAL CONGRUENCES

The homomorphism which maps each $a \in \mathbf{Z}$ into $\bar{a} \in \mathbf{Z}_m$ can be extended to a homomorphism of the polynomial domain $\mathbf{Z}[x]$ onto the domain $\mathbf{Z}_m[x]$, sending $a_0x^n + \cdots + a_n$ to $\bar{a}_0x^n + \cdots + \bar{a}_n$. Since $\mathbf{Z}_p$ is a field whereas $\mathbf{Z}_m$ is not even a domain for m composite, Theorem 2.8 and the comments following it show that $\mathbf{Z}_p[x]$ is a unique factorization domain but $\mathbf{Z}_m[x]$ is not. Of course, when a polynomial is written in a congruence as if it had coefficients in $\mathbf{Z}$, it can take many forms. Thus

$$\begin{aligned}(x-1)(x-2) &\equiv x^2 - 3x + 2 \\ &\equiv x^2 + 7x + 12 \\ &\equiv (x+4)(x+3) \pmod{5},\end{aligned}$$

but $x - \bar{1} = x + \bar{4}$ and $x - \bar{2} = x + \bar{3}$ in $\mathbf{Z}_5$, so that we do not really have two different factorizations here.

If $f(x) = \bar{a}_0x^n + \cdots + \bar{a}_n \in \mathbf{Z}_m[x]$ and $\bar{a} \in \mathbf{Z}_m$, then $f(\bar{a}) = \bar{a}_0\bar{a}^n + \cdots + \bar{a}_n$ is a well-defined element of $\mathbf{Z}_m$ called the **value** of $f(x)$ at $x = \bar{a}$. If $f(\bar{a}) = 0$, then $\bar{a}$ (or sometimes simply a itself) is called a **zero** of $f(x)$, or a **root** of the equation $f(x) = 0$ in $\mathbf{Z}_m$, or a **root** of the congruence $a_0x^n + \cdots + a_n \equiv 0 \pmod{m}$.

Theorem 3.17 *If a is a root of the congruence* $f(x) \equiv 0 \pmod{m}$, *then* $(x - \bar{a}) \mid f(x)$ *in* $\mathbf{Z}_m[x]$, *and conversely.*

Proof. As in Problem 8 of Section 1.4, there is a division theorem for monic polynomial divisors in $\mathbf{Z}_m$, and $x - \bar{a}$ is monic. Thus in $\mathbf{Z}_m$,

$$(12) \qquad f(x) = (x - \bar{a})q(x) + r(x), \qquad \text{where } r(x) = 0 \quad \text{or} \quad \partial f < \partial(x - \bar{a}).$$

Since $\partial(x - \bar{a}) = 1$, $r(x)$ must be 0 or have degree 0, that is, it must be constant. But then obviously $f(\bar{a}) = 0$ if and only if $r = 0$, and so if and only if $(x - \bar{a}) \mid f(x)$. ■

Theorem 3.18 (Lagrange's Theorem) *If p is prime, and* $f(x)$ *is of degree n over* $\mathbf{Z}_p$, *then the congruence* $f(x) \equiv 0 \pmod{p}$ (*equivalently, the equation* $f(x) = 0$ *in* $\mathbf{Z}_p$) *has at most n roots.*

Proof. To every root $\bar{a}$ corresponds a factor $x - \bar{a}$, which contributes 1 to ∂f. Factorization being unique, there can be at most n linear factors. ■

For composite modulus the corresponding assertion is in general false. For as was noted before, if ab is a nontrivial factorization of m, then

$$(x - a)(x - b) \equiv x^2 - (a + b)x \equiv x(x - a - b) \pmod{m},$$

so that $x^2 - (a + b)x \equiv 0 \pmod{m}$ has four (usually distinct) roots.

For prime modulus, there is no more reason that a congruence of degree n should have *exactly* n roots than that a polynomial over $\mathbf{Z}$ should have n zeros in $\mathbf{Z}$. There is, however, an easy way to characterize the polynomials having their full complement of zeros in $\mathbf{Z}_p$.

Theorem 3.19 *Under the hypotheses of the preceding theorem, a necessary and sufficient condition that* $f(x) \equiv 0 \pmod{p}$ *have n distinct roots is that* $f(x) \mid (x^p - x)$ *in* $\mathbf{Z}_p[x]$. *More generally, the number of distinct roots of the equation* $f(x) = 0$ *in* $\mathbf{Z}_p$ *is the degree of* $(f(x), x^p - x)$ *in* $\mathbf{Z}_p[x]$.

Proof. According to Fermat's theorem, $a^p \equiv a \pmod{p}$ for all a. Otherwise expressed, $x^p - x$ has the p zeros $\bar{0}, \bar{1}, \ldots, \overline{p - 1}$, and since factorization is unique, Theorem 3.17 shows that in $\mathbf{Z}_p$,

$$(13) \qquad x^p - x = x(x - \bar{1}) \cdots (x - \overline{p - 1}),$$

and the right-hand side contains all the linear polynomials there are in $\mathbf{Z}_p$. Hence if, in $\mathbf{Z}_p[x]$,

$$f(x) = (x - \bar{r}_1)^{s_1} \cdots (x - \bar{r}_k)^{s_k} g(x),$$

where the $\bar{r}_i$ are distinct, the s_i are positive and $g(x)$ has no linear factors at all, then

$$(f(x), x^p - x) = (x - \bar{r}_1) \cdots (x - \bar{r}_k).$$

Clearly $k = n$ if and only if $(f(x), x^p - x) = f(x)$, that is, $f(x) \mid (x^p - x)$. ■

The reader should note that a multiple factor $(x - \bar{r})^s$, where $s > 1$, accounts for only a single root, in this theorem. The theorem does not, for example, say that $f(x)$ splits completely into linear factors if and only if $f(x) \mid (x^p - x)$ in $\mathbf{Z}_p[x]$.

Corollary *If $d \mid (p - 1)$, then the congruence $x^d \equiv 1 \pmod{p}$ has d solutions.*

Theorem 3.20 (Wilson's Theorem) *If p is prime, then $(p - 1)! \equiv -1 \pmod{p}$.*

Proof. This is clearly true when $p = 2$, and for p odd it follows by comparing the coefficients of x on the two sides of equation (13). ∎

Wilson's theorem is one of a large number of theorems concerning symmetric functions of all the elements, or all the units, of $\mathbf{Z}_m$.

Returning to the general polynomial congruence

$$f(x) \equiv 0 \pmod{m}, \tag{14}$$

it is of more than passing interest that its study can be reduced to the case in which the modulus is prime, since $\mathbf{Z}_p$ is so much simpler to work in than is $\mathbf{Z}_m$. Note first that if m has the prime-power decomposition $m = p_1^{e_1} \cdots p_r^{e_r}$, then the congruence $f(x) \equiv 0 \pmod{m}$ is equivalent to the simultaneous system

$$f(x) \equiv 0 \pmod{p_1^{e_1}}, \ldots, f(x) \equiv 0 \pmod{p_r^{e_r}}. \tag{15}$$

Joseph Louis Lagrange
(1736–1813)

Lagrange and Euler are usually regarded as the two greatest mathematicians of the eighteenth century. Lagrange was born in Turin and became professor of mathematics at the artillery school there in 1755. In 1759 he became a foreign member of the Berlin Academy, and in 1766 replaced Euler there, when the latter went to Russia. Primarily an analyst, he worked in number theory only during his first 5 years in Berlin. His result on polynomial congruences was published in 1768, and his proof of Wilson's theorem in 1770. (Wilson merely conjectured it.) Laplace called his attention to the many unproved assertions of Fermat, and he provided proofs of many of them, including the four-squares theorem and the solvability of Pell's equation (cf. Chapter 8). He returned to France in 1787. By then he was depressed and nearly indifferent to mathematics; he left his copy of his masterpiece, *Mécanique analytique*, unopened for two years after it was published. During his later years he was the first professor of mathematics at the École Polytechnique, founded in 1797.

If each of these is solved individually—and there must be a solution of each, or the system has none—we arrive at a new system of the form

$$x \equiv a_{11} \quad \text{or} \quad a_{12} \quad \text{or} \quad \cdots \quad \text{or} \quad a_{1s_1} \pmod{p_1^{e_1}},$$
$$\vdots$$
$$x \equiv a_{r1} \quad \text{or} \quad a_{r2} \quad \text{or} \quad \cdots \quad \text{or} \quad a_{rs_r} \pmod{p_r^{e_r}},$$

and this evolves into the $s_1 s_2 \cdots s_r$ systems

$$x \equiv a_{1j_1} \pmod{p_1^{e_1}}, \ldots, x \equiv a_{rj_r} \pmod{p_r^{e_r}} \tag{16}$$

that result by choosing the j_i independently, $1 \leq j_1 \leq s_1, \ldots, 1 \leq j_r \leq s_r$. By the Chinese remainder theorem, each system (16) has a unique solution $(\text{mod } p_1^{e_1} \cdots p_r^{e_r})$, and we have arrived at the following result.

Theorem 3.21 *For a given polynomial $f(x) \in \mathbf{Z}[x]$, the number of solutions* (mod m) *of the congruence* (14) *is a multiplicative function of m, and for $m = p_1^{e_1} \cdots p_r^{e_r}$, the solutions of* (14) *are in 1–1 correspondence with the r-tuples whose entries are solutions of the individual congruences in* (15).

This brings us to the study of the congruence

$$f(x) \equiv 0 \pmod{p^e}. \tag{17}$$

Now every solution of (17) with $e = 2$ is also a solution with $e = 1$, but not conversely; a single residue class (mod p) splits into p residue classes (mod p^2), and certain of these may be solutions for $e = 2$ and others not; but in any case all solutions for $e = 2$ come from solutions for $e = 1$. Similarly, at higher levels, all solutions for the modulus p^{e+1} come from those for p^e. Thus it might be thought of as a problem in successive approximations: start with a solution of (17) with $e = 1$ and attempt to refine it to a solution for $e = 2$, etc. The analogue in analysis is to find better and better approximations to a solution of an equation $f(x) = 0$, given some kind of initial approximation to it. Many algorithms are known for handling the analytic problem, and one of the best known of them, Newton's method, works equally well for our present problem.

In the classical case we have a real approximation x_0 to a solution ξ of the equation $f(x) = 0$, where for simplicity we suppose that f has a Taylor expansion about the point x_0, and that $f'(x_0) \neq 0$. We wish to estimate $\xi - x_0 = h$ by an approximate value $\hat{h}$, and then repeat the procedure starting from $x_1 = x_0 + \hat{h}$. We have

$$0 = f(\xi) = f(x_0 + h) = f(x_0) + f'(x_0)h + \tfrac{1}{2}f''(x_0)h^2 + \cdots,$$

and since x_0 is supposed to be close to ξ, $|h|$ should be small and so the terms involving $h^2, h^3, \ldots$ should be comparatively negligible. Disregarding the latter terms entirely, we take $\hat{h}$ as the number satisfying

$$f(x_0) + f'(x_0)\hat{h} = 0,$$

and obtain

(18) $$x_1 = x_0 - \frac{f(x_0)}{f'(x_0)},$$

the standard formula of Newton.

In the arithmetic case, the Taylor expansion of a polynomial reduces to a finite algebraic identity,

(19) $$f(x_0 + h) = f(x_0) + f'(x_0)h + \frac{1}{2}f''(x_0)h^2 + \cdots + \frac{1}{n}f^{(n)}(x_0)h^n,$$

where $\partial f = n$ and the derivative of a polynomial $\sum_0^n a_k x^k$ can simply be defined to be the polynomial $\sum_0^n k a_k x^{k-1}$. (Thus neither the derivative nor the Taylor identity depends on real variables and limiting processes, in the case of polynomials, although admittedly no one would have thought of them without the calculus.) Now a term $c_j x^j$ in $f(x)$ leads to the corresponding term

$$\frac{j(j-1)\cdots(j-k+1)}{k!} c_j x^{j-k} = \binom{j}{k} c_j x^{j-k}$$

in $f^{(k)}(x)/k!$, so the coefficients $f^{(k)}(x_0)/k!$ are integers if $x_0 \in \mathbf{Z}$ and $f(x) \in \mathbf{Z}[x]$.

Suppose we have already found all solutions of $f(x) \equiv 0 \pmod{p^e}$, for some $e > 0$, and we would like to find those of $f(x) \equiv 0 \pmod{p^{e+1}}$. More specifically, suppose that $f(x_0) \equiv 0 \pmod{p^e}$; can we use x_0 as an approximation to one or more solutions of $f(x) \equiv 0 \pmod{p^{e+1}}$? Now x_0 is really a residue class $\bar{x}_0 \pmod{p^e}$, of course; $\bar{x}_0$ consists of all the numbers $x_0 + tp^e$, $t \in \mathbf{Z}$, and it is the union of p residue classes $\pmod{p^{e+1}}$. So the question is whether t can be chosen so that

(20) $$f(x_0 + tp^e) \equiv 0 \pmod{p^{e+1}}.$$

If in (19) we take $h = tp^e$, then all powers of h higher than the first are $0 \pmod{p^{e+1}}$ (not merely negligible, as they were in the analytic case), and we obtain

$$f(x_0 + tp^e) \equiv f(x_0) + f'(x_0)tp^e \pmod{p^{e+1}}.$$

Imposing (20) requires that t be chosen so that

$$tp^e f'(x_0) \equiv -f(x_0) \pmod{p^{e+1}},$$

or, since there is a common factor of p^e present,

$$tf'(x_0) \equiv -\frac{f(x_0)}{p^e} \pmod{p}.$$

This is a linear congruence in t, of which the number of solutions is

$$\begin{cases} 0, & \text{if } p \mid f'(x_0), \text{ but } p^{e+1} \nmid f(x_0), \\ p, & \text{if } p \mid f'(x_0) \text{ and } p^{e+1} \mid f(x_0), \\ 1, & \text{if } p \nmid f'(x_0). \end{cases}$$

In the first case, the solution x_0 at the p^e-level dies without progeny, so to speak, at the p^{e+1}-level. In the second case, x_0 was already a solution at the p^{e+1}-level, so all p residue classes (mod p^{e+1}) are solutions. In these two cases, characterized by the fact that $p \mid f'(x_0)$, x_0 is called a **singular solution** of (17). A nonsingular solution x_0 of (17), in which $p \nmid f'(x_0)$, leads to a unique solution x_1 of $f(x) \equiv 0$ (mod p^{e+1}), given by

$$x_1 \equiv x_0 - \frac{f(x_0)}{f'(x_0)} \pmod{p^{e+1}},$$

in exact analogy with Newton's formula (18).

The procedure for actually solving (17) with $e > 1$ should now be clear, *given that we know the solutions with* $e = 1$. Each of these initial solutions begets 0, 1, or p solutions at the p^2-level, each of which in turn begets 0, 1, or p solutions at the p^3-level, and so on, and all these solutions can be found by solving individual linear congruences (mod p). This is all routine, although possibly complicated; the remaining "practical" question is how to solve (17) when $e = 1$.

This is the heart of the matter, and it is genuinely difficult. Not for a specific and reasonably small prime, since one can simply test the p possibilities one after another, but for an unspecified or a very large prime. Finding an *a priori* criterion for solvability, or a method for determining the number of solutions without actually finding them, are cruder questions, and even they are difficult. We shall answer them for the general quadratic congruence $ax^2 + bx + c \equiv 0 \pmod p$ in Chapter 5, and we shall establish a criterion for the solvability of $x^n \equiv a \pmod p$ in Chapter 4, but other classes of polynomials are outside the scope of this book.

To show that we have made some progress already in this kind of question, we note that, in addition to the corollary stated along with it, Theorem 3.19 has the following consequence.

Theorem 3.22 (Euler's Criterion) *Let p be an odd prime. If $p \mid a$, the congruence*

$$x^2 \equiv a \pmod p \tag{21}$$

is (trivially) solvable. If $p \nmid a$, it is solvable or not, according as $a^{(p-1)/2} \equiv 1$ or $-1 \pmod p$.

Proof. In the case $p \nmid a$, 0 is not a root of (21), so we examine the GCD in $\mathbf{Z}_p[x]$ of $f(x)$ with $x^{p-1} - 1$, rather than with $x^p - x$. Writing

$$x^{p-1} - 1 = \{(x^2)^{(p-1)/2} - a^{(p-1)/2}\} + (a^{(p-1)/2} - 1)$$

and using the fact that $x^2 - a$ divides the quantity in braces, we see, by Theorem 3.19, that if $a^{(p-1)/2} - 1 = 0$ in $\mathbf{Z}_p$, then $x^2 - a = 0$ has two roots, and otherwise it has none. "Otherwise" means $a^{(p-1)/2} = -1$, for in $\mathbf{Z}_p$,

$$0 = a^{p-1} - 1 = (a^{(p-1)/2} - 1)(a^{(p-1)/2} + 1)$$

and $\mathbf{Z}_p$ is a domain, so that one factor on the right-hand side must be 0. ∎

For odd p, we define the **Legendre symbol** (a/p) to be 0 if $p \mid a$, and, if $p \nmid a$, to be 1 if (21) is solvable and -1 if it is not. Using this symbol, Theorem 3.22 can be phrased more simply.

Theorem 3.23 *If p is an odd prime, then for arbitrary $a \in \mathbf{Z}$,*

$$a^{(p-1)/2} \equiv (a/p) \pmod p.$$

According as $(a/p) = 1$ or -1, a is called a **quadratic residue** or **quadratic nonresidue** of p. It would probably be more expressive and less confusing (since "residue" has been used in a different way up to now) to say that a is (or is not) a square (mod p), or is a square in $\mathbf{Z}_p$, or has a square root in $\mathbf{Z}_p$, but the other language is too firmly rooted in tradition.

Note that it is only U_p and not $\mathbf{Z}_p$ which is partitioned into quadratic residues and nonresidues.

PROBLEMS

1. Suppose that $f(x) \in \mathbf{Z}[x]$ has degree n. Show that if the values $f(a)$ for $n + 1$ consecutive integers a are all divisible by p, then $p \mid f(a)$ for all $a \in \mathbf{Z}$. (Compare Problem 1 of Section 3.1.)
2. Find all solutions of the congruence

$$x^3 - 3x^2 + 27 \equiv 0 \pmod{1125}.$$

[*Answer:* $x \equiv 51, 426$, or $801 \pmod{1125}$.]
3. Show that no nonconstant polynomial over $\mathbf{Z}$ assumes only prime values for all large x. [*Hint:* Apply Taylor's theorem to $f(m + kf(m))$.]
4. Find all the nonsingular solutions of $x^2 \equiv a \pmod{2^e}$.

†5. a) Suppose that the congruence $f(x) \equiv 0 \pmod p$ has s distinct roots, all nonsingular. Show that the same is true of the congruence $f(x) \equiv 0 \pmod{p^e}$, for each $e \geq 1$.
   b) Show that if p is prime and $d \mid (p - 1)$, then the equation $x^d - 1 = 0$ has exactly d roots in $\mathbf{Z}_{p^e}$, for each $e \geq 1$.

6. If p is an odd prime, show that not all three of the sets

$$\{a_1, \ldots, a_{p-1}\}, \{b_1, \ldots, b_{p-1}\}, \{a_1b_1, \ldots, a_{p-1}b_{p-1}\}$$

can be reduced residue systems (mod p).
7. Prove that for $1 < k < p - 1$, $(p - k)!\,(k - 1)! \equiv (-1)^k \pmod p$.
8. Show that if $f(x) \in \mathbf{Z}[x]$ is not constant, then there are infinitely many primes p for which $f(x) \equiv 0 \pmod p$ has a solution. [*Hint:* Note that if $f(x) = a_0 + a_1x + \cdots + a_nx^n$, then $f(a_0x) = a_0(1 + xg(x))$ with $g(x) \in \mathbf{Z}[x]$.]
9. Let $R = \{r_1, \ldots\}$ and $N = \{n_1, \ldots\}$ be the sets of quadratic residues and nonresidues of $p > 2$, respectively; thus R and N partition U_p. Show that for all i and j,

$r_i r_j \in R$, $n_i n_j \in R$, and $r_i n_j \in N$. Conversely, show that this is the only partitioning of U_p into two sets R and N, the second being nonempty, with these multiplicative properties.

10. Show that $(-1/p) = 1$ if and only if $p \equiv 1 \pmod 4$. (In more picturesque language, -1 has a square root in the field with 5 elements, although in the enormous field $\mathbf{R}$ it has none!)
11. Prove the converse of Wilson's theorem.
12. Discuss the meaningfulness, the correctness, and the utility of the "quadratic formula"
$$x = \frac{-b \pm \sqrt{b^2 - 4ac}}{2a}$$
for solving the quadratic equation $ax^2 + bx + c = 0$ in $\mathbf{Z}_p$. Apply it in $\mathbf{Z}_7$, if possible, to solve $3x^2 - 5x + 5 = 0$; $3x^2 - 5x + 6 = 0$.
13. Say that two integers a and b are unrelated (mod m) if for every divisor $d > 1$ of m, $a \not\equiv b \pmod d$. (Unrelated numbers are of course distinct (mod m), but not vice versa.) Show that if $f(x)$ is of degree n over $\mathbf{Z}_m$, then the congruence $f(x) \equiv 0 \pmod m$ has at most n unrelated roots (mod m). Compare Theorem 3.13. [*Hint:* d may be restricted to the prime divisors of m.]
14. The polynomials in $\mathbf{Z}[x]$ assume values in $\mathbf{Z}$ for every integral value of x. But so does $x(x + 1)/2$. Show that if $f(x) \in \mathbf{Q}[x]$, then $f(a) \in \mathbf{Z}$ for all $a \in \mathbf{Z}$ if and only if $f(x)$ is of the form
$$f(x) = \sum_{k=0}^{n} c_k \binom{x}{k}, \qquad \text{where } c_k \in \mathbf{Z} \text{ for } k = 0, 1, \ldots, n.$$
Here, as usual,
$$\binom{x}{k} = \begin{cases} \dfrac{x(x-1)\cdots(x-k+1)}{k!} & \text{for } k \geq 1 \\ 1 & \text{for } k = 0. \end{cases}$$
[*Hint:* Look at $f(0), f(1), \ldots, f(n)$ when $f(x)$ has the above form.]

3.5 THE p-ADIC FIELDS

This section is a digression, aimed at showing one way in which the theory of congruences could be developed further so as to serve as a bridge between arithmetic questions and the standard machinery of algebra and analysis. Nothing said here is used later in this book.* Also, not everything said here will be within the ken of every student with background adequate to read the remainder of the book. Some of it is entirely descriptive, with no proofs at all. Much of the remainder is covered rather sketchily, since this is only a preview. It is quite likely

* But the discussion is extended a little in Section 8.1.

that the student has already encountered similar arguments in a course in calculus or general topology, when he met a rigorous treatment of the real numbers as equivalence classes of Cauchy sequences of rational numbers. The student with less background may undertake to fill in some of the details for himself, or skim the section for the general idea and perhaps refer to other books for details. Most applications are outside the scope of the present book, but valuation theory and "local" methods, of which we see the beginnings here, are central to the modern study of Diophantine equations, algebraic number theory, and algebraic geometry.

We take as starting point the question of how far the analogy can be pushed between Newton's method, which provides better and better rational approximations to a (real) root of an equation $f(x) = 0$, and the method we developed for finding solutions of $f(x) \equiv 0 \pmod{p^e}$ for higher and higher powers of p. For example, starting from the nonsingular solution $x \equiv 4 \pmod 7$ of $f(x) = x^2 - 2 \equiv 0 \pmod 7$, the method yields a unique solution at each level:

$$\begin{aligned} f(4 + 7t) &\equiv 14 + 7t \cdot 8 \equiv 0 \pmod{7^2}, & t &\equiv 5 \pmod 7, \\ f(4 + 5 \cdot 7 + t \cdot 7^2) &\equiv 1519 + 78 \cdot 7^2 t \equiv 0 \pmod{7^3}, & t &\equiv 4 \pmod 7, \\ f(4 + 5 \cdot 7 + 4 \cdot 7^2 + t \cdot 7^3) &\equiv 55223 + 470 \cdot 7^3 t \equiv 0 \pmod{7^4}, & t &\equiv 0 \pmod 7, \end{aligned}$$

and so on. The question is whether there is any sense to be made of the idea that if this were continued indefinitely, the "number"

$$\xi = 4 + 5 \cdot 7 + 4 \cdot 7^2 + 0 \cdot 7^3 + \cdots \tag{22}$$

would somehow provide a solution of the equation $x^2 - 2 = 0$, relative to the prime 7. Formally, the answer is yes, in the sense that if we square this series expansion (as if it were an absolutely convergent power series), we obtain

$$\xi^2 = 16 + 40 \cdot 7 + 57 \cdot 7^2 + 40 \cdot 7^3 + \cdots \tag{23}$$

so

$$\begin{aligned} \xi^2 - 2 &= 14 + 40 \cdot 7 + 57 \cdot 7^2 + 40 \cdot 7^3 + \cdots \\ &= (2 + 40)7 + 57 \cdot 7^2 + 40 \cdot 7^3 + \cdots \\ &= 0 + 42 \cdot 7 + 57 \cdot 7^2 + 40 \cdot 7^3 + \cdots \\ &= 0 + 0 \cdot 7 + (6 + 57)7^2 + 40 \cdot 7^3 + \cdots \\ &= 0 + 0 \cdot 7 + 0 \cdot 7^2 + (9 + 40)7^3 + \cdots \\ &= \cdots = 0. \end{aligned} \tag{24}$$

Everything goes away!

To give this meaning, we would like to arrange matters so that the partial sums $x_1 = 4$, $x_2 = 4 + 5 \cdot 7$, $x_3 = 4 + 5 \cdot 7 + 4 \cdot 7^2$, etc., converge to ξ. Lacking a ξ that we already understand, we retreat to a request for a sense in which the partial sums converge, with no mention of what the limit is to be. This suggests invoking the notion of a Cauchy sequence, or some analogue thereof. Recall that in analysis a sequence $\{a_k\}$ of rational (or real) numbers is called a Cauchy sequence if, for every $\varepsilon > 0$, there is an $n = n(\varepsilon)$ such that

$$|a_k - a_m| < \varepsilon \qquad \text{for all } k, m > n(\varepsilon).$$

In the present case, if $k > m$ and $x_k = b_0 + b_1 \cdot 7 + \cdots + b_{k-1} \cdot 7^{k-1}$, with the $b_i \in \mathbf{Z}$, then

$$x_k - x_m = b_m \cdot 7^m + \cdots + b_{k-1} \cdot 7^{k-1} = 7^m(b_m + \cdots + b_{k-1}7^{k-1-m}),$$

and the difference $x_k - x_m$ is characterized by the fact that it is divisible by a high power of 7 if k and m are both large. This suggests trying to define a new kind of absolute value—call it $|\ \ |_7$—on $\mathbf{Z}$, which is such that if $7^t \parallel x$, then $|x|_7$ is a positive real number which is small when t is large. Thus we arrive at the following definition, for an arbitrary but fixed positive prime p: **the p-adic absolute value** of $x \in \mathbf{Z}$ is

$$|x|_p = \frac{1}{p^t}, \qquad \text{where } p^t \parallel x.$$

We use the symbol $|\ \ |^*$ to stand for any one of $|\ \ |$, $|\ \ |_2$, $|\ \ |_3$, $|\ \ |_5, \ldots,$ where $|\ \ |$ is the ordinary absolute-value function of analysis. Then every instance of $|\ \ |^*$ has the following properties:

i) $|\ \ |^*$ is a mapping from $\mathbf{Z}$ to $\mathbf{Q}$;

ii) $|x|^* \geq 0$ always, with equality if and only if $x = 0$;

iii) $|xy|^* = |x|^* \cdot |y|^*$;

iv) $|x + y|^* \leq |x|^* + |y|^*$.

These are well-known properties of $|\ \ |$, of course, and the first three are obvious for $|\ \ |_p$. Something even stronger than (iv) is true for $|\ \ |_p$, namely,

v) $|x + y|_p \leq \max(|x|_p, |y|_p)$, with equality unless $|x|_p = |y|_p$.

For if $p^t \parallel x$ and $p^u \parallel y$, then $p^{\min(t,u)}$ divides $x + y$, and this is in fact the p-component of $x + y$ unless $t = u$, when it may or may not be.

It is easy to see that each $|\ \ |^*: \mathbf{Z} \to \mathbf{Q}$ can be extended in a unique way to a mapping from $\mathbf{Q}$ to $\mathbf{Q}$, namely by putting $|a/b|^* = |a|^*/|b|^*$. Any such extension is called a **valuation** on $\mathbf{Q}$.

The central fact now is that properties (i)–(iv) are the only properties of $|\ \ |$ used in the usual construction of the topological completion $\mathbf{R}$ of $\mathbf{Q}$ by means of Cauchy sequences, and hence the usual chain of reasoning extends without change to any of the valuations $|\ \ |^*: \mathbf{Q} \to \mathbf{Q}$. Briefly, it goes like this.

A sequence of rational numbers $\{a_k\}$ is a **Cauchy sequence** (relative to a fixed valuation $|\ \ |^*$, of course) if for every rational $\varepsilon > 0$, there is an $n(\varepsilon)$ such that

$$|a_k - a_m|^* < \varepsilon \qquad \text{for all } k, m > n(\varepsilon).$$

A Cauchy sequence $\{a_k\}$ is a **null sequence** if $\lim_k |a_k|^* = 0$, and two Cauchy sequences are said to be **equivalent** if they differ by a null sequence, that is, if

$\lim_k |a_k - b_k|^* = 0$. We then write $\{a_k\} \simeq \{b_k\}$. This is an equivalence relation—it is symmetric, reflexive, and transitive—and thus it partitions the set of all Cauchy sequences into equivalence classes. In case $|\ \ |^* = |\ \ |$, an equivalence class is called a real number; in case $|\ \ |^* = |\ \ |_p$, it is called a ***p*-adic number**. To cover both cases simultaneously, we shall, for the moment, refer to an equivalence class as a **limit number**, and write $\overline{\{a_k\}}$ for the class containing $\{a_k\}$. A Cauchy sequence and an infinite subsequence are always equivalent.

Define

$$\{a_k\} + \{b_k\} = \{a_k + b_k\}, \qquad \{a_k\} \cdot \{b_k\} = \{a_k b_k\}.$$

As regards addition, one sees with little effort that the sum of Cauchy sequences is again a Cauchy sequence, that addition of limit numbers is uniquely defined by

$$\overline{\{a_k\}} + \overline{\{b_k\}} = \overline{\{a_k + b_k\}},$$

and that under this definition, limit numbers form an additive commutative group, with the class of null sequences as the zero-element.

The corresponding statements are true also for multiplication, but two lemmas are useful. To show that the Cauchy sequences are closed under multiplication, it is convenient first to prove the fact that they are bounded (that is, $|a_k|^* < A$ for all k), and hence that

$$\begin{aligned} |a_k b_k - a_m b_m|^* &= |(a_k b_k - a_k b_m) + (a_k b_m - a_m b_m)|^* \\ &\leq A|b_k - b_m|^* + B|a_k - a_m|^*; \end{aligned}$$

from this point it is easy to show that $\{a_k b_k\}$ is a Cauchy sequence if $\{a_k\}$ and $\{b_k\}$ are. To show that the limit numbers, excluding the class of null sequences, form a multiplicative commutative group, it is useful to prove first that if $\{a_k\}$ is not null, there is a $\delta > 0$ and an integer n such that $|a_k|^* > \delta$ for all $k > n$. Then $\overline{\{a_k\}}^{-1}$ can be taken to be the class containing $\{b_k\}$, where $b_k = 0$ for $k \leq n$ and $b_k = 1/a_k$ for $k > n$, since $\{0, 0, \ldots, 0, 1, 1, \ldots\} \simeq \{1, 1, \ldots\}$.

Finally, the distributive law is easily proved. Hence the following theorem results.

Theorem 3.24 *The set* **R** *of real numbers forms a field, and for each prime p the set* $\mathbf{Q}_p$ *of p-adic numbers forms a field. Each of these fields contains an isomorphic copy of* **Q** *as subfield, under the correspondence* $a \leftrightarrow \{a, a, a, \ldots\}$.

Since **R** is already familiar, let us concentrate now on a p-adic field $\mathbf{Q}_p$, designating its elements by Greek letters.

Theorem 3.25 *Every nonzero p-adic number* α *has a unique "reduced" representation in the form*

$$\alpha = p^N\{d_0, d_0 + d_1 p, d_0 + d_1 p + d_2 p^2, \ldots\},$$

in which $d_k \in \mathbf{Z}$, $0 \leq d_k < p$ *for all* $k \geq 0$, $N \in \mathbf{Z}$, *and* $d_0 \neq 0$.

Proof. Suppose $\alpha = \overline{\{a_k\}}$; since $\alpha \neq 0$, only finitely many $a_k = 0$, and we can omit them without changing α. Thus we may write

$$a_k = \frac{b_k}{c_k} p^{n_k}, \qquad \text{where } b_k, c_k, n_k \in \mathbf{Z} \quad \text{and} \quad p \nmid b_k c_k.$$

Since every Cauchy sequence is bounded, $\min(n_1, n_2, \ldots)$ exists, and since $\alpha \neq 0$, $\max(n_1, n_2, \ldots)$ exists, so there is a smallest exponent N such that $n_k = N$ for infinitely many k. Delete the a_k for which $n_k < N$, and relabel the deleted sequence as $\{a_k\}$ again. Putting $m_k = n_k - N$, we see that the equivalence class α contains the sequence

$$p^N \left\{\frac{b_k}{c_k} p^{m_k}\right\}, \qquad m_k \geq 0. \tag{25}$$

For each k such that $m_k \geq k$, define $f_k = 1$; then

$$\left|\frac{b_k}{c_k} p^{m_k} - f_k p^{m_k}\right|_p = p^{-m_k} \left|\frac{b_k}{c_k} - f_k\right|_p \leq p^{-k}, \tag{26}$$

since the last absolute value written is ≤ 1. For those k for which $m_k < k$, define f_k as the smallest positive integer such that $b_k - c_k f_k \equiv 0 \pmod{p^{k-m_k}}$; f_k exists since $p \nmid c_k$, and (26) again holds, although for a different reason. Thus the sequence (25) is equivalent to $p^N\{f_k p^{m_k}\}$, and hence to a sequence $p^N\{g_k\}$, where the g_k are positive integers.

Represent each g_k in radix p:

$$g_k = d_{k0} + d_{k1}p + \cdots + d_{kn_k}p^{n_k}, \qquad 0 \leq d_{ki} < p \text{ for all } k, i.$$

Since $\{g_k\}$ is a Cauchy sequence, for each $u \geq 1$ there is an index M_u such that

$$|g_k - g_l|_p \leq p^{-u} \text{ whenever } k, l \geq M_u. \tag{27}$$

For $u = 1$, this means that $g_k \equiv g_l \pmod p$, so that $d_{k0} = d_{l0}$ if $k, l \geq M_1$, so all but finitely many of the d_{k0} have a common value, d_0. By the definition of N, $d_0 \neq 0$. Delete the finitely many exceptional g_k's, and relabel the sequence as $\{g_k\}$ again; clearly the new sequence is equivalent to the old. Call the first element of the new sequence h_1, as well as g_1.

Now (27) with $u = 2$ means that $d_0 + d_{k1}p \equiv d_0 + d_{l1}p \pmod{p^2}$, and hence that $d_{k1} = d_{l1}$ for $k, l \geq M_2$; so again, all but finitely many d_{k1} have a common value, d_1. Delete the exceptional g_k's, relabel the new sequence as $\{g_k\}$ again, and call its first element h_2. Iteration gives a sequence $\{h_k\} \simeq \{g_k\}$ such that all the h_k agree in their first digit, d_0, all but the first in their second digit, d_1, and in general all but $h_1, h_2, \ldots, h_{j-1}$ agree in their jth digit, d_{j-1}. Thus

$$\begin{aligned}
&|h_k - d_0|_p \leq p^{-1}, && k \geq 1,\\
&|h_k - (d_0 + d_1 p)|_p \leq p^{-2}, && k \geq 2,\\
&\quad \vdots\\
&|h_k - (d_0 + d_1 p + \cdots + d_{j-1}p^{j-1})|_p \leq p^{-j}, && k \geq j.\\
&\quad \vdots
\end{aligned}$$

Reintroducing the factor p^N, we obtain the result. ■

This theorem suggests introducing infinite series and convergence. We say that a series of rational numbers, $\sum_1^\infty a_k$, **converges** in $\mathbf{Q}_p$ if its partial sums $\sum_1^n a_k$ form a Cauchy sequence with respect to $|\ \ |_p$; the equivalence class defined by this sequence is then the **sum** of the series, and the series converges to its sum. In this language, the theorem says that every nonzero p-adic number α has a unique reduced series expansion,

$$\alpha = p^N(d_0 + d_1 p + d_2 p^2 + \cdots), \quad d_0 \neq 0, 0 \leq d_k < p \quad \text{for all } k. \tag{28}$$

The integer N in (28) is called the **order** of α, and the mapping $|\ \ |_p: \mathbf{Q} \to \mathbf{Q}$ can now be extended to $\mathbf{Q}_p$ by putting $|\alpha|_p = p^{-N}$. If $N \geq 0$, α is called a ***p*-adic integer**; it is easily verified that the set $\mathcal{O}_p$ of all p-adic integers is a domain, whose **units** are the integers with $|\alpha|_p = 1$, and that $\mathbf{Q}_p$ is the field of quotients of $\mathcal{O}_p$.

And now at last we can make sense of relations (22)–(24). The equation $x^2 = 2$ is solvable in $\mathbf{Q}_7$, and Newton's method provides the solutions; in (22) we were giving the first few terms of the solution ξ with $d_0 = 4$, and another, ξ', can be found with $d_0 = 3$. Equation (23) gives an expansion for ξ^2, but it is not the reduced expansion because the coefficients are not the digits $0, 1, \ldots, 6$. Equations (24) exhibit how one converts to the reduced expansion in general, and in particular they show that the reduced expansion of $\xi^2 - 2$ is

$$0 + 0 \cdot 7 + 0 \cdot 7^2 + \cdots,$$

as it should be.

We have now extended $\mathbf{Q}$ to the larger field $\mathbf{Q}_p$ in which every Cauchy sequence (with respect to $|\ \ |_p$) of rationals converges, and we have extended $|\ \ |_p$ from $\mathbf{Q}$ to $\mathbf{Q}_p$ as well. What is central, of course, is that $\mathbf{Q}_p$ is now **complete**, in the sense that repeating the whole process yields nothing new: every Cauchy sequence of p-adic numbers converges to a p-adic number. This opens the possibility of carrying over much of topology and analysis from $\mathbf{R}$ to $\mathbf{Q}_p$—compact sets, open sets, continuous functions, etc. In some respects the $\mathbf{Q}_p$ are simpler than $\mathbf{R}$ (for example, a necessary and sufficient condition for a p-adic series to converge is that its nth term tend to 0 as $n \to \infty$), and in some respects more complicated, in addition to seeming bizarre at first. Topologically, $\mathbf{Q}_p$, as a metric space with $d(\alpha, \beta) = |\alpha - \beta|_p$, splits into disjoint "spheres," each composed of all $\alpha \in \mathbf{Q}_p$ for which $|\alpha|_p$ has a given value. The set of p-adic units, for example, is an uncountable, open, compact, connected set without boundary, and every unit is at "distance" ≥ 1 from all nonunit p-adic numbers, by property (v) at the beginning of this section. Algebraically, there is no exact counterpart in $\mathbf{R}$ to the ring $\mathcal{O}_p$ of p-adic integers, for no proper subdomain of $\mathbf{R}$ has $\mathbf{R}$ as its quotient field.

$\mathbf{R}$ and all the fields $\mathbf{Q}_p$ are uncountable, no two of them are isomorphic, and none of them is algebraically closed (that is, a polynomial in one variable over one of these fields need not have a zero in that field). In a certain sense they exhaust the possibilities, for it turns out that every valuation on $\mathbf{Q}$ is either $|\ \ |$ or one of the $|\ |_p$, or a suitable power of one of these. They are not entirely independent, since for every nonzero rational number a,

$$|a| \prod_p |a|_p = 1. \tag{29}$$

The properties in the preceding paragraph are what make the p-adic fields interesting, arithmetically speaking. We pointed out earlier that a Diophantine equation $F(x_1, \ldots, x_n) = 0$ can only be solvable if every congruence

$$F(x_1, \ldots, x_n) \equiv 0 \pmod{p^k}$$

is solvable. It can be shown that for fixed p, this congruence is solvable for all $k \geq 1$ if and only if $F(x_1, \ldots, x_n) = 0$ has a solution in $\mathcal{O}_p$. So solvability in $\mathcal{O}_p$ for all p is a necessary condition for solvability in $\mathbf{Z}$, and so also is solvability in $\mathbf{R}$, obviously. These are nontrivial conditions, since $\mathbf{R}$ and the $\mathbf{Q}_p$ are not algebraically closed. Strong methods can be devised to test them, since analysis as well as arithmetic is available. They are genuinely different conditions, since the fields are nonisomorphic. And finally they are *sometimes*, taken together, sufficient as well as necessary, in a sense because they are exhaustive. When they are sufficient, one says that the Hasse principle applies. The term comes from a beautiful theorem of H. Hasse [1923], to the effect that any homogeneous quadratic equation $\sum_{i,j=1}^{n} a_{ij}x_ix_j = 0$ over $\mathbf{Z}$ has a solution different from $x_1 = \cdots = x_n = 0$ in $\mathbf{Z}$ if and only if it has a nonzero solution in $\mathbf{R}$ and in every $\mathcal{O}_p$.

PROBLEMS

1. Verify that $|\ |_p$, as a function on $\mathbf{Q}$, retains properties (ii), (iii), (v).

Problems 2–5 concern a fixed but unspecified valuation $|\ |^*$.

2. Verify that if $\{b_k\}$ is a subsequence of $\{a_k\}$, a Cauchy sequence, then $\overline{\{b_k\}} = \overline{\{a_k\}}$.
3. Verify that the set of limit numbers forms an additive group.
4. a) Verify that every Cauchy sequence is bounded.
   b) Verify that if $\{a_k\}$ is a non-null Cauchy sequence, then $|a_k|^* > \delta > 0$ for all large k, for suitable $\delta \in \mathbf{Q}$.
   c) Verify that the set of nonzero limit numbers forms a multiplicative group.
5. Verify that the distributive law holds for limit numbers.
6. Let $g > 0$ be a composite integer. For each nonzero $x \in \mathbf{Z}$, there is a unique highest power of g which divides x: $g^t \mid x$, $g^{t+1} \nmid x$. Put $|x|_g = g^{-t}$. Which of properties (i)–(v) hold in this case? Find two non-null Cauchy sequences of integers, relative to $|\ |_6$, whose product is null.
7. a) Find the reduced expansion in $\mathbf{Q}_5$ of 127; of -2; of 5/16; of 3/5.
   b) Find the first few terms in the reduced expansions in $\mathbf{Q}_5$ of both solutions of $x^2 = 1$; of one solution of $x^2 = -1$.

8. Let p be an odd prime. The problem is to investigate the solvability of the equation $x^2 = a$ in $\mathbf{Q}_p$, for $a \in \mathbf{Z}$, $a \neq 0$. Show that
   a) attention can be restricted to the case $p^2 \nmid a$;
   b) if $(a/p) = 1$, the equation has two distinct solutions;
   c) if $(a/p) = -1$, or if $p \parallel a$, the equation has no solution.
9. Verify that $\mathcal{O}_p$ is a domain, and that its units are the $\alpha \in \mathcal{O}_p$ with $|\alpha|_p = 1$.
10. Verify that (29) holds for $a \in \mathbf{Q}$.
11. Show that the sequence of coefficients in the reduced expansion of a nonzero rational number, as an element of $\mathbf{Q}_p$, is eventually periodic, and that if the coefficients in an expansion are eventually periodic, the expansion is that of an $a \in \mathbf{Q}$.
12. Verify that a p-adic series $\sum_1^\infty \alpha_k$ converges if and only if $|\alpha_k|_p \to 0$ as $k \to \infty$.
13. Show that no p-adic series converges conditionally: if $\sum_1^\infty \alpha_k$ converges to α, then every series resulting from this one by changing the order of terms also converges to α.

NOTES AND REFERENCES

Section 3.1

The lower bound in Case I of Fermat's problem is due to Brillhart, Tonascia, and Weinberger [1971].

Section 3.2

What we have called the order of a (mod m) was referred to in the older literature as the **exponent to which a belongs** (mod m).

Concerning the values assumed by the φ-function, two tantalizing problems have remained unsolved for many years. Carmichael [1907] gave a fallacious proof that if the equation $\varphi(x) = n$ has at least one solution, it has at least two; and Lehmer [1932] conjectured that if $\varphi(n) \mid (n - 1)$, then n is prime. No proofs have been found.

Section 3.3

The Chinese monk Yih-hing knew Theorem 3.16 in the seventh century. Theorem 3.14 was known even earlier, to Sun-tse, in the first century.

The Chinese remainder theorem is concerned with the intersection of arithmetic progressions. Questions regarding the union of progressions can be quite difficult. A set of congruences $x \equiv a_i \pmod{m_i}$, $i = 1, \ldots, n$ is called a covering set if the moduli are distinct and every x in $\mathbf{Z}$ satisfies one of them. (Example: $\{a, m\} = \{0, 2\}, \{0, 3\}, \{1, 4\}, \{1, 6\}, \{11, 12\}$.) Two long-unsolved problems: Can all m_i be odd? Can the minimum m_i be chosen arbitrarily?

Readers with the necessary algebraic background may prefer an alternative development of Theorems 3.14 and 3.15 which also yields the multiplicativity of the φ-function. The map $\rho: x \mapsto ((x)_{m_1}, \ldots, (x)_{m_n})$ is clearly a homomorphism from $\mathbf{Z}$ into $\mathbf{Z}_{m_1} \times \cdots \times \mathbf{Z}_{m_n}$ (considered as additive groups), so $\mathbf{Z}/\ker \rho$ is mapped 1–1 into $\mathbf{Z}_{m_1} \times \cdots \times \mathbf{Z}_{m_n}$. Here the kernel is the ideal generated by $[m_1, \ldots, m_n]$, and if $m_1, \ldots, m_n$ are relatively prime in pairs, the kernel is $(m_1 \cdots m_n)$, so ρ is then an isomorphism of $\mathbf{Z}_{m_1 \cdots m_n}$ into

$\mathbf{Z}_{m_1} \times \cdots \times \mathbf{Z}_{m_n}$. But both of these additive groups have order $m_1 \cdots m_n$, so the map is onto. Hence ρ defines a ring isomorphism of $\mathbf{Z}_{m_1 \cdots m_n}$ onto $\mathbf{Z}_{m_1} \otimes \cdots \otimes \mathbf{Z}_{m_n}$; this gives Theorem 3.15 and hence 3.14. The multiplicativity of the φ-function becomes the assertion that the groups of units in these Rings must have the same order.

Section 3.5

The valuation-theoretic approach to problems in arithmetic and analysis originated in the development by E. Kummer and L. Kronecker of the theory of algebraic numbers, and in that by K. Weierstrass of the theory of analytic functions of a complex variable. The work of these men was published over the latter half of the nineteenth century, but p-adic numbers as such were introduced by K. Hensel only in 1897, in a paper on algebraic number theory. He published a book giving a p-adic treatment of algebraic numbers in 1908, and a much more elementary book on number theory in 1913.

For detailed development of the contents of this section, see Artin [1959], Bachman [1964], Borevich and Shafarevich [1966], Hasse [1963], Hensel [1913], Lewis [1969], and Mahler [1961] and [1973].

The theorem attributed to Hasse is usually called the Hasse-Minkowski Theorem, since H. Minkowski had earlier proved an equivalent theorem, not using the language of p-adic numbers. The proofs are in Borevich and Shafarevich [1966], Hasse [1923], and Minkowski [1890].

4

Primitive Roots and the Group U_m

4.1 PRIMITIVE ROOTS

Each residue class $a \in U_m$ generates a cyclic group, which we have designated by $\langle a \rangle$, consisting of all the distinct powers $a, a^2, \ldots, a^t$, where $\text{ord}_m\, a = t$. It can happen that $U_m = \langle g \rangle$, for suitable m and g, and in that case g is called a **primitive root** of m. (g is not necessarily unique.) Obviously U_m then has the simplest possible structure, and it is important to know when it happens.

Since U_m is of order $\varphi(m)$, a primitive root can also be characterized as an integer g, or as a residue class $\bar{g}$ (mod m), such that $\text{ord}_m\, g = \varphi(m)$. If m has a primitive root, then Euler's theorem actually gives the smallest exponent r such that $a^r \equiv 1 \pmod m$ for all $a \in U_m$.

For arbitrary $a \in U_m$, we need to know the connection between the order of a and the orders of powers of a.

Theorem 4.1 *If* $a \in U_m$ *and* $\text{ord}_m\, a = t$, *then* $\text{ord}_m(a^n) = t/(n, t)$.

Proof. Among the sequence of powers $a^n, a^{2n}, \ldots$, the relation $a^{kn} = 1$ holds for the first time when kn is first a multiple of t. Now $t \mid kn$ if and only if

$$\frac{t}{d} \,\Big|\, k\frac{n}{d}, \qquad \text{where } d = (n, t),$$

and since t/d is relatively prime to n/d, this requires that t/d divide k, and the smallest such k is obviously t/d itself. ■

We now proceed in stages, first considering the simplest case in which $m = p$, a prime.

Theorem 4.2 *If* $t \nmid (p - 1)$, *no integer has order* t (mod p). *If* $t \mid (p - 1)$, *the number of* $a \in U_m$ *of order* t *is either* 0 *or* $\varphi(t)$.

Proof. We already know (Theorem 3.11) that the first sentence is true. Suppose $t \mid (p - 1)$, and that $\text{ord}_p\, a = t$. Then $x^t \equiv 1 \pmod p$ has the distinct solutions $x \equiv a, a^2, \ldots, a^t$ (in accordance with the corollary to Theorem 3.19), and it can have no others, by Lagrange's theorem. Thus every element of order t is congruent to one of $a, a^2, \ldots, a^t$, and by the preceding theorem, $\text{ord}_p(a^n) = t$ if and only if $(n, t) = 1$; there are $\varphi(t)$ such n with $1 \le n \le t$. ■

Theorem 4.3 *If $t \mid (p - 1)$, the number of $a \in U_p$ with* $\text{ord}_p\, a = t$ *is $\varphi(t)$. In particular, U_p is cyclic and there are $\varphi(p - 1)$ primitive roots of p.*

Proof. For each divisor t of $p - 1$, let $\psi(t)$ be the number mentioned in the theorem. Every $a \in U_p$ has a unique order, and there are $p - 1$ such a's, so

$$\sum_{t \mid (p-1)} \psi(t) = p - 1.$$

On the other hand, by Theorem 3.12,

$$\sum_{t \mid (p-1)} \varphi(t) = p - 1,$$

and by Theorem 4.2, $\psi(t) = 0$ or $\varphi(t)$ for each t. Obviously, these three facts are compatible only if $\psi(t) = \varphi(t)$ for each t. ■

The next simplest case is $m = p^n$, $n > 1$. It turns out that 2 behaves differently from the other primes, and we set it aside temporarily. Note that in the next theorem a is an integer, not a residue class, since otherwise z might not be well-defined.

Theorem 4.4 *Suppose that p is prime, and that $p \nmid a$. Let* $\text{ord}_p\, a = t$ *and let p^z be the p-component of $a^t - 1$, that is, $p^z \parallel (a^t - 1)$. Then if $p > 2$ or $z > 1$,*

$$t_n = \text{ord}_{p^n}\, a = \begin{cases} t & \textit{for } n \le z \\ tp^{n-z} & \textit{for } n \ge z. \end{cases}$$

Proof. a) Suppose that $n \le z$. Then

$$a^t \equiv 1 \pmod{p^n}$$

so $t_n \mid t$. But $a^{t_n} \equiv 1 \pmod{p^n}$ implies $a^{t_n} \equiv 1 \pmod{p}$, so $t \mid t_n$. Hence $t_n = t$.

b) By definition, $z > 0$. Suppose $n > z$, and put $a^t = 1 + up^z$, where $p \nmid u$. Thus for $k = z$ and $u_k = u$,

$$\text{(1)} \qquad a^{tp^{k-z}} = 1 + u_k p^k, \qquad \text{where } p \nmid u_k.$$

For any $k \ge z$ for which this relation holds, taking the pth power of both sides gives

$$\text{(2)} \qquad a^{tp^{k-z+1}} = 1 + \binom{p}{1} u_k p^k + \cdots + \binom{p}{p-1} (u_k p^k)^{p-1} + (u_k p^k)^p.$$

Clearly the "interior" binomial coefficients

$$\binom{p}{s} = \frac{p(p-1)\cdots(p-s+1)}{s!}, \qquad 0 < s < p$$

are divisible by p, since p occurs in the numerator but not in the denominator. Thus the p-components of the successive terms after the first, on the right hand side of (2), are at least

$$p^{k+1}, \quad p^{2k+1}, \quad \ldots, \quad p^{(p-1)k+1}, \quad p^{kp}.$$

Here all the exponents after the first are strictly larger than $k + 1$. This is obvious for all but the last, and the inequality $kp > k + 1$ is equivalent to $k(p - 1) > 1$, which follows from the hypothesis that either $p > 2$ or $k \geq z > 1$. Hence there is a $v_k \in \mathbf{Z}$ such that

$$\begin{aligned} a^{tp^{k-z+1}} &= 1 + p^{k+1}(u_k + pv_k) \\ &= 1 + p^{k+1}u_{k+1}, \quad p \nmid u_{k+1}. \end{aligned}$$

Hence (1) holds for all $k \geq z$, by induction.

In particular, (1) implies that $t_n \mid tp^{n-z}$. Suppose $t_n = t'p^{n-r}$, where $t' \mid t$ and $r \geq z$. Now $a^{t_n} \equiv 1 \pmod{p^n}$ implies $a^{t_n} \equiv 1 \pmod{p}$, so $t \mid t_n$; since $(t, p) = 1$, we have $t \mid t'$, so $t = t'$. Hence

$$a^{tp^{n-r}} \equiv 1 \pmod{p^n},$$

and (1) shows that $r \leq z$. Thus $r = z$, and $t_n = tp^{n-z}$. ■

We can use Theorem 4.4 to construct primitive roots of p^n when p is odd. Let $g \in \mathbf{Z}$ be a primitive root of p and suppose first, in the notation of Theorem 4.4, that $z = 1$, so that $p^2 \nmid (g^{p-1} - 1)$. Then for $n \geq 1$,

$$\operatorname{ord}_{p^n} g = (p - 1)p^{n-1} = \varphi(p^n),$$

so g is also a primitive root of p^n. On the other hand, if $z > 1$, consider the number $g_1 = g + p$, which is again a primitive root of p. Let $p^{z_1} \parallel (g_1^{p-1} - 1)$. We have

$$\begin{aligned} g_1^{p-1} - 1 = (g + p)^{p-1} - 1 &\equiv g^{p-1} + (p - 1)g^{p-2}p - 1 \\ &\equiv (p - 1)g^{p-2}p \not\equiv 0 \pmod{p^2}, \end{aligned}$$

so $z_1 = 1$, and the preceding argument shows that g_1 is a primitive root of p^n for all $n \geq 1$.

Theorem 4.5 *Each positive power of an odd prime has a primitive root.*

The case $p = 2$ is not as simple, since Theorem 4.4 definitely fails when $p = 2, z = 1$. For 3 is a primitive root of 2 for which $z = 1$, but it is not a primitive root of 2^3; in fact, there are no primitive roots of 8, since $\varphi(8) = 4$ while $1^2 \equiv 3^2 \equiv 5^2 \equiv 7^2 \pmod 8$. On the other hand, we can take $a = 5$, $t = 1$, $z = 2$ in Theorem 4.4, and obtain

$$\operatorname{ord}_{2^n} 5 = 2^{n-2} = \tfrac{1}{2}\varphi(2^n). \tag{3}$$

Theorem 4.6 *Both 2 and 2^2 have the primitive root -1. For $n \geq 3$, 2^n does not have primitive roots. On the other hand, the powers $5, 5^2, 5^3, \ldots, 5^{2^{n-2}}$ constitute half of a reduced residue system* (mod 2^n), *namely all the integers* $\equiv 1$

(mod 4). *The missing residue classes are represented by* $-5, -5^2, \ldots, -5^{2^{n-2}}$. *In group-theoretic language,* U_{2^n} *is not a cyclic group but has two generators,* $\overline{-1}$ *and* $\overline{5}$, *of orders* 2 *and* $\frac{1}{2}\varphi(2^n)$.

Proof. Since $a^2 \equiv 1 \pmod 8$ for all odd a, the relation

$$a^{2^{k-2}} \equiv 1 \pmod{2^k}$$

holds for $k = 3$, and for any $k \geq 3$ for which it holds,

$$a^{2^{k-1}} = (1 + 2^k u)^2 = 1 + 2^{k+1}(u + 2^{k-1}u^2) \equiv 1 \pmod{2^{k+1}},$$

so it holds for all $k \geq 3$. Thus always $(\operatorname{ord}_{2^k} a) \mid 2^{k-2}$. The third sentence of the theorem follows from (3) and the fact that all powers of 5 are $\equiv 1 \pmod 4$, together with the fact that there are exactly 2^{n-2} positive integers less than 2^n and $\equiv 1 \pmod 4$. Similarly, the numbers $-5, -5^2, \ldots, -5^{2^{n-2}}$ are distinct (mod 2^n), and they are all $\equiv -1 \pmod 4$, so they must be congruent in some order to $3, 7, 11, \ldots, 2^n - 1$. ∎

Theorem 4.7 *The numbers having primitive roots are* 2, 4, p^n, *and* $2p^n$, *where* $n \in \mathbf{Z}^+$ *and* p *runs over the odd primes.*

Proof. If $g \in \mathbf{Z}$ is a primitive root of p^n, then so is $g + p^n$, and one of these two primitive roots is odd—say g_2. Since g_2 is a primitive root,

$$\text{if } d \mid \varphi(p^n), \quad \text{then } g_2^d \equiv 1 \pmod{p^n} \text{ if and only if } d = \varphi(p^n).$$

But $\varphi(2p^n) = \varphi(p^n)$, and since g_2 is odd, $2 \mid (g_2^d - 1)$ for all d, so

$$\text{if } d \mid \varphi(2p^n), \quad \text{then } g_2^d \equiv 1 \pmod{2p^n} \text{ if and only if } d = \varphi(2p^n).$$

This means that g_2 is a primitive root of $2p^n$.

What is left is to show that m does not have a primitive root if at least two of the prime-power factors in $m = \prod p_i^{e_i}$ are such that $\varphi(p_i^{e_i}) > 1$. Put $M = [\varphi(p_1^{e_1}), \varphi(p_2^{e_2}), \ldots]$. Since

$$a^{\varphi(p_i^{e_i})} \equiv 1 \pmod{p_i^{e_i}}, \qquad i = 1, 2, \ldots$$

also

$$a^M \equiv 1 \pmod{p_i^{e_i}}, \qquad i = 1, 2, \ldots$$

and hence

$$a^M \equiv 1 \pmod m.$$

But if $\varphi(x) > 1$ then $\varphi(x)$ is even, so the LCM in the exponent is strictly smaller than the product of the entries, and that product is $\varphi(m)$, by Theorem 3.7. ∎

The reader may have noticed that the proof that primes have primitive roots was not terribly helpful for computations—it gave no direct algorithm for actually finding one. For a single prime this is a finite problem, so it can be solved by trial and error: reduce $2, 2^2, 2^3, \ldots$ to their least positive remainders (mod p), to find $\operatorname{ord}_p 2$; if $\operatorname{ord}_p 2 < p - 1$, choose an integer a which is not congruent to a power of 2 (for see Theorem 4.1) and find its order, etc. This search can be made somewhat more efficient by exploiting the following fact, which we have already used implicitly: *if* $\text{ord}_p a = t$ *and* $\text{ord}_p b = u$ *and* $(t, u) = 1$, *then* $v = \text{ord}_p ab = tu$. For clearly $(ab)^{tu} \equiv 1 \pmod p$, so $v \mid tu$. If $v = t_1 u_1$, where $t_1 \mid t$ and $u_1 \mid u$, then $\text{ord}_p a^u = t$, and $1 \equiv (ab)^{t_1 u} \equiv (a^u)^{t_1} \pmod p$, so $t \mid t_1$. Similarly, $u \mid u_1$. Suitably modified, this principle enables us to go from elements a and b of arbitrary orders t and u to an element of order $[t, u]$. For example, since

$$\text{ord}_{13} 10 = 6 \qquad \text{and} \qquad \text{ord}_{13} 8 = 4,$$

we see that $10^2 \cdot 8$ or 7 is a primitive root of 13.

In fact, this idea can be made to yield a new proof that primes have primitive roots. Suppose that q is prime and that q^f is the q-component of $p - 1$, with $f > 0$. Then by the Corollary to Theorem 3.19, the congruences

$$x^{q^{f-1}} \equiv 1 \pmod p \qquad \text{and} \qquad y^{q^f} \equiv 1 \pmod p$$

have q^{f-1} and q^f solutions, respectively, and a y which is not an x has order $q^f \pmod p$. For each q such that $q \mid (p - 1)$, take one such y, and multiply these y's together; the product is a primitive root of p.

Instead of testing all the powers $a, a^2, a^3, \ldots$ to determine $\text{ord}_p a$ and then recognizing a primitive root when some a has order $p - 1$, it may be more efficient to apply the following criterion, which bears an obvious affinity to the converse of Fermat's theorem mentioned in Problem 11 of Section 3.2:

Theorem 4.8 *If* $p \nmid a$ *and for every prime divisor* q *of* $p - 1$,

$$a^{(p-1)/q} \not\equiv 1 \pmod p,$$

then a *is a primitive root of* p.

New kinds of difficulties arise as soon as one asks a question involving primitive roots of various primes simultaneously. In the *Disquisitiones*, Gauss gave a primitive root and certain indices, for each prime power less than 100, and he always chose $g = 10$ when it is a primitive root, for computational simplicity. (And for another reason, to be found in the problems following.) One third of the primes < 100 allow $g = 10$, but Gauss refrained from speculating, on the basis of such scanty evidence, about whether there are infinitely many primes of this kind. More data became available in 1839, when C. G. J. Jacobi published his *Canon Arithmeticus*. This was a two-way table for solving $y \equiv g^x \pmod{p^e}$ for either variable, for each prime power < 1000, again using 10 as base when possible. In the introduction, Jacobi pointed out that of the 365 primes < 2500, fully 148 have 10 as primitive root. Much larger tables exist now, of course, and they continue to support the more general conjecture, commonly attributed to E. Artin in 1927, that every nonsquare integer $a \neq 0, -1$ is a primitive root of infinitely many primes. In fact Artin made a quantitative conjecture concerning the fraction of the primes $p \leq x$, for large x, for which such an a is a primitive root of p. A modified version of this conjecture (the need for which was shown by computations by D. H. Lehmer) was formulated by H. Heilbronn, and this was proved by Hooley [1967] under the assumption that another unproved conjecture (the so-called extended Riemann hypothesis) is true. The formula is complicated and varies with the nature of a, but for both $a = 2$ and $a = 10$ it predicts that, of the $\pi(x)$ primes up to x, about 37.396% will have a as primitive root. For $x = 50{,}000$, the actual ratio is 0.37456 for $a = 2$ and 0.37924 for $a = 10$.

There are also fascinating unsolved questions about how the $\varphi(p - 1)$ primitive roots are distributed among the integers $1, 2, \ldots, p - 1$. If $g(p)$ denotes the smallest positive primitive root of p, then it is known that for $\varepsilon > 0$ and $p \to \infty$, $\lim g(p)/p^{1/4+\varepsilon} = 0$ and $\lim g(p)/\log p \neq 0$; numerical evidence suggests that $g(p) < c \log^3 p$ for some constant $c > 0$, but not even the weaker inequality $g(p) < cp^\varepsilon$ can be proved. Almost nothing is known about the size of the smallest positive prime $P(p)$ which is a primitive root of p, although again numerical evidence suggests that $P(p) < c' \log^3 p$.

PROBLEMS

†1. Show that if p is prime, n is positive and $a \equiv b \pmod{p^n}$, then $a^{p^k} \equiv b^{p^k} \pmod{p^{n+k}}$. (This generalizes the first step in the proof given for Theorem 4.6.)

2. Let g be a primitive root of p, an odd prime. Show that $-g$ is also a primitive root of p if and only if $p \equiv 1 \pmod 4$.

3. Show that if $p = 2^m + 1$ and $(a/p) = -1$, then a is a primitive root of p.

4. Show that if p is an odd prime and $\text{ord}_p\, a = t > 1$, then

$$\sum_{k=1}^{t-1} a^k \equiv -1 \pmod{p}.$$

5. Compute the orders of 2 and 7 (mod 73), and from this information find a primitive root of 73.

6. For what primes p is $a^{17} \equiv a \pmod{p}$ for all a?

7. Show that if $\text{ord}_p\, a = 3$, then $\text{ord}_p\, (a + 1) = 6$. [*Hint:* Factor $a^3 - 1$.]

8. Under what circumstances do the kth powers of the elements of a reduced residue system (mod p) again constitute a reduced residue system (mod p)?

9. Show that if m has primitive roots, there are $\varphi(\varphi(m))$ of them, and their product is congruent to 1 (mod m) if $m > 6$.

10. Find all the primitive roots of 25.

11. Show that if g is a primitive root of p^n for all $n \geq 1$, then for each such n the roots of the congruence

$$x^{p-1} \equiv 1 \pmod{p^n}$$

are $g^{kp^{n-1}}$, $k = 1, 2, \ldots, p - 1$.

12. Prove, by induction on n or otherwise, that if p is an odd prime and $p \nmid b$ and $a^p \equiv b^p \pmod{p^{n+1}}$, then $a \equiv b \pmod{p^n}$.
13. Show that a primitive root of p^e is also a primitive root of p^f for $1 \le f < e$.
14. Prove that 10 is a primitive root of infinitely many prime powers.
15. Suppose that $m > 1$ and $(m, 10) = 1$. Let $a_1, a_2, \ldots$ and $b_1, b_2, \ldots$ be defined by the following conditions:

$$\begin{aligned} 10 &= ma_1 + b_1, & 0 &\le b_1 < m, \\ 10b_{k-1} &= ma_k + b_k, & 0 &\le b_k < m, \quad \text{for } k \ge 2. \end{aligned}$$

a) Show that $a_1, a_2, \ldots$ are the successive digits in the decimal expansion of $1/m$:

$$\frac{1}{m} = 0.a_1a_2\ldots.$$

b) Show that $b_k \equiv 10^k \pmod{m}$, for $k \ge 1$.
c) Show that $a_1, a_2, \ldots$ is periodic from the beginning: for some h, $a_1 = a_h$, $a_2 = a_{h+1}, \ldots$.
d) Find the exact length of the period (i.e., the minimal value of $h > 1$), in terms of concepts occurring in the text. [*Hint:* You may find it convenient to prove and use the fact that for $k \ge 2$, $b_{k-1}/m = 0.a_ka_{k+1}\ldots$.]

16. Let p be an odd prime and let r and n be positive integers. Show that the sum of the rth powers of the elements of a reduced residue system $\pmod{p^n}$ is always divisible by p^{n-1}, and is divisible by p^n if $(p - 1) \nmid r$.
17. Define the function $\lambda(m)$ as follows: $\lambda(1) = 1$; $\lambda(p^e) = \varphi(p^e)$ if p is an odd prime;

$$\lambda(2^e) = \begin{cases} \varphi(2^e) & \text{if } e = 1 \text{ or } 2, \\ \frac{1}{2}\varphi(2^e) & \text{if } e > 2; \end{cases}$$

and for distinct primes $p_1, \ldots, p_r$ and nonnegative $e_1, \ldots, e_r$,

$$\lambda(p_1^{e_1} \cdots p_r^{e_p}) = [\lambda(p_1^{e_1}), \ldots, \lambda(p_r^{e_r})].$$

a) Show that if $(a, m) = 1$, then $a^{\lambda(m)} \equiv 1 \pmod{m}$.
b) Show that for each $m > 1$ there is an a such that $\operatorname{ord}_m a = \lambda(m)$.
c) If $m > 4$ and $(a, m) = 1$, show that $a^{\varphi(m)/2} \equiv (a/p) \pmod{m}$ if $m = p^e$ or $2p^e$, while $a^{\varphi(m)/2} \equiv 1 \pmod{m}$ otherwise.

18. A famous unproved conjecture due to E. Catalan (1844) is that 8 and 9 are the only consecutive positive integers which are powers of smaller integers; more briefly, the conditions $a^x - b^y = 1$, $x > 1$, $y > 1$ imply $a = y = 3$, $b = x = 2$. Show that at any rate this is the only solution under the additional restriction that a and b be prime. [*Hint:* When $b = 2$, consider two cases. If $a \equiv 3 \pmod 4$, then since $y > 1$, x is even; factor $a^x - 1$. In the other case, if $2^z \parallel a - 1$ then $z \ge 2$, so Theorem 4.4 applies, yielding $2^{y-z} = x$ and hence $a^x = 2^y + 1 \le x(a - 1) + 1$, which is impossible for $a > 1$ and $x > 1$. For the case $a = 2$, see Section 3.1, Problem 9.] Tijdemann [1976] has shown that there are only finitely many solutions for arbitrary $a, b > 0$, and in fact that for any solution, $a^x < 10^{10^{500}}$.

19. Consider the following three assertions:

    i) $m > 1$ has a primitive root;
    ii) the only solutions of $x^2 = 1$ in U_m are $x = \pm 1$;
    iii) $\prod_{a \in U_m} a \equiv -1 \pmod{m}$.

    Show directly (without using Theorem 4.7) that (i) implies (ii), that (i) implies (iii), and (independently) that (ii) implies (iii). Show with the help of Theorem 4.7 that (ii) implies (i). Show finally that if it could be proved directly that (ii) is necessary and sufficient for (i), then Theorem 4.7 would be a corollary. [*Hints:* Recall Theorem 3.21. In dealing with (iii), it may be useful to pair each $a \in U_m$ with its inverse, and to consider which elements are their own inverses.]

4.2 THE STRUCTURE OF U_m

For the moment, suppose that $m = p_1^{e_1} \cdots p_r^{e_r}$. Recall that in Theorem 3.15 we established a ring isomorphism between $\mathbf{Z}_m$ and $\mathbf{Z}_{p_1^{e_1}} \oplus \cdots \oplus \mathbf{Z}_{p_r^{e_r}}$ by means of the 1–1 correspondence

$$(x)_m \leftrightarrow [(x)_{p_1^{e_1}}, \ldots, (x)_{p_r^{e_r}}].$$

Now if $x \in \mathbf{Z}$, then clearly $(x)_m \in U_m$ (that is, $(x, m) = 1$) if and only if $(x)_{p_i^{e_i}} \in U_{p_i^{e_i}}$ for $1 \leq i \leq r$. Thus we have the following result, which merely describes the image of U_m under the isomorphism.

Theorem 4.9 *If $m = p_1^{e_1} \cdots p_r^{e_r}$, where the p_i are arbitrary distinct primes and the e_i are positive, then U_m is isomorphic to $U_{p_1^{e_1}} \times \cdots \times U_{p_r^{e_r}}$, the group of r-tuples under componentwise multiplication.*

We can go a step further, using what we know about the individual factors U_{p^e}. But now 2 is exceptional so we change notation slightly. We continue to use $\langle a \rangle$ for the cyclic group generated by a, when a is an element of a specified group.

Theorem 4.10 *Suppose $m > 1$ has the prime-power decomposition $m = 2^e p_1^{e_1} \cdots p_r^{e_r}$, where $e \geq 0$, $r \geq 0$, and if $r > 0$ then $p_1, \ldots, p_r$ are distinct odd primes and $e_1, \ldots, e_r$ are positive. Let $g_1, \ldots, g_r$ be primitive roots of $p_1, \ldots, p_r$, if $r > 0$. Then U_m is isomorphic to the product of multiplicative cyclic groups:*

$$\langle (-1)_{2^2} \rangle \times \langle (5)_{2^e} \rangle \times \langle (g_1)_{p_1^{e_1}} \rangle \times \cdots \times \langle (g_r)_{p_r^{e_r}} \rangle,$$

where the first two factors are to be omitted if $e = 0$ or 1 and the second factor is to be omitted if $e = 2$. Differently expressed, to every integer a there corresponds exactly one collection of exponents $[\eta, \varepsilon, \varepsilon_1, \ldots, \varepsilon_r]$ such that

$$\begin{aligned} a &\equiv (-1)^\eta 5^\varepsilon \pmod{2^e}, & 0 \leq \eta < 2,\quad 0 \leq \varepsilon < \tfrac{1}{2}\varphi(2^e), \\ a &\equiv g_1^{\varepsilon_1} \pmod{p_1^{e_1}}, & 0 \leq \varepsilon_1 < \varphi(p_1^{e_1}), \\ &\vdots \\ a &\equiv g_r^{\varepsilon_r} \pmod{p_r^{e_r}}, & 0 \leq \varepsilon_r < \varphi(p_r^{e_r}), \end{aligned} \tag{4}$$

with the same omissions as before.

Now fix m and, simply to be definite, assume for the moment that $r > 0$ and $8 \mid m$, so that all the indicated congruences actually occur in (4), with $p_1, \ldots, p_r$ in fixed order. Then the theorem says that for fixed primitive roots $g_1, \ldots, g_r$, there is a 1–1 correspondence between $a \in U_m$ and $(r + 1)$-tuples $\{\eta, \varepsilon, \varepsilon_1, \ldots, \varepsilon_r\}$ with $\eta \in \mathbf{Z}_2$, $\varepsilon \in \mathbf{Z}_{2^{e-2}}$, and $\varepsilon_i \in \mathbf{Z}_{\varphi(p_i^{e_i})}$ for $1 \le i \le r$. Moreover, if $a \leftrightarrow \{\eta, \varepsilon, \varepsilon_1, \ldots, \varepsilon_r\}$ and $a' \leftrightarrow \{\eta', \varepsilon', \varepsilon_1', \ldots, \varepsilon_r'\}$, then clearly

$$aa' \leftrightarrow \{\eta + \eta', \varepsilon + \varepsilon', \varepsilon_1 + \varepsilon_1', \ldots, \varepsilon_r + \varepsilon_r'\}.$$

In other words, the correspondence

$$a \leftrightarrow \{\eta, \varepsilon, \varepsilon_1, \ldots, \varepsilon_r\}$$

defines an isomorphism between U_m and the product of groups

$$\mathbf{Z}_2 \times \mathbf{Z}_{\varphi(2^e)/2} \times \mathbf{Z}_{\varphi(p_1^{e_1})} \times \cdots \times \mathbf{Z}_{\varphi(p_r^{e_r})},$$

in which the factors are regarded merely as additive cyclic groups, of the orders indicated by their subscripts. The vector of exponents on the right-hand side in the above correspondence is called an **index vector** of a; given the primitive roots $g_1, \ldots, g_r$, it is unique in the above product of groups.

In the remainder of this chapter we designate by q a modulus for which U_q is cyclic, and by g one of its primitive roots. For such modulus, the index vector of a has only a single entry, and this is then called the **index** of a. It bears much the same relation to a as $\log x$ bears to x when $x \in \mathbf{R}^+$, since the defining relation $a \equiv g^{\text{ind}\, a} \pmod q$ easily yields

$$\begin{aligned} \text{ind}(ab) &\equiv \text{ind}\, a + \text{ind}\, b \pmod{\varphi(q)} \\ \text{ind}\, a^n &\equiv n\, \text{ind}\, a \pmod{\varphi(q)}. \end{aligned} \tag{5}$$

The usefulness of indices will become apparent in the remaining sections of this chapter. We end the present section with a numerical example.

The problem is to determine the primitive roots, and a table of indices with respect to one of them, for $q = 41$. Here q is prime and $\text{ord}_q\, a \mid 40$ for all $a \in U_q$; there are $\varphi(40) = 16$ primitive roots. According to Theorem 4.1, the primitive roots are the elements of U_q which are neither squares nor fifth powers, since $40 = 2^2 \cdot 5$. The easy way to compute the successive squares $1^2, 2^2, \ldots, 40^2$ is to use the relation $(n + 1)^2 = n^2 + (2n + 1)$ to obtain them by addition rather than multiplication:

3 5 7 9 11 13 15 17 19 21
$a^2 \pmod{41}$: 1 4 9 16 25 36 8 23 40 18 39

23 25 27 29 31 33 35 37 39 41 43
21 5 32 20 10 2 37 33 31 ¦ 31 33.

There is no point in continuing, since $x^2 \equiv (41 - x)^2 \pmod{41}$; the quadratic residues of q are

1, 2, 4, 5, 8, 9, 10, 16, 18, 20, 21, 23, 25, 31, 32, 33, 36, 37, 39, 40.

Of the remaining 20 elements of U_q, 16 are primitive roots, so 4 are fifth powers but not squares. A short computation shows that, mod 41,

$$\begin{array}{llll} 3^5 \equiv 38, & 6^5 \equiv 27, & 7^5 \equiv 38, & 11^5 \equiv 3, \\ 12^5 \equiv 3, & 13^5 \equiv 38, & 14^5 \equiv 27, & 15^5 \equiv 14, \end{array}$$

so these fifth powers are 3, 14, 27, 38, while the remaining numbers,

$$6, 7, 11, 12, 13, 15, 17, 19, 22, 24, 26, 28, 29, 30, 34, 35,$$

are primitive roots.

Let $g = 6$. The successive powers of g are computed recursively, using $g^{n+1} \equiv g \cdot g^n \pmod q$, to give the following table:

a	6	36	11	25	27	39	29	10	19	32	28	4	24
ind a	1	2	3	4	5	6	7	8	9	10	11	12	13

a	21	3	18	26	33	34	40	25	5	30	16	14	2
ind a	14	15	16	17	18	19	20	21	22	23	24	25	26

a	12	31	22	9	13	37	17	20	38	23	15	8	7	1
ind a	27	28	29	30	31	32	33	34	35	36	37	38	39	40

Here each entry in the first row, after the 6, comes from multiplying the preceding entry by 6 and reducing (mod 41), and each entry in the second row gives the exponent of the power of 6 which yields the entry above it.

Once a table of this kind is available, congruences of the form $ax^n \equiv b \pmod q$ become much easier to solve. For example, the linear congruence

$$16x \equiv 37 \pmod{41}$$

is equivalent to

$$\text{ind } 16 + \text{ind } x \equiv \text{ind } 37 \pmod{40},$$

so $\text{ind } x \equiv 32 - 24 \equiv 8 \pmod{40}$ and $x \equiv 10 \pmod{41}$.

Similarly, from $3x^2 \equiv 17 \pmod{41}$ we obtain

$$2 \text{ ind } x \equiv \text{ind } 17 - \text{ind } 3 \equiv 33 - 15 \equiv 18 \pmod{40},$$

or $\text{ind } x \equiv 9$ or $29 \pmod{40}$, yielding $x \equiv 19$ or $22 \pmod{41}$. On the other hand, the congruence $3x^2 \equiv 10 \pmod{41}$ is not solvable, since the congruence $2 \text{ ind } x \equiv \text{ind } 10 - \text{ind } 3 \equiv 8 - 15 \pmod{40}$ has no solution.

A short table of indices is to be found at the back of this book.

PROBLEMS

1. Find all the primitive roots of 37; of 29; of 81; of 26. Think first—then compute.
2. Given that 10 is a primitive root of 23, construct a table of indices and use it to find all the other primitive roots, and to solve the congruences
   a) $17x \equiv 10 \pmod{23}$; b) $17x^2 \equiv 10 \pmod{23}$.
3. Develop a method for solving the congruence
$$Ax^2 + Bx + C \equiv 0 \pmod p$$
   by use of indices, when p is an odd prime not dividing A. (Make the leading coefficient 1 and then, after suitable modification if necessary, complete the square.) Apply your method to
   a) $15x^2 - 4x + 9 \equiv 0 \pmod{41}$;
   b) $15x^2 - 4x + 23 \equiv 0 \pmod{41}$.
4. Modify Theorem 4.10 as follows. Let $m = 2^e p_1^{e_1} \cdots p_r^{e_r}$, and let g_i be a primitive root of the odd prime p_i, for $1 \le i \le r$. Considering U_m as a subset of $\mathbf{Z}_{2^e} \oplus \mathbf{Z}_{p_1^{e_1}} \oplus \cdots \oplus \mathbf{Z}_{p_r^{e_r}}$, it contains the elements
$$\gamma_0 = [-1, 1, \ldots, 1],\ \gamma_1 = [1, g_1, 1, \ldots, 1], \ldots, \gamma_r = [1, \ldots, 1, g_r],$$
   in each of which all entries but one are 1; for $e \ge 3$ also define $\gamma_{-1} = [5, 1, \ldots, 1]$. Then for $e \ge 3$, the elements $\gamma_{-1}, \gamma_0, \gamma_1, \ldots, \gamma_r$ constitute a **basis** for the multiplicative group U_m, in the sense that each $a \in U_m$ has a unique representation in the form
$$a = \gamma_{-1}^{x_{-1}} \gamma_0^{x_0} \gamma_1^{x_1} \cdots \gamma_r^{x_r},$$
   with $0 \le x_{-1} < 2^{e-2}$, $0 \le x_0 < 2$, $0 \le x_i < \varphi(p_i^{e_i})$ for $1 \le i \le r$. If $e = 2$, then $\gamma_0, \gamma_1, \ldots, \gamma_r$ constitute a basis; if $e = 0$ or 1, omit γ_0. Find a basis for U_{36}, for U_{120}.
5. Let p be an odd prime and suppose $n > 0$, $x > 1$. Abbreviate $x^{p^{n-1}}$ as y (so that $y \equiv x \pmod p$), and put
$$f(x) = \frac{y^p - 1}{y - 1} = y^{p-1} + \cdots + y + 1$$
$$= (y-1)^{p-1} + \binom{p}{1}(y-1)^{p-2} + \cdots + \binom{p}{2}(y-1) + p.$$
   Show that
   a) $p^2 \nmid f(x)$. [Two cases: $p \mid f(x)$ or $p \nmid f(x)$.]
   b) $f(x) > p$.
   c) $(f(x), x) = 1$.
   d) If q is prime, $q \ne p$ and $q \mid f(x)$, then $q \equiv 1 \pmod{p^n}$. [Recall Problem 3 of Section 2.1.]

   Deduce that there exists a prime $q \equiv 1 \pmod{p^n}$, and then by taking $x = q_1 \cdots q_r$, where each $q_i \equiv 1 \pmod{p^n}$, that there are infinitely many such primes. (Cf. Problem 21, Section 3.2, for the case $p = 2$. The present result is generalized in Problem 11 of Section 6.2.)

6. Calculate a table of index vectors for the case $m = 40$. What corresponds to relations (5) in this case? Use the table as an aid in solving the following congruences or showing them to be unsolvable:

   a) $13x^3 \equiv 21 \pmod{40}$ b) $13x^2 \equiv 21 \pmod{40}$

   c) $13x^2 \equiv 37 \pmod{40}$ d) $13x^2 \equiv 2 \pmod{40}$

   e) $13x^2 \equiv 4 \pmod{40}$.

4.3 *n*th-POWER RESIDUES

Generalizing the notion of quadratic residue, we say that $a \in U_m$ is an ***n*th-power residue** (or nonresidue) of m if the congruence

$$x^n \equiv a \pmod{m} \tag{6}$$

is (or is not) solvable, that is, if the equation $x^n = a$ is (or is not) solvable in $\mathbf{Z}_m$, and hence in U_m. Thus the whole subject is really concerned with the group U_m. This is reflected by the fact that the next theorem and its proof extend immediately to an arbitrary finite abelian group, in place of U_m.

Theorem 4.11 *If a and b are both nth-power residues of m, then the congruences $x^n \equiv a \pmod{m}$ and $x^n \equiv b \pmod{m}$ have the same number of solutions.*

Proof. Let $x_1, \ldots, x_k$ be all the solutions of (6), and suppose that $y^n \equiv b \pmod{m}$. Then $(yx_1x_j^{-1})^n \equiv b \pmod{m}$ for $j = 1, \ldots, k$, and clearly $yx_1x_j^{-1} \not\equiv yx_1x_l^{-1} \pmod{m}$ if $j \neq l$. By symmetry, each congruence in the theorem has at least as many solutions as the other. ■

Since $x_ix_j^{-1}$ is a solution of

$$u^n \equiv 1 \pmod{m}, \tag{7}$$

the proof just given suggests the following theorem and its proof.

Theorem 4.12 *The set of nth-power residues* (mod m) *forms a subgroup $U_m^{(n)}$ of U_m. Every solution of* (6) *is of the form ux, where x is a fixed solution of* (6) *and u ranges over the solutions of* (7). *(In algebraic language, the map $x \mapsto x^n$ is a homomorphism of U_m onto $U_m^{(n)}$ whose kernel is the group of solutions of* (7), *and the solutions of* (6) *form a coset of this group in U_m.)*

In studying nth-power residues we may, by virtue of the Chinese remainder theorem (as in Theorem 3.21), restrict attention to the case of prime-power modulus. For odd primes we have a criterion for a to be an nth-power residue, given in the next theorem.

Theorem 4.13 *Let q be a number having a primitive root, and suppose that $a \in U_q$. Then a is an nth-power residue of q if and only if*

$$a^{\varphi(q)/d} \equiv 1 \pmod{q}, \qquad \textit{where } d = (n, \varphi(q)). \tag{8}$$

The number of nth-power residues of q is $\varphi(q)/d$, and each of them is the nth power of exactly d integers (mod q).

Proof. Taking indices in (6), we obtain

$$n \text{ ind } x \equiv \text{ind } a \pmod{\varphi(q)},$$

which is solvable if and only if $(n, \varphi(q)) \mid \text{ind } a$. The latter condition is equivalent to ind $a \equiv 0 \pmod{d}$, and hence to

$$\frac{\varphi(q)}{d} \text{ ind } a \equiv 0 \pmod{\varphi(q)},$$

and hence to (8). Finally, if g is a primitive root of q, then the $\varphi(q)/d$ numbers g^d, $g^{2d}, \ldots, g^{(\varphi(q)/d)d}$ are distinct (mod q) and satisfy (8), and there are $\varphi(q)/d$ of them. Theorem 4.11 now yields the final assertion of the theorem. ■

Note that this theorem generalizes Euler's criterion, Theorem 3.22 (where $n = 2$, $q = p$), and also the Corollary to Theorem 3.19. It covers the case of odd prime-power moduli but fails for moduli $2^e \geq 8$, when the situation is more complicated.

Theorem 4.14 *Suppose that $e \geq 3$. Then every odd a is an nth-power residue of 2^e if n is odd. If n is even, a is an nth-power residue of 2^e if and only if* $a \equiv 1 \pmod{(2^e, 4n)}$. *In either case, the number of nth-power residues* (mod 2^e) *is*

$$\frac{2}{(n, 2)} \cdot \frac{2^{e-2}}{(n, 2^{e-2})}.$$

Remark. The theorem is also valid for $e = 2$; see Problem 4.

Proof. Let $d = (2^{e-2}, n)$. We use the unique representation

$$a \equiv (-1)^{\alpha} 5^{\beta} \pmod{2^e}, \qquad 0 \leq \alpha < 2, \quad 0 \leq \beta < 2^{e-2}$$

as in (4), and ask for ξ, η such that

$$((-1)^{\xi} 5^{\eta})^n \equiv a \pmod{2^e}.$$

This holds if and only if $n\xi \equiv \alpha \pmod{2}$ and $n\eta = \beta \pmod{2^{e-2}}$. The first of these congruences is solvable if and only if $(n, 2) \mid \alpha$, and there are then $(n, 2)$ solutions ξ (mod 2). The second congruence is solvable if and only if $d \mid \beta$, which is the case if and only if

$$(5^{\beta})^{2^{e-2}/d} \equiv 1 \pmod{2^e},$$

or equivalently, by the definition of β,

$$a^{2^{e-2}/d} \equiv (-1)^{\alpha \cdot 2^{e-2}/d} \pmod{2^e}.$$

But the exponent on -1 here is always even if $(n, 2) \mid \alpha$, since in that case if $\alpha = 1$, then $d = 1$, so we obtain the following as a necessary and sufficient condition for a to be an nth-power residue of 2^e:

$$(9) \qquad (n, 2) \mid \alpha \quad \text{and} \quad a^{2^{e-2}/d} \equiv 1 \pmod{2^e}.$$

If n is odd, then $(n, 2) = 1$ and $d = 1$, so (9) always holds, since a is odd. Suppose then that $2^\nu \parallel n$, where $\nu > 0$; then $d = 2^{\min(e-2,\nu)}$. Now the first relation in (9) holds if and only if $\alpha = 0$, so that $a \in \langle 5 \rangle$, or equivalently

$$(10a) \qquad a \equiv 1 \pmod 4.$$

The second relation in (9) becomes

$$(10b) \qquad a \equiv 1 \pmod{2^e} \qquad \text{if } e \leq \nu + 2,$$

$$(10c) \qquad a^{2^{e-2-\nu}} \equiv 1 \pmod{2^e} \qquad \text{if } e \geq \nu + 2.$$

Suppose first that $e \geq \nu + 2$. Then (10a) and (10c) imply that a is an nth-power residue of 2^e, and hence that it is an nth-power residue of $2^{\nu+2}$; this in turn implies that (10a) and (10c) hold for $e = \nu + 2$. On the other hand,

$$a \equiv 1 \pmod{2^{\nu+2}} \quad \text{implies} \quad a^{2^{e-2-\nu}} \equiv 1 \pmod{2^e} \qquad \text{for } e > \nu + 2,$$

by Problem 1 of Section 4.1. Thus (10c) is equivalent to

$$a \equiv 1 \pmod{2^{\nu+2}},$$

which combines with (10b) to give

$$a \equiv 1 \pmod{2^{\min(e,\nu+2)}}.$$

This is the same as $a \equiv 1 \bigl(\text{mod } (2^e, 4n)\bigr)$, and this obviously subsumes (10a).

When the congruence $n\eta \equiv \beta \pmod{2^{e-2}}$ is solvable it has d solutions $\pmod{2^{e-2}}$, so there are $(n, 2) \cdot d$ pairs $\{\xi, \eta\}$ which lead to distinct values of $x \equiv (-1)^\xi 5^\eta \pmod{2^e}$ for which $x^n \equiv a \pmod{2^e}$, when a is an nth-power residue. Since U_{2^e} has 2^{e-1} elements, the number of nth-power residues must be

$$\frac{2^{e-1}}{(n, 2)d}. \quad \blacksquare$$

It is easy to ask exceedingly difficult questions about nth-power residues if the latter are regarded as integers rather than as residue classes, although such questions are slightly more tractable than the corresponding ones about primitive roots. For example, the smallest positive quadratic nonresidue of p is known to be $< cp^{1/(4\sqrt{e})+\varepsilon}$ (for a weaker result, see Problem 12 of Section 6.11), and some theorems are known about the regularity of distribution of the residues among $1, 2, \ldots, p - 1$, and, contrariwise, the occurrence of clumps. But these are terminal theorems, not fraught with implications elsewhere. Of incomparably greater importance for the development of number theory was a relationship first noticed by Euler (1744), clearly formulated by Legendre (1785), and proved by Gauss (1796), between the solvability of $x^2 \equiv p \pmod{q}$ and that of $x^2 \equiv q \pmod{p}$, where p and q are distinct primes. This was the famous law of quadratic reciprocity, the principal subject of the next chapter.

PROBLEMS

1. Let $p > 3$ be prime. How many solutions are there of the congruence $x^3 \equiv 1 \pmod{p}$? How many cubic residues of p are there?
2. Show that if p is an odd prime, there are $\frac{1}{2}(p - 1)$ quadratic residues of p, and the same number of nonresidues.
3. Is 5 a cubic residue of 18? What are all the cubic residues of 18?
4. Show that Theorem 4.14 remains valid for $e = 2$.
5. Give a simpler formulation of Theorem 4.14 for the case $n = 2$.
6. Let p be any prime and suppose $p \nmid na$. Show that the number of solutions of $x^n \equiv a \pmod{p^e}$ is the same for all $e \geq 1$.
7. Show that the map $x \mapsto x^n$ is a homomorphism of U_m onto a subgroup. Under what circumstances is it an isomorphism of U_m with itself?
8. Suppose $(a, m) = 1$. Under what circumstances is the congruence $x^n \equiv a \pmod{m}$ solvable? How many solutions are there?

4.4 AN APPLICATION TO FERMAT'S EQUATION

A simple way of attempting to show that the equation

$$x^n + y^n = z^n \tag{11}$$

has no nonzero solutions for $n \geq 3$ is to try to show that the infinitely many congruences

$$x^n + y^n \equiv z^n \pmod{p}, \qquad p = 2, 3, 5, \ldots$$

impose absurd conditions on the variables. For example, in the case $n = 3$ the congruence

$$x^3 + y^3 \equiv z^3 \pmod{7}$$

implies that $7 \mid xyz$. For if $7 \nmid u$, then $u^6 \equiv 1 \pmod{7}$, so that $u^3 \equiv \pm 1 \pmod{7}$, and for no choice of signs is $\pm 1 \pm 1 \equiv \pm 1 \pmod{7}$. If we could find infinitely many primes p such that

$$x^3 + y^3 \equiv z^3 \pmod{p} \quad \text{implies} \quad p \mid xyz,$$

then clearly equation (11) could have no nonzero solutions for $n = 3$. We shall show that this cannot be done, either for $n = 3$ or for larger n. The proof depends on the following combinatorial lemma.

Theorem 4.15 *If the numbers* 1, 2, ..., N *are distributed into* m *disjoint classes, and if** $N > m!e$, *then at least one class contains the difference of two of its elements.*

Proof. Suppose that the numbers 1, 2, ..., N have been put into m disjoint classes so that no class contains the difference of any two of its elements. Let a class having the largest number of elements be called C_1; then if C_1 is composed of $x_1, \ldots, x_{n_1}$, we have $N \leq n_1 m$. If the names are so chosen that $x_1 < x_2 < \cdots < x_{n_1}$, the $n_1 - 1$ differences

$$x_2 - x_1, \quad x_3 - x_1, \quad \ldots, \quad x_{n_1} - x_1 \tag{12}$$

are also integers between 1 and N, inclusive, and by assumption they lie in the remaining $m - 1$ classes. Let C_2 be a class in which the largest number of differences (12) lie. If C_2 contains the n_2 differences

$$x_\alpha - x_1, \quad x_\beta - x_1, \quad \ldots,$$

then clearly $n_1 - 1 \leq n_2(m - 1)$. Now the $n_2 - 1$ differences

$$x_\beta - x_\alpha, \quad x_\gamma - x_\alpha, \quad \ldots \tag{13}$$

do not lie in either C_1 or C_2, so they must be distributed among the remaining $m - 2$ classes. If n_3 is the largest number of differences (13) in any single class, then $n_2 - 1 \leq n_3(m - 2)$. Continuing in this way, we have

$$n_\mu - 1 \leq n_{\mu+1}(m - \mu), \tag{14}$$

for $\mu = 1, 2, \ldots, m_1$, where m_1 is such that $n_{m_1} = 1$. From (14), we have

$$\frac{n_\mu}{(m-\mu)!} \leq \frac{1}{(m-\mu)!} + \frac{n_{\mu+1}}{(m-\mu-1)!}, \qquad \mu = 1, 2, \ldots, m_1$$

and adding all these inequalities gives

$$\frac{n_1}{(m-1)!} \leq \frac{1}{(m-1)!} + \frac{1}{(m-2)!} + \cdots + \frac{1}{(m-m_1)!} < e.$$

Hence $N \leq n_1 m < m!e$, and the proof is complete. ■

Theorem 4.16 *There are only finitely many primes* p *for which every solution of the congruence*

$$x^n + y^n \equiv z^n \pmod{p} \tag{15}$$

---

* Here, contrary to our convention, the number $e = 2.718\cdots$ is not an integer.

is such that $p \mid xyz$. *More precisely, if* $p > n!e + 1$, *then* (15) *has solutions such that* $p \nmid xyz$.

Proof. First suppose that $n \mid (p - 1)$, so that $p - 1 = nr$ for suitable r. Let g be a primitive root of p, and let s_m be the smallest positive residue (mod p) of g^m. Then the numbers $s_1, \ldots, s_{p-1}$ are the integers $1, 2, \ldots, p - 1$ in some order. We now classify the numbers s_m according to the residue classes of their subscripts (mod n), so that for each t with $0 \le t \le n - 1$, the numbers

$$s_t, \quad s_{t+n}, \quad \ldots, \quad s_{t+(r-1)n}$$

form a single class, there being n classes altogether. By Theorem 4.15, if $p - 1 > n!e$, then some class contains three elements, say $s_{t+jn}, s_{t+kn}, s_{t+ln}$, such that

$$s_{t+jn} - s_{t+kn} = s_{t+ln}.$$

But then

$$g^{t+jn} \equiv g^{t+kn} + g^{t+ln} \pmod{p},$$

whence

$$g^{jn} \equiv g^{kn} + g^{ln} \pmod{p},$$

and the numbers $x = g^k$, $y = g^l$, $z = g^j$ give the desired solution of (15).

If $n \nmid (p - 1)$, let $d = (n, p - 1)$. Then by what we have just proved, the congruence

$$x^d + y^d \equiv z^d \pmod{p}$$

is solvable with $p \nmid xyz$ if $p - 1 > d!e$. But by Theorem 4.13, any dth power is an nth power residue of p, since

$$(u^d)^{(p-1)/(p-1,n)} = (u^d)^{(p-1)/d} = u^{p-1} \equiv 1 \pmod{p}.$$

Thus there exist x_1, y_1, and z_1 such that

$$x_1^n \equiv x^d, \quad y_1^n \equiv y^d, \quad z_1^n \equiv z^d \pmod{p},$$

and hence

$$x_1^n + t_1^n \equiv z_1^n \pmod{p}. \quad \blacksquare$$

PROBLEMS

1. Show that if $x^3 + y^3 \equiv z^3 \pmod{9}$, then $3 \mid xyz$. Use the result of this section, together with the method of Section 3.4, to show that this is an atypical phenomenon: for fixed n, the congruence

$$x^n + y^n \equiv z^n \pmod{p^a}$$

has a solution such that $p \nmid xyz$, if p is sufficiently large and $a \ge 1$.

2. Show that no integer $\equiv 5 \pmod{7}$ is of the form $x_1^6 + x_2^6 + x_3^6 + x_4^6$, with all $x_i \in \mathbf{Z}$.
3. Show that no integer of the form $4^a(8k + 7)$ is a sum of three squares.

NOTES AND REFERENCES

Section 4.1

In 1769 Lambert stated, in substance, that every prime has a primitive root, although the name primitive root was only introduced by Euler in 1773, when he gave a faulty proof of Lambert's assertion. Legendre gave the first correct proof in 1785, and in fact he proved Theorem 4.3, in a different way from that presented here (which is due to Gauss). Indeed, Gauss developed the whole theory much as it is presented in this and the following two sections, except for the statements explicitly mentioning groups, and the referenced modern work.

For two different heuristic derivations of the formula for the frequency of primes with a as a primitive root, see Western and Miller [1968] and Goldstein [1971]. For a fairly elementary proof that $g(p) < \sqrt{p}\,(\log p)^{17}$, see Erdös [1945]. For other work on the distribution of primitive roots, see LeVeque [1974], vol. 4.

Section 4.2

Extensive tables of primitive roots and indices are available. See, for example, Western and Miller [1968], which gives the least primitive root, and indices, for all $p < 50{,}000$, or Hauptman, Vegh, and Fisher [1970], which gives all primitive roots for all $p < 5000$.

Section 4.3

The bound for the smallest positive quadratic residue of p is due to Burgess [1957]. For related literature, see LeVeque [1974], vol. 4.

Section 4.4

The main theorem was first proved by Dickson [1909]. The proof given here is due to Schur [1917]. Dickson's proof is more difficult, but gives the better lower bound $p > cn^4$, for suitable c, in Theorem 4.16.

5

Quadratic Residues

5.1 INTRODUCTION

The theorems of the preceding chapter concerning nth power residues lie rather close to the surface. Further theorems of any generality concerning polynomial congruences of degree larger than 2 are substantially more difficult, and we shall not continue in that direction. But for quadratic congruences there is still a fundamental question that can be answered by elementary means, namely, what are the primes of which a given integer a is a quadratic residue? To contrast this properly with what has been proved thus far, recall the Legendre symbol (a/p), which was given the value 1 or -1 according as a is or is not a quadratic residue of p. For fixed p, we could as well write $(\bar{a}/p)$, since $(a/p) = (b/p)$ if $a \equiv b \pmod{p}$. Up to now, we have considered $(\bar{a}/p)$ as a function of its first entry, and have characterized and counted the (finitely many) solutions of $(\bar{a}/p) = 1$, for fixed p. The new question is altogether different, for if $a \in \mathbf{Z}$ is fixed and p is variable, there is no *a priori* reason to expect any structure in the set of solutions p of $(a/p) = 1$, and at this point we do not even have a method for finding all solutions, since there are infinitely many primes to be tested. It turns out that there *is* some structure, and that the solutions *can* be foretold, and that this is all connected with another unexpected phenomenon: if p and q are distinct odd primes, there is a simple connection between the values of (p/q) and (q/p). Thus, magically, an infinite problem turns into a finite one! (The same kind of thing happens with nth power residues, but the matter cannot be explained in elementary terms.) The reader who wants to have the pleasure of discovering for himself one of the most beautiful, subtle, and fecund theorems in elementary number theory should not fail to carry out the calculations suggested in the first problem at the end of this section.

The prime 2 plays a rather special role in the theory of quadratic residues, not so much because of an intrinsic difference between 2 and the other primes (which does exist, as we saw in connection with primitive roots) as because when $f(x) = x^2 - a$, $2 \mid f'(x)$ for all x, so that all solutions are singular, in the language of Section 3.4. For cubic congruences, 3 must be treated separately. At any rate, throughout this chapter we shall use the symbol p to represent an *odd* prime.

Moreover, we shall frequently use the terms residue and nonresidue without the modifying adjective "quadratic," since these are the only kind discussed in this chapter.

We begin by showing that questions about quadratic residues of composite moduli can be reduced to the case of prime moduli.

Theorem 5.1 *A number $a \in U_m$ is a quadratic residue of m if and only if it is a residue of all odd prime divisors of m and is congruent to* 1 (mod 4) *if* $2^2 \parallel m$ *and to* 1 (mod 8) *if* $8 \mid m$.

Proof. Let $m = 2^e p_1^{e_1} \cdots p_r^{e_r}$. Then the congruence $x^2 \equiv a \pmod{m}$ is equivalent to the system

$$x^2 \equiv a \pmod{2^e}, \ldots, \quad x^2 \equiv a \pmod{p_r^{e_r}},$$

as in Theorem 3.21. Clearly, if a is a quadratic residue of a power of any prime, it is a residue of the prime itself. Contrariwise, if a is a quadratic residue of the odd prime p, then

$$a^{(p-1)/2} \equiv 1 \pmod{p}$$

by Theorem 4.13, so

$$a^{p^{e-1}(p-1)/2} \equiv 1 \pmod{p^e}$$

by Problem 1 of Section 4.1, so a is a residue of p^e, by Theorem 4.13 again. As for the modulus 4, 1 is a residue and -1 is not. For the modulus 2^e ($e \geq 3$), Theorem 4.14 with $n = 2$ gives $a \equiv 1 \pmod{8}$ as the necessary and sufficient condition for a to be a residue. ∎

The above proof can be retraced step-by-step to obtain the number of solutions of $x^2 \equiv a \pmod{m}$ when it is solvable. Theorem 4.13 says that there are then 2 solutions for $m = p^e$. The number of solutions of $x^2 \equiv a \pmod{2^e}$ is clearly 1 when $e = 1$, and 2 when $e = 2$. When $e \geq 3$ the number of residues is 2^{e-3}, by Theorem 4.14, so by Theorem 4.11 the congruence $x^2 \equiv a \pmod{2^e}$ has $\varphi(2^e)/2^{e-3} = 4$ solutions when it is solvable. Thus we have the following special case of Theorem 3.21:

Theorem 5.2 *If $a \in U_m$ and the congruence $x^2 \equiv a \pmod{m}$ is solvable, it has 2^{r+u} solutions, where r is the number of distinct odd prime divisors of m and u is* 0, 1, *or* 2 *according as* $4 \nmid m$, $2^2 \parallel m$, *or* $8 \mid m$.

It is possible now to give a complete analysis of the congruence $ax^2 + bx + c \equiv 0 \pmod{m}$, but it is tedious and not sufficiently rewarding. The easy case is that in which $(m, 2a) = 1$, so that the congruence is equivalent to $(2ax + b)^2 \equiv b^2 - 4ac \pmod{m}$, but aside from this there are numerous special cases. We shall not pursue the matter further.

PROBLEMS

1. Construct a table of values of (p/q) for all odd p, $q \leq 23$, such as this:

$q \backslash p$	3	5	7	$\cdots$	23
3	0	-1	1		-1
5	-1	0			
$\vdots$					
23					0

   Classify these p and q (mod 4), and hunt for the rule which distinguishes between the cases $(p/q) = (q/p)$ and $(p/q) = -(q/p)$.

2. Show that if $p \equiv 3 \pmod 4$, the solutions of $x^2 \equiv a \pmod p$, supposing it is solvable, are $x \equiv \pm a^{(p+1)/4} \pmod p$.
3. Decide whether 37 is a residue of 2772, and, if so, how many solutions the congruence $x^2 \equiv 37 \pmod{2772}$ has.
4. Show that the product of the quadratic residues of p is congruent to 1 or $-1 \pmod p$, according as $p \equiv -1$ or $1 \pmod 4$. [*Hint:* Use a primitive root.]
5. Suppose that $(a, 10) = 1$. Show that if the last n digits of a^2 are known ($a \in \mathbf{Z}$), then there are exactly 2, 4, or 8 possibilities for the last n digits of a, according as $n = 1$, $n = 2$, or $n \geq 3$.
6. Show that the only 3-digit integers a whose squares end in a are 376 and 625.

5.2 QUADRATIC RESIDUES OF PRIMES, AND THE LEGENDRE SYMBOL

As was seen in Section 5.1, the quadratic residues of powers of 2 can be given explicitly, and the quadratic residues of powers of an odd prime are identical with those of the prime itself. Consequently, there remains only the investigation of quadratic residues of odd primes. It will be recalled that the Legendre symbol has been defined as follows:

$$(a/p) = \begin{cases} 1, & \text{if } a \text{ is a quadratic residue of } p, \\ -1, & \text{if } a \text{ is a quadratic nonresidue of } p, \\ 0, & \text{if } p \mid a. \end{cases}$$

Theorem 5.3 *The Legendre symbol (a/p) has the following properties:*

a) $(ab/p) = (a/p)(b/p)$. *Thus the product of two residues or two nonresidues is a residue; the product of a residue and a nonresidue is a nonresidue.*

b) *If* $a \equiv b \pmod p$, *then* $(a/p) = (b/p)$.

Adrien Marie Legendre
(1752–1833)

It was Legendre's fate to be eclipsed repeatedly by younger mathematicians. He invented the method of least squares in 1806, but Gauss revealed in 1809 that he had done the same in 1795. He labored for 40 years on elliptic integrals, and then Abel and Jacobi revolutionized the subject in the 1820s with the introduction of elliptic functions. He conjectured the prime number theorem and the law of quadratic reciprocity, but could not prove either. Still, he created much beautiful mathematics, including the determination of the number of representations of an integer as a sum of two squares, and the exact conditions under which the equation $ax^2 + by^2 + cz^2 = 0$ holds for some $x, y, z \neq 0, 0, 0$. He also wrote an elementary geometry text which, in 39 editions of the English translations, replaced Euclid's *Elements* in American schools.

c) $(a^2/p) = 1$ *if* $p \nmid a$.
d) $(-1/p) = (-1)^{(p-1)/2}$. *Thus* -1 *is a residue of* p *if* $p \equiv 1 \pmod 4$, *but not if* $p \equiv -1 \pmod 4$.

Proof. By Theorem 3.23, $(a/p) \equiv a^{(p-1)/2} \pmod p$. Hence

$$(ab/p) \equiv (ab)^{(p-1)/2} = a^{(p-1)/2}b^{(p-1)/2} \equiv (a/p)(b/p) \pmod p,$$

and since (a/p) assumes only the values 0 and ± 1, it follows that $(ab/p) = (a/p)(b/p)$. Property (d) also follows immediately from this congruence. Properties (b) and (c) are obvious. ∎

It follows from Theorem 5.3 that in investigating the Legendre symbol (a/p), there will be no loss in generality in assuming that a is a positive prime. For example, Theorem 5.3 shows that

$$\begin{aligned}(-48/31) &= (-1/31)(48/31) = (-1/31)(3/31)(16/31)\\ &= (-1/31)(3/31)\\ &= (30/31)(3/31) = (2/31)(3/31)(5/31)(3/31)\\ &= (2/31)(5/31),\end{aligned}$$

so that $(-48/31)$ can be evaluated either from

$$(-48/31) = (-1)^{(31-1)/2}(3/31) = -(3/31)$$

or from

$$(-48/31) = (2/31)(5/31).$$

In general, (a/p) can be written as the product of Legendre symbols, in which the first entries are the distinct prime divisors of a which divide a to an odd power.

Although it will be used only in the case where a is prime, the following theorem is valid for all a's for which $p \nmid a$.

Theorem 5.4 (Gauss's Lemma) *If μ is the number of elements of the set $a, 2a, \ldots, \frac{1}{2}(p-1)a$ whose numerically smallest remainders* (mod p) *are negative, then*

$$(a/p) = (-1)^{\mu}.$$

Example. If $a = 3$, $p = 31$, the numerically smallest remainders (mod 31) of $3 \cdot 1, 3 \cdot 2, \ldots, 3 \cdot 15$ are $3, 6, 9, 12, 15, -13, -10, -7, -4, -1, 2, 5, 8, 11, 14$; thus $\mu = 5$, $(3/31) = -1$, and from the above numerical example, $(-48/31) = 1$.

Proof. Replace the numbers of the set $a, 2a, \ldots, \frac{1}{2}(p-1)a$ by their numerically smallest remainders (mod p); denote the positive ones by $r_1, r_2, \ldots$ and the negative ones by $-r_1', -r_2', \ldots$. Clearly no two r_i's are equal, and no two r_i''s are equal. If $m_1 a \equiv r_i$ and $m_2 a \equiv -r_j' \pmod p$, then $r_i = r_j'$ would imply $a(m_1 + m_2) \equiv 0 \pmod p$, which implies $m_1 + m_2 \equiv 0 \pmod p$, and this is impossible because the m's are strictly between 0 and $p/2$. Hence the $(p-1)/2$ numbers r_i, r_i' are distinct integers between 1 and $(p-1)/2$ inclusive, and are therefore exactly the numbers $1, 2, \ldots, (p-1)/2$ in some order. Hence,

$$a \cdot 2a \cdots \frac{p-1}{2} a \equiv (-1)^{\mu} \frac{p-1}{2}! \pmod p,$$

$$a^{(p-1)/2} \equiv (-1)^{\mu} \pmod p.$$

Since also $a^{(p-1)/2} \equiv (a/p) \pmod p$, it follows that $(a/p) \equiv (-1)^{\mu} \pmod p$, and finally, $(a/p) = (-1)^{\mu}$. ■

In distinction to Euler's criterion, Gauss's lemma can be used to characterize the primes of which a given integer a is a quadratic residue. For example, if $a = 2$, then μ is the number of numbers $2m$, with $1 \le m \le (p-1)/2$, which are greater than $p/2$; clearly, this is true if and only if $m > p/4$. Thus if we write $[x]$ to stand for the largest integer not exceeding x, it follows that

$$\mu = \frac{p-1}{2} - \left[\frac{p}{4}\right].$$

If now

$$\begin{aligned}
p &= 8k+1, \quad \text{then } \mu = 4k - [2k + \tfrac{1}{4}] = 4k - 2k \equiv 0 \pmod 2,\\
p &= 8k+3, \quad \text{then } \mu = 4k + 1 - [2k + \tfrac{3}{4}] = 4k + 1 - 2k \equiv 1 \pmod 2,\\
p &= 8k+5, \quad \text{then } \mu = 4k + 2 - [2k + 1 + \tfrac{1}{4}] = 2k + 1 \equiv 1 \pmod 2,\\
p &= 8k+7, \quad \text{then } \mu = 4k + 3 - [2k + 1 + \tfrac{3}{4}] = 2k + 2 \equiv 0 \pmod 2.
\end{aligned}$$

Since it happens that the quantity $(p^2 - 1)/8$ satisfies exactly the same congruences as μ above, this result can be stated in the following form.

Theorem 5.5 *2 is a quadratic residue of p if $p \equiv \pm 1 \pmod 8$, but not if $p \equiv \pm 3 \pmod 8$. Briefly, $(2/p) = (-1)^{(p^2-1)/8}$.*

As an application of Theorem 5.5, we have

Theorem 5.6

a) *2 is a primitive root of the prime $p = 4q + 1$ if q is an odd prime.*

b) *2 is a primitive root of $p = 2q + 1$ if q is a prime of the form $4k + 1$.*

c) *-2 is a primitive root of $p = 2q + 1$ if q is a prime of the form $4k - 1$.*

Proof. a) If $\text{ord}_p\, 2 = t$, then $t \mid (p - 1)$, which is equivalent to saying that $t \mid 4q$. Aside from 4, every proper divisor of $4q$ is also a divisor of $2q$, and if $2^4 \equiv 1 \pmod p$, then p is 5 and q is not prime. Hence it suffices to show that $2^{2q} \not\equiv 1 \pmod p$. But $2^{2q} = 2^{(p-1)/2} \equiv (2/p) \pmod p$, and $(2/p) = -1$ since $p \equiv 5 \pmod 8$. Parts (b) and (c) can be proved in a similar fashion. ■

Part (a) shows that 2 is a primitive root of 13, 29, 53, . . . ; part (b) shows that 2 is a primitive root of 11, 59, 83, . . . , and part (c) that -2 is a primitive root of 7, 23, 47, ... The conjecture that 2 is a primitive root of infinitely many primes would follow from Theorem 5.5 if it could be shown that there are infinitely many primes p of the kinds described in (a) and (b). Referring to (a), this requires a proof that the function $4x + 1$ assumes prime values for infinitely many prime arguments. Unfortunately, there is no nonconstant rational function known to have this property except x itself. If one could prove that the function $x + 2$ has it, one would have proved a conjecture which is one of the outstanding problems in additive number theory: that there are infinitely many "twin primes," such as 17 and 19, or 101 and 103. (See Section 6.12.)

PROBLEMS

1. Apply Gauss's lemma to determine the primes of which -2 is a quadratic residue, and show that your result is consistent with Theorem 5.3, parts (a) and (d), and Theorem 5.5.
2. Complete the proof of Theorem 5.6.
3. Suppose $p \nmid a$. Show that if $p \equiv 1 \pmod 4$, then both or neither of a and $-a$ are residues of p, while if $p \equiv -1 \pmod 4$, exactly one is a residue.
4. a) Prove part (a) of Theorem 5.3 directly from the definition of quadratic residue, without recourse to Euler's criterion. [*Hint:* The case $(a/p) = (b/p) = 1$ is easy. For the remainder, consider the set ar, where a is fixed and r ranges over the residues of p.]

   b) Deduce part (d) of Theorem 5.3 from part (a) and Wilson's theorem.
5. For which prime powers p^e are there integers x and y such that $p \nmid xy$ and $x^2 + y^2 \equiv 0 \pmod{p^e}$?
6. Show that if p and $q = 2p + 1$ are both odd primes, then -4 is a primitive root of q.

7. Prove that 2 is a nonresidue of any integer $\equiv \pm 3 \pmod 8$, without using Gauss's lemma. [*Hint:* Suppose that $n > 5$ is the smallest such integer of which 2 is a residue. Show that there is an odd integer b such that $0 < b < n$ and $b^2 \equiv 2 \pmod n$, so that $b^2 = 2 + mn$. Think about m.]
8. Show that if $p \nmid m$, then $\sum_{a=1}^{p} (ma/p) = 0$.
9. Show that the Diophantine equation $y^2 = x^3 + 7$ has no solution. [*Hint:* Show that if x, y were a solution, x would be odd, say $x = 2z + 1$, and factor $y^2 + 1$ as a polynomial in z.]
10. a) Generalize Gauss's lemma as follows: Put $v = (p-1)/2$, and let $S = \{a_1, \ldots, a_v\}$ be any set of v elements of U_p with the property that it is disjoint from the set $-S = \{-a_1, \ldots, -a_v\}$. For $a \in U_p$, let μ be the number of elements of $aS = \{aa_1, \ldots, aa_v\}$ not belonging to S. Then $(a/p) = (-1)^\mu$.
    b) Show that $\{2, 4, 6, \ldots, 2v\}$ and $\{g, g^2, g^3, \ldots, g^v\}$ have the property described in (a), if g is a primitive root of p.

5.3 THE LAW OF QUADRATIC RECIPROCITY

Returning to the problem of evaluating (a/p), we see that since the symbol is a multiplicative function of the first entry, and since the values of $(-1/p)$, $(2/p)$, and (a^2/p) are known, there remains only the problem of evaluating (q/p) for positive odd primes $q \neq p$. This—and a great deal more—can be done with the help of the following fundamental theorem.

Theorem 5.7 (Quadratic reciprocity law) *If p and q are distinct positive odd primes, then $(p/q) = (q/p)$ unless both p and q are of the form $4k - 1$, in which case $(p/q) = -(q/p)$. More briefly,*

$$(p/q)(q/p) = (-1)^{\frac{p-1}{2}\cdot\frac{q-1}{2}}$$

Proof. By Gauss's lemma, the numbers μ and ν in the equations

$$(q/p) = (-1)^\mu, \qquad (p/q) = (-1)^\nu$$

are the numbers of the multiples

$$q, 2q, \ldots, \frac{p-1}{2} q,$$

and

$$p, 2p, \ldots, \frac{q-1}{2} p$$

whose absolutely smallest remainders (mod p) and (mod q), respectively, are negative, and we need only show that

$$\mu + \nu \equiv \frac{p-1}{2}\cdot\frac{q-1}{2} \pmod 2.$$

Consider one of the multiples xq, with $1 \leq x \leq (p-1)/2$. If y is chosen so that

$$-\frac{p}{2} < qx - py < \frac{p}{2},$$

then clearly $qx - py$ is the numerically smallest remainder of $qx \pmod{p}$. From this inequality we find that y must lie in an interval of unit length:

$$\frac{qx}{p} - \frac{1}{2} < y < \frac{qx}{p} + \frac{1}{2}.$$

Thus y is unique and nonnegative; if $y = 0$, then $qx - py = qx > 0$, and there is no contribution to μ in this case. Moreover, we see that for $x \leq (p-1)/2$,

$$\frac{qx}{p} + \frac{1}{2} \leq \frac{q}{2} - \frac{q}{2p} + \frac{1}{2} < \frac{q+1}{2},$$

so that we may, without loss, restrict y to the interval $0 < y \leq (q-1)/2$. The number μ denotes therefore the number of combinations of x and y from the sequences

$$\mathcal{P} \qquad 1, 2, \ldots, \frac{p-1}{2}$$

and

$$\mathcal{Q} \qquad 1, 2, \ldots, \frac{q-1}{2},$$

respectively, for which

$$0 > qx - py > -\frac{p}{2}.$$

Similarly, ν is the number of pairs x and y from the sequences $\mathcal{P}$ and $\mathcal{Q}$, respectively, for which

$$0 > py - qx > -\frac{q}{2}.$$

For any other pair x and y from $\mathcal{P}$ and $\mathcal{Q}$ respectively, either

$$py - qx > \frac{p}{2}$$

or

$$py - qx < -\frac{q}{2};$$

let there be λ of the former and ρ of the latter. Then clearly

$$\frac{p-1}{2} \cdot \frac{q-1}{2} = \mu + \nu + \lambda + \rho.$$

Finally, as x and y run through $\mathcal{P}$ and $\mathcal{Q}$ respectively, the numbers

$$x' = \frac{p+1}{2} - x \quad \text{and} \quad y' = \frac{q+1}{2} - y$$

run through the same sequences, but in the opposite order. And if $py - qx > p/2$, then

$$\begin{aligned} py' - qx' &= p\left(\frac{q+1}{2} - y\right) - q\left(\frac{p+1}{2} - x\right) \\ &= \frac{p-q}{2} - (py - qx) < \frac{p-q}{2} - \frac{p}{2} = -\frac{q}{2}. \end{aligned}$$

Conversely, if $py - qx < -q/2$, then $py' - qx' > p/2$. Hence $\lambda = \rho$, and

$$\frac{p-1}{2} \cdot \frac{q-1}{2} = \mu + \nu + 2\lambda \equiv \mu + \nu \pmod 2. \quad \blacksquare$$

By combining the law of quadratic reciprocity with the properties of the Legendre symbol mentioned in Theorem 5.3, it is easy to evaluate (q/p) if p and q do not lie beyond the extent of the available tables of factorizations of integers. For example, 2819 and 4177 are both primes and $4177 \equiv 1 \pmod 4$, so that

$$\begin{aligned} (2819/4177) &= (4177/2819) = (1358/2819) = (2 \cdot 7 \cdot 97/2819) \\ &= (2/2819)(7/2819)(97/2819) \\ &= -1 \cdot -(2819/7)(2819/97) = (5/7)(6/97) \\ &= (7/5)(2/97)(97/3) \\ &= (2/5)(1/3) = -1, \end{aligned}$$

and so 2819 is not a quadratic residue of 4177.

Moreover, the quadratic reciprocity law can be used to determine the primes p of which a given prime q is a quadratic residue. This result, which is contained in the next theorem, has sometimes been taken as the quadratic reciprocity law, rather than Theorem 5.7. (Each can be deduced from the other.)

Theorem 5.8 *Let q be a fixed positive odd prime, and let p range over the odd positive primes $\neq q$. Every such p has a unique representation in exactly one of the two forms*

(1) $$p = 4qk \pm a, \quad \text{with } k \in \mathbf{Z}, \quad 0 < a < 4q, \quad a \equiv 1 \pmod 4.$$

When (1) *holds,*

(2) $$(q/p) = (a/q).$$

Thus the p for which $(q/p) = 1$ are exactly those $p \equiv \pm a \pmod{4q}$, for all a such that

(3) $$0 < a < 4q, \quad a \equiv 1 \pmod 4, \quad (a/q) = 1.$$

The a's satisfying (3) *are given by the smallest positive remainders* (mod $4q$) *of the odd squares* $1^2, 3^2, \ldots, (q-2)^2$.

Proof. By the division theorem, there are unique k', a' such that

$$p = 4qk' + a', \qquad 1 \le a' < 4q,$$

and clearly a' is odd. If $a' \equiv 1 \pmod 4$, (1) holds with the plus sign and with $k = k'$, $a = a'$. If $a' \equiv -1 \pmod 4$, (1) holds with the minus sign and $k = k' + 1$, $a = 4q - a'$. Any other value of k than k' and $k' + 1$ would yield $|a| > 4q$.

To verify (2), first suppose that the plus sign is correct in (1). Then $p \equiv 1 \pmod 4$, and $p \equiv a \pmod q$, so $(q/p) = (p/q) = (a/q)$. If the minus sign is correct, then $p \equiv -1 \pmod 4$ and $p \equiv -a \pmod q$, so either

$$q \equiv -1 \pmod 4, \quad \text{and then } (q/p) = -(p/q) = -(-a/q) = (a/q),$$

or

$$q \equiv 1 \pmod 4, \qquad \text{and then } (q/p) = (p/q) = (-a/q) = (a/q).$$

Finally, if $(a/q) = 1$, there is a b such that

$$a \equiv b^2 \pmod q \qquad \text{and} \qquad 1 \le b \le q - 1,$$

whence also

$$a \equiv (q-b)^2 \pmod q \qquad \text{and} \qquad 1 \le q - b \le q - 1.$$

Since either b or $q - b$ is odd—say b'—we have

$$a \equiv b'^2 \pmod q, \qquad 1 \le b' \le q - 2, \qquad b' \equiv 1 \pmod 2.$$

But then also

$$a \equiv 1 \equiv b'^2 \pmod 4,$$

so that

$$a \equiv b'^2 \pmod{4q},$$

as asserted. ■

To illustrate, take $q = 3$. Then the only integer satisfying the conditions (3) is 1, so that 3 is a quadratic residue of primes $12k \pm 1$. Every other odd number is of one of the forms $12k \pm 3$ or $12k \pm 5$, and no prime except 3 occurs in the progressions $12k \pm 3$. Hence $(3/p)$ is completely determined by the equations

$$(3/p) = \begin{cases} 1 & \text{if } p \equiv \pm 1 \pmod{12}, \\ -1 & \text{if } p \equiv \pm 5 \pmod{12}. \end{cases}$$

Similarly, taking $q = 17$ we consider the squares

$$1^2, 3^2, 5^2, 7^2, 9^2, 11^2, 13^2, 15^2,$$

which reduce (mod 68) to

$$1, 9, 25, 49, 13, 53, 33, 21.$$

We have that 17 is a quadratic residue of primes of the forms

$$68k \pm 1, 9, 13, 21, 25, 33, 49, \text{ and } 53,$$

and a nonresidue of primes of the forms

$$68k \pm 5, 29, 37, 41, 45, 57, 61, \text{ and } 65;$$

17 itself is the only prime of the forms $68k \pm 17$.

In general, out of the $2q$ progressions $4qk \pm a$, exactly $q - 1 = \frac{1}{2}\varphi(4q)$ contain only primes of which q is a residue, $q - 1$ contain only primes of which q is a nonresidue, and two (either $4qk \pm q$ or $4qk \pm 3q$, according as $q \equiv 1$ or $3 \pmod 4$) contain no primes besides q itself.

Determining the primes of which a composite number is a quadratic residue is somewhat more complicated. To illustrate, consider the problem of finding the primes p for which $(6/p) = 1$. This requires that either $(2/p) = (3/p) = 1$ or $(2/p) = (3/p) = -1$, so that either

$$p \equiv \pm 1 \pmod 8 \qquad \text{and} \qquad p \equiv \pm 1 \pmod{12}$$

or

$$p \equiv \pm 3 \pmod 8 \qquad \text{and} \qquad p \equiv \pm 5 \pmod{12},$$

all combinations of signs being allowed. Thus we have the following pairs of congruences, each pair to be solved simultaneously:

$$\begin{array}{lll} p \equiv 1 \pmod 8 & p \equiv -1 \pmod 8 & p \equiv 1 \pmod 8 \\ p \equiv 1 \pmod{12} & p \equiv -1 \pmod{12} & p \equiv -1 \pmod{12} \\[1ex] p \equiv -1 \pmod 8 & p \equiv 3 \pmod 8 & p \equiv -3 \pmod 8 \\ p \equiv 1 \pmod{12} & p \equiv 5 \pmod{12} & p \equiv -5 \pmod{12} \\[1ex] p \equiv 3 \pmod 8 & p \equiv -3 \pmod 8 & \\ p \equiv -5 \pmod{12} & p \equiv 5 \pmod{12}. & \end{array}$$

Four of these pairs are internally inconsistent, while the others have the solutions $p \equiv \pm 1, \pm 5 \pmod{24}$. For these primes, $(6/p) = 1$, while for primes $p \equiv \pm 7, \pm 11 \pmod{24}$, $(6/p) = -1$. Only eight residue classes (mod 24) contain primes, since $\varphi(24) = 8$.

The law of quadratic reciprocity has an interesting early history. Euler first stated it, in full generality, in 1744/46, in a form essentially the same as Theorem 5.8: q is a quadratic residue of p if and only if one of $\pm p$ is a residue of $4q$. In 1783 (the year of Euler's death) a second version appeared in his *Opuscula Analytica*: $(-1)^{(p-1)/2}p$ is a residue of q if and only if q is a residue of p. This is an easy variant of Theorem 5.7. Legendre introduced his symbol in an article in 1785, and at the same time stated the reciprocity law without using the symbol. He gave the elegant second formulation in Theorem 5.7 in his book of 1798. Euler gave a faulty proof, in a second paper in 1783, of a special case of the theorem, and Legendre gave a proof, but with a gap in it, in 1785. Gauss discovered the theorem empirically (and in complete ignorance of earlier work) just before his eighteenth birthday, in 1795; he wrote, "for a whole year this theorem tormented me and absorbed my best efforts until at last I obtained a proof." He published this (a difficult induction) and a second proof five years later in the *Disquisitiones*; his first proof depending on Theorem 5.4 appeared in 1808, and all told he gave eight proofs. He recounted what he had learned by then of the earlier literature in the *Disquisitiones*, but Legendre felt strongly that he was not given sufficient credit for discovering the theorem, though he agreed that Gauss had provided the first proof, and he regarded Gauss as an enemy from that time on. Apparently neither of them was aware of either of the general statements given by Euler—rather astonishing, considering that they both knew of his faulty 1783 proof, and that the two 1783 papers were published in the same volume!

Long before any general results were known, Fermat had characterized the primes of which 2, -2, 3, and -3 are residues; proofs were supplied by Euler for ± 3 in 1760, and by Lagrange for ± 2 in 1775.

The theorem has proved to be of the highest importance, throughout a large portion of number theory. Besides being a powerful tool for solving problems, it has been a fruitful source of new problems. As was mentioned earlier, there are reciprocity theorems for nth power residues, but they cannot be obtained—nor even easily phrased—through purely rational considerations. Indeed, Gauss studied the Gaussian integers $\mathbf{Z}[i]$, and thereby initiated algebraic number theory, in an unsuccessful attempt to prove the quartic reciprocity law, which he had discovered empirically.

PROBLEMS

1. Evaluate the Legendre symbols $(503/773)$ and $(501/773)$.
2. Characterize the primes of which 5 is a quadratic residue; those of which 10 is a quadratic residue.
3. Show that if $p = 4m + 1$ and $d \mid m$, then $(d/p) = 1$. [*Hint:* Let q be a prime divisor of m, and consider separately the cases $q = 2$ and $q > 2$.]
4. Show that if n is a value assumed by the polynomial $x^2 - ay^2$ for some $x, y \in \mathbf{Z}$, then for every prime divisor p of n, either $p \mid x$ or $(a/p) = 1$.
5. Decide which of the following congruences are solvable:

   a) $x^2 \equiv 2455 \pmod{4993}$,

   b) $1709x^2 \equiv 2455 \pmod{4993}$,

   c) $x^2 \equiv 245 \pmod{27496}$,

   d) $x^2 \equiv 5473 \pmod{27496}$.

6. Show that 7 is a primitive root of any prime of the form $2^{4n} + 1$ with $n > 0$. [*Hint:* Show first that it suffices to prove that $(7/p) = -1$, and then show that any *prime* of the specified form is congruent to 3 or 5 (mod 7). Note that $2^4 \equiv 2 \pmod 7$.]
7. Theorem 5.8 has been deduced from Theorem 5.7 (and preceding results) in the text. Carry out the deduction in the opposite direction.
8. If q is an odd prime, the primes p for which $(q/p) = 1$ are known, by Theorem 5.8, to lie in certain residue classes (mod $4q$). Show that when $q \equiv 1 \pmod 4$, these p's can be described more briefly by congruences (mod $2q$). For example, 17 is a residue of exactly the primes of the forms $34k \pm 1$, 9, 13, and 15. [*Hint:* Show that $u^2 \equiv -1 \pmod{2q}$ is solvable and choose one solution of it; pair each odd b, $1 \le b \le q - 2$, with odd $a \equiv bu \pmod{2q}$, $1 \le a \le q - 2$, so that $a^2 = -b^2 + (2l + 1)2q$.]
9. Show that for $p > 3$, the congruence $x^2 \equiv -3 \pmod p$ is solvable if and only if $p \equiv 1 \pmod 6$. Deduce that there are infinitely many primes of the form $6k + 1$. (See Problem 4, Section 1.4, for the case $6k - 1$.)
10. Deduce Euler's second form of the law of quadratic reciprocity from Theorem 5.7, and vice versa.

5.4 THE JACOBI SYMBOL

As was pointed out at the end of the proof of the law of quadratic reciprocity, it is necessary to have available rather extensive factorization tables if one is to evaluate Legendre symbols with large entries. Partly to obviate such a list, and partly for theoretical purposes, it has been found convenient to extend the definition of the Legendre symbol (a/p) so as to give meaning to (a/b) when b is not a prime. This is done in the following way: put $(a/1) = 1$, and if b is greater than 1 and odd, put

$$(a/b) = (a/p_1)(a/p_2)\cdots(a/p_r), \tag{4}$$

where $p_1p_2\cdots p_r$ is the prime factorization of b, and the symbols on the right in (4) are Legendre symbols. Then the symbol on the left in (4) is called a *Jacobi symbol*; like the Legendre symbol, it is undefined for even second entry. As we shall see, others of its properties are also similar to those of the Legendre symbol, but there is one crucial point at which the similarity breaks down: it may happen that $(a/b) = 1$ even when a is not a quadratic residue of b. For it is clearly necessary that each of the Legendre symbols (a/p_i) have the value 1 in order for a to be a residue of b, while $(a/b) = 1$ if an even number of the factors in (4) are -1 while the remainder are $+1$. On the other hand, a is certainly not a quadratic residue of b if $(a/b) = -1$, since then some $(a/p_i) = -1$.

The following theorem lists properties of the Jacobi symbol which were proved for the Legendre symbol in Theorems 5.3, 5.5, and 5.7, together with one (the second) which is peculiar to the extended function.

Theorem 5.9 *The Jacobi symbol has these properties, for arguments for which it is defined:*

a) $(a_1a_2/b) = (a_1/b)(a_2/b)$.

b) $(a/b_1b_2) = (a/b_1)(a/b_2)$.

c) *If* $a_1 \equiv a_2 \pmod{b}$, *then* $(a_1/b) = (a_2/b)$.

d) $(-1/b) = (-1)^{(b-1)/2}$.

e) $(2/b) = (-1)^{(b^2-1)/8}$.

f) *If* $(a, b) = 1$, *then* $(a/b)(b/a) = (-1)^{\frac{1}{2}(a-1)\cdot\frac{1}{2}(b-1)}$.

Proof.

a) Put $b = p_1 \cdots p_r$. Then

$$(a_1a_2/b) = (a_1a_2/p_1)\cdots(a_1a_2/p_r),$$

and since these are Legendre symbols,

$$(a_1a_2/b) = (a_1/p_1)\cdots(a_1/p_r)(a_2/p_1)\cdots(a_2/p_r) = (a_1/b)(a_2/b).$$

b) Put $b_1 = p_1 \cdots p_r$ and $b_2 = p_1' \cdots p_s'$. Then

$$\begin{aligned}(a/b_1b_2) &= (a/p_1 \cdots p_r p_1' \cdots p_s')\\ &= ((a/p_1)\cdots(a/p_r))((a/p_1')\cdots a/p_s'))\\ &= (a/b_1)(a/b_2).\end{aligned}$$

c) If $a_1 \equiv a_2 \pmod{b}$ and $b = p_1 \cdots p_r$, then $a_1 \equiv a_2 \pmod{p_i}$ for $i = 1, \ldots, r$. Hence $(a_1/p_i) = (a_2/p_i)$, and

$$(a_1/b) = (a_1/p_1)\cdots(a_1/p_r) = (a_2/p_1)\cdots(a_2/p_r) = (a_2/b).$$

d) Put $b = p_1 \cdots p_r$. Then

$$(-1/b) = \prod_{i=1}^{r}(-1/p_i) = \prod_{i=1}^{r}(-1)^{(p_i-1)/2}$$

or

$$(-1/b) = (-1)^{\sum_{i=1}^{r}\frac{p_i-1}{2}}. \tag{5}$$

But if m and n are odd, then

$$\begin{aligned}(m-1)(n-1) &\equiv 0 \pmod{4},\\ mn - 1 &\equiv m + n - 2 \pmod{4},\\ \frac{mn-1}{2} &\equiv \frac{m-1}{2} + \frac{n-1}{2} \pmod{2}.\end{aligned}$$

Repeated application of this fact shows that

$$\sum_{i=1}^{r}\frac{p_i - 1}{2} \equiv \frac{p_1 \cdots p_r - 1}{2} \pmod{2},$$

so that $(-1/b) = (-1)^{(b-1)/2}$, by (5).

e) The proof of this is the same as that just given for (d), except that, using the fact that $m^2 \equiv 1 \pmod 8$ if m is odd, we deduce from the congruence

$$(m^2 - 1)(n^2 - 1) \equiv 0 \pmod{64}$$

that

$$\frac{m^2 - 1}{8} + \frac{n^2 - 1}{8} \equiv \frac{(mn)^2 - 1}{8} \pmod 2.$$

f) Put $a = p_1 \cdots p_r$, $b = p'_1 \cdots p'_s$. Then since $(a, b) = 1$,

$$\begin{aligned}(a/b)(b/a) &= \prod_{i=1}^{s} (a/p'_i) \prod_{j=1}^{r} (b/p_j) \\ &= \prod_{j=1}^{r} \prod_{i=1}^{s} (p_j/p'_i) \cdot \prod_{j=1}^{r} \prod_{i=1}^{s} (p'_i/p_j) \\ &= \prod_{j=1}^{r} \prod_{i=1}^{s} (p_j/p'_i)(p'_i/p_j) \\ &= (-1)^{\sum_{j=1}^{r} \sum_{i=1}^{s} \frac{p_j - 1}{2} \cdot \frac{p'_i - 1}{2}} \\ &= (-1)^{\sum_{j=1}^{r} \frac{p_j - 1}{2} \cdot \sum_{i=1}^{s} \frac{p'_i - 1}{2}} \\ &= (-1)^{\frac{1}{2}(a-1) \cdot \frac{1}{2}(b-1)}. \quad \blacksquare\end{aligned}$$

Because the laws of operation and combination are the same for the two types, Jacobi symbols can be used (and according to the same rules) in evaluating Legendre symbols, even though they do not give complete information about the

Carl G. J. Jacobi (1804–1851)

Mathematics in Germany was at low ebb when Jacobi was a student at Potsdam and Berlin, and he was mostly self-educated, through reading the works of Euler and Lagrange. He became a splendid teacher, and did much to revive German mathematics, in Königsberg and Berlin. His first love was the theory of elliptic functions, but he also wrote in other branches of analysis and in geometry and mechanics. Interested in the history of mathematics, Jacobi was a prime mover in the publication of Euler's collected works (still not completed!). He and Dirichlet were close friends; they independently sired two quite different kinds of analytic number theory. Although his friends predicted he would work himself to death, he died instead of smallpox.

quadratic character of a modulo b; all that is required is that one begin with a Legendre symbol. This means that the first entry in each symbol does not have to be factored, except that powers of 2 must be removed. Thus, using the numerical example considered earlier, we have

$$\begin{aligned}(2819/4177)\,(4177/2819) &= (1358/2819) = (2/2819)(679/2819)\\ &= -(679/2819) = (2819/679) = (103/679)\\ &= -(679/103) = -(61/103) = -(103/61)\\ &= -(42/61) = -(2/61)(21/61) = (61/21)\\ &= (19/21) = (21/19) = (2/19) = -1,\end{aligned}$$

and we can again conclude that 2819 is a nonresidue of 4177.

The Jacobi symbol is not useful merely for evaluating Legendre symbols. We now give another application, for which we would have to invoke Dirichlet's theorem on primes in a progression if we used only symbols with prime second entries.

It is clear that if a given integer a is congruent to 1 (mod p) for every prime p, then $a = 1$, since $p \mid (a - 1)$ implies $p \leq |a| + 1$ unless $a - 1 = 0$. Here we have a trivial instance of the following principle: if an assertion involving a congruence holds for every prime modulus p, then the statement with the congruence replaced by the corresponding equation may be implied. (This is similar to the Hasse principle mentioned at the end of Chapter 3 but is even more demanding, since only prime moduli are considered, rather than prime powers.) From this point of view, it is natural to ask whether it is true that if, for fixed integers a and n, a is an nth power modulo p for every p, then a must be an nth power. (Saying that a is an nth power (mod p) means, of course, that a is congruent to the nth power of some integer; in other words, that $p \mid a$ or else a is an nth power residue of p.) Unfortunately, this is not quite the case: if the congruence $x^n \equiv a \pmod{p}$ is solvable for every p, then $a = b^n$ for some b if $8 \nmid n$; but if $8 \mid n$, either $a = b^n$ or $a = 2^{n/2}b^n$. Powers of 2 higher than the second cause difficulty here, just as they did in the study of primitive roots. (Cf. Problem 1 at the end of this section.)

At the present time, the theorem just stated cannot be proved in a simple way, except in the special case $n = 2$, which we now treat.

Theorem 5.10 *The congruence $x^2 \equiv a \pmod{p}$ is solvable for every prime p if and only if $a = b^2$ for some $b \in \mathbf{Z}$.*

Proof. If $a = b^2$, the congruence $x^2 \equiv a \pmod{p}$ has the solution $x \equiv b \pmod{p}$ for every p.

Suppose, on the other hand, that a is not a square. Then it suffices to prove that there exists an odd positive integer P such that $(a/P) = -1$, for then P must have a prime factor p such that $(a/p) = -1$. We consider three cases, which together exhaust all instances in which a is not a square.

I. $a = \pm 2^k b$, where k and b are positive odd integers. In this case choose P, as is possible by the Chinese remainder theorem, so that

$$P \equiv 5 \pmod 8,$$
$$P \equiv 1 \pmod b.$$

Then $(\pm 2/P) = -1$, $(2^{k-1}/P) = 1$, and $(b/P) = (P/b) = (1/b) = 1$, so

$$(a/P) = -1 \cdot 1 \cdot 1 = -1.$$

II. $a = \pm 2^{2h} q^k b$, where q, k, and b are positive odd integers, q is prime, and $q \nmid b$. Choose P so that

$$P \equiv 1 \pmod{4b},$$
$$P \equiv n \pmod q,$$

where n is any nonresidue of q. Then $(\pm 1/P) = 1$, $(2^{2h}/P) = 1$, $(b/P) = (P/b) = 1$, and $(q^k/P) = (q/P) = (P/q) = (n/q) = -1$, so $(a/P) = -1$.

III. $a = -b^2$. Choose $P \equiv 3 \pmod 4$ so that $(P, b) = 1$. Then

$$(a/P) = (-1/P) = -1. \quad \blacksquare$$

PROBLEMS

1. Show that the congruence

$$x^{2^k} \equiv 2^{2^{k-1}} \pmod p$$

has a solution for every prime p, if $k \geq 3$. [*Hint:* Verify the factorization

$$x^{2^k} - 2^{2^{k-1}} = (x^2 - 2)(x^2 + 2)((x - 1)^2 + 1)((x + 1)^2 + 1) \times (x^{2^3} + 2^{2^2}) \cdots (x^{2^{k-1}} + 2^{2^{k-2}}),$$

and show that every p divides one of the first three factors for suitable x.]

2. Show that if the congruence $x^n \equiv a \pmod{p^k}$ is solvable for every prime power modulus, then a is an nth power. [*Hint:* Consider the moduli p^{e+1}, where $p^e \parallel a$ and e is positive.]
3. Evaluate $(751/919)$, both with and without the use of Jacobi symbols. The entries are primes.
4. Show that Theorem 5.10 remains true if the phrase "for every prime p" is replaced by "for all but finitely many primes p."

5.5 FACTORIZATION OF LARGE INTEGERS

The closely related problems of deciding whether a given large integer is prime or composite, and of factoring it if it is composite, have engaged number theorists since the time of Mersenne. Originally the interest lay partly in determining new perfect numbers (see Chapter 6) and partly in finding a prime-generating formula, a noble theoretical goal but one that has seemed unattainable, to most mathematicians, for many years. But Mersenne's and Fermat's lists of alleged primes raised a new kind of challenge, to find efficient ways to determine the multiplicative structure of interestingly large numbers. "Interestingly large" means too large to attack unimaginatively with the well-known methods of the day, and yet small enough to offer some hope of success if one is sufficiently clever. Euler took up the challenge, and his efforts led him to the law of quadratic reciprocity and the study of quadratic forms, so it would be a mistake to dismiss this kind of problem as "mere computation." Still, the theoretical by-products were sufficiently limited that computational number theory was somewhat out of fashion for many years, until high-speed computers not only changed by orders of magnitude the size of feasible computations, but breathed new life into the panmathematical problem of finding better algorithms.

In the eighteenth century the table of primes extended only to 10^5, and for Euler, doing all calculations by hand (or mentally), a number of 8 or 9 digits was interestingly large. When the factor tables were extended to 10^7 in the nineteenth century, numbers of 12 or 13 digits could usually be handled, and numbers of special types (divisors of $a^n \pm b^n$) having up to 18 or 20 digits might succumb. Nowadays numbers of 20 digits usually can be factored or shown to be prime by comparatively amateurish programs, if some computer time is available, and at least one program exists by which any 40-digit number can be factored in 50 minutes on a suitable computer. Of course, partial or complete factorizations of some much larger integers are known. Since several of the techniques depend on ideas closely connected with the present and the preceding chapter, we digress a moment to talk about some of them. Throughout, N is the integer whose multiplicative structure is to be determined.

The most rudimentary method is to use the fact that if N is not prime, its smallest prime divisor p is $\leq \sqrt{N}$ (since $p \leq N/p$), so it suffices to try dividing N successively by 2, 3, 5, 7, . . . up to the last prime $\leq \sqrt{N}$. An apparent difficulty is that one must first know all these primes, but that is rather easy to accomplish with the sieve of Eratosthenes, which will be developed in the next chapter. In any case, it is not really necessary; using trial divisors which are not prime is wasteful but not otherwise harmful, and the prime number theorem says it is not terribly wasteful, unless N is huge. For the approximate equality $\pi(x) \approx x/\log x$ means that about one integer out of every $n \log 10$ is prime, in the range $1, 2, \ldots, 10^n$, so that, for example, about one in 11 is prime, up to 10^5. But the integers up to 10^5 not divisible by any of 2, 3, or 5 contain all these primes (except 2, 3, and 5 themselves), they are much easier to generate, and they constitute only slightly more than one in 4 of the integers up to 10^5, so there is not much waste. Generating the integers not divisible by 2, 3, 5, 7, or 11 is a little more trouble, but leaves fewer than 2 out of 11 integers. In any case, this method requires on the order of $c\sqrt{N}$

divisions (an uncomfortably large number), and makes no use of any special property of N. The latter feature might be called generality if N has no special attributes, and insensitivity if it has and they are ignored.

Fermat (1643) suggested using the fact that there is a 1–1 correspondence between factorizations $N = uv$ $(u > v \geq 1)$ of the odd nonsquare integer N and representations $N = x^2 - y^2 = (x - y)(x + y)$, with $x, y > 0$. For then $x > y$ and hence

$$\begin{aligned} u &= x + y, \\ v &= x - y, \end{aligned} \qquad \text{and} \qquad \begin{aligned} x &= (u + v)/2, \\ y &= (u - v)/2. \end{aligned}$$

As in Problem 12 of Section 1.4, we can use the relation $(x + 1)^2 - x^2 = 2x + 1$ to compute $(x + 1)^2 - N$ when $x^2 - N$ is already known, taking as initial value of x the smallest integer $m > \sqrt{N}$. Thus for $N = 1501$, we have $m = 39$, $39^2 - N = 20$, $40^2 - N = 20 + (2 \cdot 39 + 1) = 99$, and continuing, we obtain

$$\begin{array}{ccccccccccc} & 79 & 81 & 83 & 85 & 87 & 89 & 91 & 93 & 95 & 97 \\ 20 & 99 & 180 & 263 & 348 & 435 & 524 & 615 & 708 & 803 & 900 = 30^2 \end{array}$$

Thus $49^2 - N = 30^2$, so $N = (49 - 30)(49 + 30) = 19 \cdot 79$, and the procedure could be repeated for the factors, if the structure of either were unknown.

Note that it is necessary to examine each of the numbers 20, 99, 180, . . . to see whether it is a square. But one can rely mostly on the two terminal digits, 20, 99, 80, . . . , since squares must end in 00, 25, e1, e4, e9, or o6, where e is an even digit and o is an odd digit. (See Problem 11, Section 1.4.) In the present case this leaves only 524 and 900 to test directly for squareness.

The device worked well in this instance, but if N had happened to be prime the computation would have continued to the trivial factorization $x - y = 1$, $x + y = N$, requiring the examination of all values of $x^2 - N$ for $\sqrt{N} < x \leq (N + 1)/2$. This is worse than $c\sqrt{N}$ divisions! In general we cannot test all possible values of x, but must find a way to exclude most values *a priori*, even though this involves computing the remaining squares by multiplication rather than addition.

One way is this. Choose an odd prime p; then for each $u \pmod p$ there is a unique $v \pmod p$ such that $uv \equiv N \pmod p$, and each pair u, v gives one allowable sum $u + v \equiv 2x \pmod p$, and so one allowable $x \pmod p$. [*Example:* $N = 1501 \equiv 3 \pmod 7$, so

$$\{u, v\} = \{1, 3\}, \{2, 5\}, \text{ or } \{4, 6\},$$

so $2x \equiv 4, 0$, or $10 \pmod 7$, so $x \equiv 2, 0$, or $5 \pmod 7$.] It can be shown that there are at most $(p + 1)/2$ allowable residue classes for $x \pmod p$, which means that almost half the integers are excluded. Using k different primes leaves only about one out of every 2^k of the x's to be tested, from the interval $(\sqrt{N}, (N + 1)/2)$, and this is feasible if k is large enough. With the example $N = 1501$, we have $N \equiv 1 \pmod 3$, $1 \pmod 5$, $3 \pmod 7$, $5 \pmod{11}$, and $6 \pmod{13}$, which implies

$$
\begin{aligned}
x &\equiv 1 \text{ or } 2 \pmod 3,\\
x &\equiv 0, 1, \text{ or } 4 \pmod 5,\\
x &\equiv 0, 2, \text{ or } 5 \pmod 7,\\
x &\equiv 3, 4, 5, 6, 7, \text{ or } 8 \pmod{11},\\
x &\equiv 3, 4, 6, 7, 9, \text{ or } 10 \pmod{13}.
\end{aligned} \tag{6}
$$

The first three lines easily yield $x \equiv 40, 44, 49, 56, 61, 70, 79, 86, 89, 91, 104, 110, 119, 121, 124, 131, 139$, or $140 \pmod{105}$, where we have chosen elements from the complete residue system $39, 40, \ldots, 143 \pmod{105}$ because we search for x with $39 \le x \le 751$. The first two possible values of x, namely 40 and 44, would be eliminated by invoking the congruences modulo 11 and 13 in (6), and the first remaining value, $x = 49$, gives the factorization! This is mere luck, naturally, but the method really does work: for example, a 12-digit factor of $10^{25} - 1$ was shown to be prime in this way, in 1897, before electric desk calculators were developed.

Note that all the relationships such as "$N \equiv 3 \pmod 7$ implies $x \equiv 0, 2$, or $5 \pmod 7$" can be established and recorded permanently, for all p up to some convenient number, and then each new N merely requires a new application of the Chinese remainder theorem to a set of restrictions such as those in (6).

Another type of exclusion, devised by Legendre, depends on the law of quadratic reciprocity in the form of Theorem 5.8 for its effective application. Suppose that N can be written in the form $N = a^2 - db^2$, where d is square-free. We may suppose that $(d, N) = (b, N) = 1$, since otherwise we know a divisor of N. Then the equation gives $(ab^{-1})^2 \equiv d \pmod N$, so d is a quadratic residue of N, and hence $(d/p) = 1$ for every prime divisor p of N. This eliminates the primes in about half the residue classes $\pmod{4d}$, and thus about half the primes can be excluded as trial divisors. If a sufficient number of different d's can be found, the number of trial divisors will be reduced to manageable size. Before computers, of course, the work was facilitated by a table giving allowable residue classes for each d up to some convenient limit.

In our example $N = 1501$, the calculation tabulated earlier which gave $N = 49^2 - 30^2$ also yields $N = 39^2 - 5 \cdot 2^2 = 40^2 - 11 \cdot 3^2$, and examination of the squares a little smaller than N yields $N = 31^2 + 2^2 \cdot 3^3 \cdot 5 = 34^2 + 3 \cdot 5 \cdot 23 = 36^2 + 5 \cdot 41$. Thus each prime divisor p of N is such that $5, 11, -3 \cdot 5$, $-3 \cdot 5 \cdot 23$, and $-5 \cdot 41$ are residues of p, and hence so are -3 (since $5 \cdot -3 \cdot 5$ is), 23, and -41. These conditions require that

$$
\begin{gathered}
p \equiv \pm 1 \pmod{10}, \quad p \equiv \pm 1, 5, 7, 9, \text{ or } 19 \pmod{44}, \quad p \equiv \pm 1 \pmod 6,\\
p \equiv \pm 1, 7, 9, 11, 13, 15, 19, 25, 29, 41, \text{ or } 43 \pmod{92}, \text{ etc.}
\end{gathered}
$$

The restrictions modulo 10 and 44 already exclude every $p < \sqrt{N}$ except $p = 19$.

So far we have been discussing methods for finding factors of N, but of course if there is good reason to believe that N is prime it might be more useful to try to prove that fact. If N is prime, then

$$a^{N-1} \equiv 1 \pmod{N} \tag{7}$$

for every a prime to N, but the converse is not true. (See Problem 23, Section 3.2.) But if there is an a with $\operatorname{ord}_N a = N - 1$, then surely N is prime, and this will be the case if $a^{N-1} \equiv 1 \pmod{N}$ while $a^{(N-1)/p} \not\equiv 1 \pmod{N}$ for all $p \mid (N-1)$. This is a useful test for primality only for special N, since it requires knowledge of all prime divisors of $N - 1$. (See Problem 11, Section 3.2, for an application.) But (7) is a rather strong necessary condition for primality which is easy to apply and it surely would be among the first things to try, for some small values of a, if it were suspected that N is prime. If N meets the test for a certain a, the following theorem gives an exclusionary principle if anything at all is known about the factorization of $N - 1$.

Theorem 5.11 *Let $N - 1 = kq^n$, where $k > 0$, $n > 0$, and q is prime, and suppose (7) holds for some a. Then for every prime factor p of N, either $p \mid (a^{(N-1)/q} - 1)$ or $p \equiv 1 \pmod{q^n}$. In particular, if $(a^{(N-1)/q} - 1, N) = 1$, then every prime factor of N is $\equiv 1 \pmod{q^n}$.*

Proof. Suppose $\operatorname{ord}_p a = t$. Then $t \mid N - 1$, so $t \mid kq^n$. If $a^{(N-1)/q} \not\equiv 1 \pmod{p}$, then $t \nmid kq^{n-1}$. In that case $q^n \mid t$, and the theorem follows since $t \mid p - 1$. ∎

If the factored portion of $N - 1$ exceeds the unfactored complementary portion, this gives a simple test for primality:

Theorem 5.12 *Suppose $N - 1 = FC$, where the prime factors of F are known, and $F > C > 0$. If (7) holds while $(a^{(N-1)/q} - 1, N) = 1$ for every prime divisor of q of F, then N is prime.*

Proof. If $p \mid N$ and $q^n \parallel F$, then by the preceding theorem, $p \equiv 1 \pmod{q^n}$. Hence $p \equiv 1 \pmod{F}$, and consequently $p > F$. Hence

$$p^2 > F^2 \geq (C + 1)F > N.$$

But if every prime divisor of N is $> \sqrt{N}$, then N is prime. ∎

Consider, for example, the number

$$N = 9{,}999{,}999{,}900{,}000{,}001 = \frac{10^{24} + 1}{10^8 + 1}.$$

We have

$$N - 1 = \frac{10^{24} - 10^8}{10^8 + 1} = 10^8(10^8 - 1),$$

and we put $F = 10^8$, $a = 7$. A tedious but straightforward calculation shows that

$$7^{(N-1)/10} \equiv 8383924385890424 \pmod{N};$$

put $r = 838 \cdots 24$. It is easy to check that $(r^2 - 1, N) = (7^{(N-1)/5} - 1, N) = 1$, that $r^5 \equiv -1 \pmod{N}$, and that $r^{10} \equiv 7^{N-1} \equiv 1 \pmod{N}$, so N is prime.

Of course, if the factorization of $N - 1$ is not known, its decomposition may be regarded as a new subproblem. For example, $N - 1$ may split easily into several small primes and one large factor N_1 which one could again attempt to show is prime, and the whole process might repeat.

There are primality tests based on completely different principles, of course (see Problem 12, Section 1.4, for example), and many devices for factorization in addition to those described in this section. On the theoretical side, much attention has been given to finding optimal algorithms (with respect to machine time and storage space), for doing these and other arithmetic calculations. Factoring by testing for divisibility by the primes $< \sqrt{N}$ requires about $cN^{1/2}$ operations, but algorithms recently devised require only $cN^{1/4+\varepsilon}$ operations.

PROBLEMS

1. Factor $N = 5917$ by each of the methods described in this section.
2. Show that if $a^{N-1} \equiv 1 \pmod{N}$ and $N - 1 = mp$ with $p > \sqrt{N}$, and if $a^m \not\equiv 1 \pmod{N}$, then N is prime. [*Hint:* If q is the smallest prime divisor of N and $\text{ord}_q\, a = t$, show that $t \mid mp$ but $t \nmid m$.]
3. Prove that if p is an odd prime and $a > 1$, then every prime divisor q of $(a^p - 1)/(a - 1)$ is such that $q \equiv 1 \pmod{2p}$ and $(a/q) = 1$. Factor $2^{23} - 1$ and $3^{13} - 1$.
4. Use of the representations $N = a^2 - db^2$ with both positive and negative d suggests the utility of a theorem similar to Theorem 5.8 for the solutions p of the equation $(-q/p) = 1$, where q is a positive odd prime. Prove the theorem in the form in which Euler stated it: Let r run over the smallest positive residues (mod $4q$) of $1^2, 3^2, 5^2, \ldots,$ and let n run over the remaining numbers of the form $4k + 1, 0 < n < 4q$. Then for every odd prime p,

if $p = 4qk + r$	then q is a residue, and	$-q$ is a residue of p,
$p = 4qk - r$	residue	nonresidue,
$p = 4qk + n$	nonresidue	nonresidue,
$p = 4qk - n$	nonresidue	residue.

5. Show that the congruences in (6) are typical: there are always $\frac{1}{4}\{p + (N/p)\}$ allowable values of $x \pmod{p}$.
6. The method of exclusion of quadratic nonresidues can be used for other problems in addition to factoring, for example, for actually solving the congruence $x^2 \equiv a \pmod{p}$ for large p. (See Problem 2 of Section 5.1 for the case $p \equiv 3 \pmod{4}$.) Show that the problem can be rephrased as that of finding y such that $0 \le y < p/2$ and $a + py$ is a square; thus $a + py$ must be a quadratic residue of any prime q, and this restricts y. Use this idea to solve $x^2 \equiv 2 \pmod{1901}$ with a calculator, or $x^2 \equiv 2 \pmod{181}$ by hand.

NOTES AND REFERENCES

Section 5.3

There is an account of the history of the law of quadratic reciprocity in Bachmann [1902]. The story of the Euler-Legendre-Gauss dispute, given there in detail, is based on an article by Kronecker [1875]. Bachmann also lists the 50 proofs that had been published by the end of the nineteenth century, with some indication of methods; many more have been published since then.

Gauss's original proof was by induction, somewhat along the lines indicated in Problem 7, Section 5.2. Euler's proof was inadequate in that he assumed that the solutions p of $(q/p) = 1$ must be those in certain arithmetic progressions, and he merely determined which progressions. The lacuna in Legendre's argument was more interesting —he had already conjectured, in 1788, that there are infinitely many primes in all progressions $ak + b$ with $(a, b) = 1$, and he based his proof on this conjecture. He gave an erroneous proof of the conjecture in 1808, 29 years before Dirichlet's proof appeared.

We have devoted more space to the early history of quadratic reciprocity than to that of most other topics because of a curious gap in Dickson's *History of the Theory of Numbers*, a usually reliable friend: the law of quadratic reciprocity is nowhere stated, and no proofs of it are discussed. (A chapter on the subject was written, but it was unsatisfactory and was not published.)

The proof we have given here is a variant by Frobenius [1914] of an 1872 simplification by C. Zeller of Gauss's fifth proof. The Frobenius article gives a comparative analysis of several proofs.

For a readable and relatively elementary exposition of the cubic reciprocity law, see Ireland and Rosen [1972]

Section 5.4

The conditions under which $x^n \equiv a \pmod{p}$ is solvable for every p are due to Trost [1934]; the theorem was rediscovered by Ankeny and Rogers [1951].

Section 5.5

Theorem 5.11 was discovered by H. C. Pocklington in 1914, but was first put to use by Lehmer [1927]. Theorem 5.12 and other types of converses of Fermat's theorem were given by Robinson [1957]. For more recent results see Brillhart, Lehmer, and Selfridge [1975], and Guy [1975]; the latter has an extensive bibliography.

There is a very readable discussion of the algorithms presented here, and others, in Knuth [1969].

Primality tests depending upon the factorization of $N + 1$ rather than $N - 1$ are also known. These, together with a clear analysis of the time-cost of solving various number-theoretic problems on a computer, are to be found in the expository article of Lehmer [1969].

Pollard [1974] has shown how to factor a number N in $cN^{1/4+\varepsilon}$ steps, and how to decide whether N is prime in $c'N^{1/8+\varepsilon}$ steps. But c and c' are too large for the algorithms to be practical, for N of the size being tested at the present time.

Gauss suggested the method in Problem 6 for solving $x^2 \equiv a \pmod{p}$. For a newer algorithm, see Shanks [1972].

6

Number-Theoretic Functions and the Distribution of Primes

6.1 INTRODUCTION

A number-theoretic function is any function which is defined for positive integral argument or arguments. Euler's φ-function is such, as are $n!$, n^2, e^n, etc. The functions which are interesting from the point of view of number theory are, of course, those like φ whose value depends in some way on the arithmetic nature of the argument, and not simply on its size. But then the behavior of the function is likely to be highly irregular, and it may be a difficult matter to describe how rapidly the function values grow as the argument increases. One way to proceed is to find upper or lower bounds which are monotonic elementary functions such as $\log \log n$ or $e^{\sqrt{n}}$; such bounds gloss over the fine structure of the function, but they are much easier to work with, and suffice for some purposes. Not only are the techniques for finding bounds for number-theoretic functions likely to be new to the reader, but the whole subject of "estimative analysis," with its manifold devices, may be unfamiliar. For this reason, a number of problems have been included in this chapter as much for the sake of the manipulative practice they provide as for the immediate importance of the results. If further justification is needed, other than that these kinds of estimations form a steady diet in much of advanced number theory, it might be found in the fact that facility in approximating complicated expressions by simpler ones is of central importance in many other branches of mathematics besides number theory, and many of the techniques developed here should prove useful elsewhere.

It should be emphasized that almost no result in this chapter is the best of its kind that is known; essentially all the estimates that will be obtained can be and have been improved upon. But the improved versions depend on more intricate or more sophisticated proofs which are not suitable here; moreover, the results to be obtained are frequently just as useful as more precise estimates.

Two of the most interesting number-theoretic functions are $\tau(n)$, the number of positive divisors of n, and $\sigma(n)$, the sum of these divisors. These functions have been treated extensively in the literature, partly because of their simplicity and partly because they occur in a natural way in the investigation of many other problems. For this reason we shall pause briefly to demonstrate some of their fundamental properties. Recall that, as noted in Chapter 3, a number-theoretic function which is not identically zero is said to be **multiplicative** if $f(mn) = f(m)f(n)$ whenever $(m, n) = 1$.

Theorem 6.1 *The functions σ and τ are multiplicative.*

Proof. Assume that $(m, n) = 1$. Then by the unique factorization theorem, every divisor* of mn can be represented uniquely as the product of a divisor of m and a divisor of n, and conversely, every such product is a divisor of mn. Clearly this implies that τ is multiplicative, and that

$$\sum_{d \mid m} d \cdot \sum_{d' \mid n} d' = \sum_{d'' \mid mn} d'',$$

so that also $\sigma(m)\sigma(n) = \sigma(mn)$. ■

If f is any multiplicative function and the prime power factorization of n is

$$n = \prod_{i=1}^{r} p_i^{e_i},$$

then

$$f(n) = f\left(\prod_{i=1}^{r} p_i^{e_i}\right) = \prod_{i=1}^{r} f(p_i^{e_i}),$$

and so the function is completely determined when its value is known for every prime-power argument. In the cases at hand, we have

$$\tau(p^e) = e + 1$$

and

$$\sigma(p^e) = 1 + p + \cdots + p^e = \frac{p^{e+1} - 1}{p - 1}.$$

Thus we have proved

Theorem 6.2 *If $n = p_1^{e_1} \cdots p_r^{e_r}$, then*

$$\tau(n) = \prod_{i=1}^{r} (e_i + 1) \quad \text{and} \quad \sigma(n) = \prod_{i=1}^{r} \frac{p_i^{e_i+1} - 1}{p_i - 1}.$$

There is another way of proving the multiplicativity of σ and τ which uses a basic property of all multiplicative functions:

Theorem 6.3 *If f is multiplicative and F is the function defined by the equation*

$$F(n) = \sum_{d \mid n} f(d),$$

then F is also multiplicative.

---

* Here and throughout this chapter, variables occurring as arguments of number-theoretic functions are understood to be positive. The same applies to their divisors.

Remark. The multiplicativity of σ and τ follows immediately from the relations

$$\sigma(n) = \sum_{d \mid n} d, \qquad \tau(n) = \sum_{d \mid n} 1,$$

since the functions f_1 and f_2 defined by the equations

$$f_1(n) = n \qquad \text{and} \qquad f_2(n) = 1 \qquad \text{for all } n$$

are obviously multiplicative.

Proof. Let $(m, n) = 1$. Then every divisor d of mn can be written uniquely as the product of a divisor d_1 of m and a divisor d_2 of n, and $(d_1, d_2) = 1$. Hence

$$F(mn) = \sum_{d \mid mn} f(d) = \sum_{\substack{d_1 \mid m \\ d_2 \mid n}} f(d_1 d_2) = \sum_{\substack{d_1 \mid m \\ d_2 \mid n}} f(d_1) f(d_2)$$

$$= \sum_{d_1 \mid m} f(d_1) \cdot \sum_{d_2 \mid n} f(d_2) = F(m)F(n). \quad \blacksquare$$

We shall see in the next section that the converse of Theorem 6.3 also holds.

A problem that was of great interest to the Greeks was that of determining all the **perfect numbers**, that is, numbers such as 6 which are equal to the sum of their proper divisors. In our notation this amounts to asking for all solutions of the equation

$$\sigma(n) = 2n.$$

It was known as early as Euclid's time that every number of the form

$$n = 2^{p-1}(2^p - 1),$$

in which both p and $2^p - 1$ are primes, is perfect. This is easy to verify:

$$\sigma(n) = \frac{2^p - 1}{2 - 1} \cdot \frac{(2^p - 1)^2 - 1}{(2^p - 1) - 1} = (2^p - 1) \cdot 2^p = 2n.$$

It happens that a partial converse also holds: every *even* perfect number n is of the Euclid type. To see this we put $n = 2^{k-1} \cdot n'$, where $k \geq 2$. Then

$$\sigma(n) = \sigma(2^{k-1})\sigma(n') = (2^k - 1)\sigma(n'),$$

so that if n is perfect, it must be that

$$(2^k - 1)\sigma(n') = 2n = 2^k n'.$$

This implies that $(2^k - 1) \mid n'$, so we put $n' = (2^k - 1)n''$ and obtain

$$\sigma(n') = 2^k n''.$$

Since n' and n'' are divisors of n' whose sum is

$$n'' + (2^k - 1)n'' = 2^k n'' = \sigma(n'),$$

it must be that they are the only divisors of n', so that n' must be prime, and hence

$n'' = 1$, $n' = 2^k - 1$. Thus $n = 2^{k-1}(2^k - 1)$, where $2^k - 1$ is prime; this can happen only if k itself is prime.

There are two problems connected with perfect numbers which have not yet been solved. One is whether there are any odd perfect numbers; various necessary conditions are known for an odd number to be perfect, which show that any such number must be extremely large, but no conclusive results have been obtained. The other question, of course, is about the primes p for which $2^p - 1$ is prime. As was pointed out earlier, these Mersenne primes continue to occur, although with decreasing frequency, as far as computations have been pushed. There is no reason to suppose that there are only finitely many, but no proof that there are infinitely many.

Aside from φ, σ, and τ, the function with which we shall be most concerned in this chapter is $\pi(x)$, already defined in Chapter 1 as the number of primes not exceeding x. (We now drop the restriction that all variables are integer-valued. The last letters of the alphabet may, as usual, be real-valued.) It was shown there that $\pi(x)$ increases indefinitely with x, that is, that *there are infinitely many primes*. We now give another proof, which depends on the unique factorization theorem.

Assume that there are only k primes, say $p_1, \ldots, p_k$. By the unique factorization theorem, every integer larger than 1 can be written uniquely as the product of a square-free number (that is, an integer which is the product of distinct primes) and a square. But with only k primes at our disposal, there are only

$$\binom{k}{0} + \binom{k}{1} + \cdots + \binom{k}{k} = 2^k$$

square-free numbers, and there are not more than $\sqrt{n}$ perfect squares less than or equal to n. This means that there are at most $2^k\sqrt{n}$ positive integers not exceeding n, which is obviously false if $n > 2^k\sqrt{n}$, that is, if $\sqrt{n} > 2^k$. Actually, this argument proves a little more, namely that

$$2^{\pi(n)} > \sqrt{n}, \qquad \text{or} \qquad \pi(n) > \frac{\log n}{2 \log 2}.$$

For later use in this chapter we now prove a general combinatorial theorem of very wide applicability. (The product representation for the φ-function, for example, is a special case.) The result is sometimes called the **principle of cross-classification**; it also occurs conspicuously in elementary probability theory.

Theorem 6.4 *Let S be a set of N distinct elements, and let $S_1, \ldots, S_r$ be arbitrary subsets of S containing $N_1, \ldots, N_r$ elements, respectively. For $1 \le i < j < \cdots < l \le r$, let $S_{ij\ldots l}$ be the intersection of $S_i, S_j, \ldots, S_l$, that is, the set of all elements of S common to $S_i, S_j, \ldots, S_l$, and let $N_{ij\ldots l}$ be the* *number of elements of* $S_{ij\cdots l}$. *Then the number of elements of S not in any of* $S_1, \ldots, S_r$ *is**

$$K = N - \sum_{1 \le i \le r} N_i + \sum_{1 \le i < j \le r} N_{ij} - \sum_{1 \le i < j < k \le r} N_{ijk} + \cdots + (-1)^r N_{12\cdots r}.$$

Remark. To obtain the product formula for the φ-function, take S to be the set of integers $1, \ldots, n$, and for $1 \le k \le r$, take S_k to be the set of elements of S which are divisible by p_k, where $n = p_1^{e_1} \cdots p_r^{e_r}$. If $d \mid n$, the number of integers $s \le n$ such that $d \mid s$ is n/d; hence

$$\varphi(n) = n - \sum_{1 \le i \le r} \frac{n}{p_i} + \sum_{1 \le i < j \le r} \frac{n}{p_i p_j} - \cdots = n \prod_{p \mid n} \left(1 - \frac{1}{p}\right).$$

Proof. Let a certain element s of S belong to exactly m of the sets $S_1, \ldots, S_r$. If $m = 0$, s is counted just once in the above expression for K, namely in N itself. If $0 < m \le r$, then s is counted once, or $\binom{m}{0}$ times, in N, $\binom{m}{1}$ times in the terms N_i, $\binom{m}{2}$ times in the terms N_{ij}, etc. Hence the net contribution to K arising from the element s is

$$\binom{m}{0} - \binom{m}{1} + \binom{m}{2} - \cdots + (-1)^m \binom{m}{m} = (1 - 1)^m = 0. \quad \blacksquare$$

PROBLEMS

1. Find an expression for $\sigma_k(n)$, the sum of the kth powers of the divisors of n.
2. Prove that

$$\sum_{d \mid n} \tau^3(d) = \left(\sum_{d \mid n} \tau(d)\right)^2.$$

[*Hint:* Show that both sides are multiplicative functions, so that it suffices to consider the case $n = p^e$. Cf. Problem 7 of Section 1.3.]

3. Show that, if $\sigma(n)$ is odd, then n is a square or the double of a square.
4. Show that the number of ordered pairs of positive integers whose LCM is n is $\tau(n^2)$. [First understand the case $n = p^e$.]
5. Show that

$$\sum_{n=1}^{\infty} \frac{\tau(n)}{n^s} = \left(\sum_{n=1}^{\infty} \frac{1}{n^s}\right)^2$$

---

* In this identity and elsewhere, an empty sum (one having no terms) is to be given the value 0. Similarly, an empty product, such as occurs in Problem 7 below when $n = d$, is given the value 1.

if $s > 1$. [The series involved converge absolutely, and therefore can be rearranged in any order.]

6. a) Show that the sum of the odd divisors of n is

$$-\sum_{d \mid n} (-1)^{n/d} d.$$

[*Hint:* Let d_1 be an odd divisor of n, and find the total contribution to this sum from all divisors of n of the form $2^k d_1$.]

b) Show that if n is even, then

$$\sum_{d \mid n} (-1)^{n/d} d = 2\sigma(n/2) - \sigma(n).$$

7. Show that, if $d \mid n$ and $(n, r) = 1$, then the number of solutions (mod n) of

$$x \equiv r \pmod{d}, \qquad (x, n) = 1$$

is

$$\frac{\varphi(n)}{\varphi(d)} = \frac{n}{d} \prod_{\substack{p \mid n \\ p \nmid d}} \left(1 - \frac{1}{p}\right).$$

[*Hint:* Take S of Theorem 6.4 to be the n/d numbers

$$x = r + td, \qquad 1 \le t \le n/d.$$

If $p \mid d$, then $p \nmid x$. Let the subsets consist of those elements of S divisible by the various primes which divide n but not d.]

8. Let $\omega(n)$ be the number of distinct primes dividing n. Show that ω is an **additive** function, in the sense that it satisfies the condition $f(mn) = f(m) + f(n)$ whenever $(m, n) = 1$.

9. Show that, in the notation of the preceding problem,

$$\varphi(n) \ge n \prod_{k=2}^{\omega(n)+1} \left(1 - \frac{1}{k}\right) = \frac{n}{\omega(n) + 1}.$$

Show also that $2^{\omega(n)} \le \tau(n) \le n$, and conclude that

$$\varphi(n) > \frac{cn}{\log n} \qquad \text{for} \quad n \ge 2,$$

for a suitable constant $c > 0$.

10. Let S be an arbitrary finite set, with subsets S_i, S_{ij}, etc., as in Theorem 6.4, and let f be a complex-valued function defined on S. Show that

$$\sum_{\substack{x \in S \\ x \notin \cup S_i}} f(x) = \sum_{x \in S} f(x) - \sum_{i=1}^{r} \sum_{x \in S_i} f(x) + \sum_{i<j} \sum_{x \in S_{ij}} f(x) - \cdots.$$

In particular, suppose that S consists of N real numbers, ordered so that $x_1 \le x_2 \le \cdots \le x_N$, let S_i be the subset $\{x_1, \ldots, x_i\}$, for $i = 1, \ldots, N$, and put $f(x_1) = x_1$ and $f(x_j) = x_j - x_{j-1}$ for $j = 2, \ldots, N$. What does the above identity give in this case? Deduce the result in Problem 4(a) of Section 2.4.

11. a) Let $p_1, \ldots, p_k$ be distinct prime divisors of the positive integer n. Show that there are $2^k n/p_1 \cdots p_k$ positive integers $m \leq n$ such that for each $j \leq k$, $m \equiv 0$ or $-1 \pmod{p_j}$.
    b) Let $\theta(n)$ be the number of positive integers $m \leq n$ such that
$$(m, n) = (m + 1, n) = 1.$$
    Show that
$$\theta(n) = n \prod_{p \mid n} \left(1 - \frac{2}{p}\right).$$
12. Let $d_1, \ldots, d_t$ be a set of positive divisors of $n = p_1^{e_1} \cdots p_r^{e_r}$, with the property that $d_i \nmid d_j$ if $i \neq j$. Show that
$$t \leq \frac{\tau(n)}{\max_i (e_i + 1)}.$$
    [*Hint:* A divisor $d = p_1^{b_1} \cdots p_r^{b_r}$ is completely determined by the r-tuple $\{b_1, \ldots, b_r\}$. No two of the r-tuples $\{b_1, \ldots, b_{r-1}, 0\}$ can be identical.]
13. Show that if f is multiplicative and k is an integer such that $f(k) \neq 0$, then the function $F_k(n) = f(kn)/f(k)$ is multiplicative in n.

6.2 THE MÖBIUS FUNCTION

As we saw in Theorem 6.3, if f is any multiplicative function and F is its sum function, so that
$$F(n) = \sum_{d \mid n} f(d),$$
then F is also multiplicative. We now ask whether the converse is true—whether the multiplicativity of F implies that of f. To this end we attempt to express $f(n)$ as a sum, over the divisors of n, of terms involving $F(d)$. Assuming that F is multiplicative, and that the converse in question is valid, we can restrict attention to $f(p^e)$. Since
$$f(p^e) = F(p^e) - F(p^{e-1}),$$
we can write
$$f(p^e) = \sum_{\alpha=0}^{e} \mu(p^{e-\alpha}) F(p^\alpha) = \sum_{d \mid p^e} \mu\left(\frac{p^e}{d}\right) F(d),$$
if we define the function μ in the following way:
$$\begin{aligned} \mu(1) &= 1, \\ \mu(p) &= -1, \\ \mu(p^e) &= 0 \qquad \text{for } e > 1. \end{aligned}$$
If we now require in addition that μ be multiplicative, then $\mu(n)$ is defined for all positive integers n, and it is easily seen that
$$\mu(n) = \begin{cases} 1 & \text{if } n = 1, \\ 0 & \text{if } n \text{ is divisible by a square larger than } 1, \\ (-1)^v & \text{if } n = p_1 \cdots p_v, \text{ where the } p_i \text{ are distinct primes.} \end{cases}$$

This function μ is commonly called the **Möbius function**; it plays an important role in the theory of numbers. On the basis of the heuristic argument above, it is reasonable to conjecture that, for any n,

$$f(n) = \sum_{d \mid n} \mu\left(\frac{n}{d}\right) F(d),$$

and that from this formula one might be able to deduce the multiplicativity of f from that of F. We now substantiate these conjectures.

Theorem 6.5
$$\sum_{d \mid n} \mu(d) = \begin{cases} 1 & \text{if } n = 1, \\ 0 & \text{if } n > 1. \end{cases}$$

Proof. By Theorem 6.3, the function

$$M(n) = \sum_{d \mid n} \mu(d)$$

is multiplicative, and since

$$M(p^e) = \begin{cases} 1 & \text{if } e = 0, \\ 1 - 1 + 0 + \cdots + 0 & \text{if } e \geq 1, \end{cases}$$

we see that $M(n) = 0$ if n is divisible by any prime, that is, if $n > 1$. ∎

Theorem 6.6 (Möbius inversion formula) *If f is any number-theoretic function (not necessarily multiplicative) and*

$$F(n) = \sum_{d \mid n} f(d),$$

then

$$f(n) = \sum_{d \mid n} F(d)\mu\left(\frac{n}{d}\right) = \sum_{d \mid n} F\left(\frac{n}{d}\right)\mu(d) = \sum_{d_1 d_2 = n} \mu(d_1)F(d_2).$$

Proof. We have

$$\sum_{d \mid n} \mu(d)F\left(\frac{n}{d}\right) = \sum_{d_1 d_2 = n} \mu(d_1)F(d_2) = \sum_{d_1 d_2 = n} \mu(d_1) \sum_{d \mid d_2} f(d)$$

$$= \sum_{d_1 d \mid n} \mu(d_1)f(d) = \sum_{d \mid n} f(d) \sum_{d_1 \mid \frac{n}{d}} \mu(d_1).$$

But by Theorem 6.5, the coefficient of $f(d)$ in the final sum is zero unless $n/d = 1$ (that is, unless $d = n$), when it is 1, so that this last sum is equal to $f(n)$. ∎

As an example of Theorem 6.6, we have

Theorem 6.7
$$\varphi(n) = n \sum_{d \mid n} \frac{\mu(d)}{d}.$$

Proof. This follows immediately from Theorem 3.12:

$$\sum_{d \mid n} \varphi(d) = n.$$

It can also be obtained directly from the product representation of $\varphi(n)$, by multiplying out:

$$\varphi(n) = n \prod_{p \mid n} \left(1 - \frac{1}{p}\right) = n \left(1 + \sum_{\substack{p_1 \cdots p_\nu \mid n \\ p_1 < \cdots < p_\nu}} \frac{(-1)^\nu}{p_1 \cdots p_\nu}\right) = n \sum_{d \mid n} \frac{\mu(d)}{d}. \quad ■$$

Theorem 6.8 *If*

$$F(n) = \sum_{d \mid n} f(d)$$

and F is multiplicative, so is f.

Proof. If $(m, n) = 1$, then

$$\begin{aligned} f(mn) &= \sum_{\substack{d_1 \mid m \\ d_2 \mid n}} F(d_1 d_2) \mu\left(\frac{mn}{d_1 d_2}\right) \\ &= \sum_{\substack{d_1 \mid m \\ d_2 \mid n}} F(d_1) F(d_2) \mu\left(\frac{m}{d_1}\right) \mu\left(\frac{n}{d_2}\right) \\ &= \sum_{d_1 \mid m} F(d_1) \mu\left(\frac{m}{d_1}\right) \sum_{d_2 \mid n} F(d_2) \mu\left(\frac{n}{d_2}\right) \\ &= f(m) f(n). \quad ■ \end{aligned}$$

PROBLEMS

1. Prove that $\sum_{d \mid n} \sigma(d) \mu(n/d) = n$, for every $n \in \mathbf{Z}^+$.
2. Show that

$$\sum_{d^2 \mid n} \mu(d) = |\mu(n)|.$$

[*Hint:* Show that if $n = n_1^2 n_2$, where n_2 is square-free, then $d^2 \mid n$ if and only if $d \mid n_1$.]

3. Let

$$\Lambda(n) = \begin{cases} \log p & \text{if } n \text{ is a power of any prime } p, \\ 0 & \text{otherwise.} \end{cases}$$

Show that $\log n = \sum_{d \mid n} \Lambda(d)$, and deduce that

$$\sum_{d \mid n} \mu(d) \log d = -\Lambda(n).$$

4. Show that if f is multiplicative, then

$$\sum_{d \mid n} \mu(d) f(d) = \prod_{p \mid n} (1 - f(p)).$$

[*Hint:* Show that the function on the left is multiplicative.]

5. What is the error in the following argument? Given $F(n) = \sum_{d \mid n} f(d)$, we obtain $f(n) = \sum_{d \mid n} \mu(n/d) F(d)$ by the inversion theorem. Writing $g(d)$ for $\mu(n/d)f(d)$, we have $f(n) = \sum_{d \mid n} g(d)$, and hence $g(n) = \sum_{d \mid n} \mu(d) f(n/d)$. But $g(n) = \mu(1)F(n)$, so $F(n) = \sum_{d \mid n} \mu(d) f(n/d)$, contrary to the definition of $F(n)$.

6. a) In Theorem 6.6 it is tacitly assumed that $f(n)$ and $F(n)$ are defined on $\mathbf{Z}^+$, and that $F(n) = \sum_{d \mid n} f(d)$ for all such n. Show that if $f(n)$ and $F(n)$ are defined for all positive divisors n of a fixed $m \in \mathbf{Z}^+$, and if the above equation holds for all such n, then the inversion formula applies for such n.
   b) In the proof of Theorem 4.3, Theorem 4.2 was used. In the notation occurring there, prove that if $d \mid p - 1$, then $\sum_{c \mid d} \psi(c) = d$, and deduce Theorem 4.3 directly from (a) above and Theorem 6.7. Why can't Theorem 6.6 be used as it is stated, for this purpose?

7. Show that for $s > 1$,

$$\sum_{n=1}^{\infty} \frac{1}{n^s} \cdot \sum_{n=1}^{\infty} \frac{\mu(n)}{n^s} = 1.$$

8. Let $J_k(n)$ be the number of ordered sets of k not necessarily distinct positive integers, none of which exceeds n and whose GCD is prime to n. Show, in the order indicated, that

   a) $\sum_{d \mid n} J_k(d) = n^k$,

   b) J_k is multiplicative,

   c) $J_k(n) = n^k \prod_{p \mid n} \left(1 - \dfrac{1}{p^k}\right)$.

9. Let f be any number-theoretic function of two variables. Show that if F is defined by the equation

$$F(m, n) = \sum_{\substack{d_1 \mid m \\ d_2 \mid n}} f(d_1, d_2),$$

then

$$f(m, n) = \sum_{\substack{d_1 \mid m \\ d_2 \mid n}} \mu(d_1)\mu(d_2) F\left(\frac{m}{d_1}, \frac{n}{d_2}\right).$$

10. a) If θ is any multiplicative function, then the function θ' defined by the equation

$$\sum_{d \mid n} \theta(d)\theta'\left(\frac{n}{d}\right) = \begin{cases} 1 & \text{if } n = 1, \\ 0 & \text{if } n > 1, \end{cases}$$

exists and is also multiplicative. In this notation, find μ' and τ'.

b) If θ and θ' have the relation specified in (a) and if

$$F(n) = \sum_{d \mid n} f(d)\theta\left(\frac{n}{d}\right),$$

then

$$f(n) = \sum_{d \mid n} F(d)\theta'\left(\frac{n}{d}\right).$$

11. A complex number $z = \cos\alpha + i\sin\alpha = e^{i\alpha}$ is called an **nth root of unity** if $z^n - 1 = 0$, and is a **primitive** nth root of unity if in addition $z^k - 1 \neq 0$ for $1 \leq k < n$. The **nth cyclotomic polynomial**, $\Phi_n(x)$, is the monic polynomial of which the zeros are the distinct primitive nth roots of unity. Take $n > 1$.

a) Show that $\zeta = e^{2\pi i/n}$ is a primitive nth root of unity.

b) Show that

$$\Phi_n(x) = \prod_{\substack{1 \leq j \leq n \\ (j,n)=1}} (x - \zeta^j),$$

and that $x^n - 1 = \prod_{d \mid n} \Phi_d(x)$.

c) By applying the inversion formula to $\log \Phi_n(x)$, or otherwise, show that

$$\Phi_n(x) = \prod_{d \mid n} (x^{n/d} - 1)^{\mu(d)}.$$

In particular, conclude that $\Phi_n \in \mathbf{Z}[x]$.

d) Suppose that p is prime, $p \nmid n$, and $\Phi_n(a) \equiv 0 \pmod{p}$. Show that

i) $p \nmid a$;

ii) $t \mid n$, where $t = \operatorname{ord}_p a$;

iii) if $a^{n/d} \equiv 1 \pmod{p}$, then $d \,\Big|\, \dfrac{n}{t}$;

iv) if $p^s \,\|\, (a^t - 1)$, then $p^s \,\|\, (a^{mt} - 1)$ when $p \nmid m$;

v) if $n/t > 1$, then $p \nmid \Phi_n(a)$ [use (c)];

vi) the congruence $\Phi_n(x) \equiv 0 \pmod{p}$ is solvable if and only if $p \equiv 1 \pmod{n}$.

e) Show that there are infinitely many primes $p \equiv 1 \pmod{n}$. [*Hint:* Consider $\Phi_n(np_1 \cdots p_r y)$, where the p_i are $\equiv 1 \pmod{n}$.]

6.3 THE FUNCTION [x]

Another function which is of importance in number theory is the function $[x]$, introduced in the preceding chapter to represent the largest integer not exceeding x. In other words, for each real x, $[x]$ is the unique integer such that

$$x - 1 < [x] \leq x < [x] + 1.$$

For later purposes we list some of the properties of $[x]$.

a) $x = [x] + ((x))$, where $0 \leq ((x)) < 1$. $((x))$ is called the **fractional part** of x.

b) $[x + n] = [x] + n$, if n is an integer.

c) $[x] + [-x] = \begin{cases} 0 & \text{if } x \text{ is an integer,} \\ -1 & \text{otherwise.} \end{cases}$

d) $[x_1] + [x_2] \le [x_1 + x_2]$.

e) $[x/n] = [[x]/n]$ if n is a positive integer.

f) $0 \le [x] - 2[x/2] \le 1$. (Equivalently, $[x] - 2[x/2]$ assumes only the values 0 and 1.)

g) The number of integers m for which $x_1 < m \le x_2$ is $[x_2] - [x_1]$.

h) The number of positive multiples of m which do not exceed x is $[x/m]$, when $m > 0$, $x > 0$.

i) The least nonnegative residue of a, modulo m, is $m\left(\left(\dfrac{a}{m}\right)\right)$.

These properties may easily be proved using the definition of $[x]$ and the first property above. Also, the graph of $y = [x]$ is sometimes useful.

Another quantity closely related to $[x]$ is the nearest integer to x, which is $[x + \frac{1}{2}]$. Sometimes the quantity $-[-x]$ occurs; it is the smallest integer not less than x.

In order to simplify the notation, summation signs will sometimes be used with a real number x as upper limit. In these cases, it is understood that the summation variable takes values up to $[x]$; in other words,

$$\sum_{k=a}^{x} f(k) = \sum_{k=a}^{[x]} f(k).$$

The following relation between the greatest integer function and the factorial function will be of importance later.

Theorem 6.9 *If n is a positive integer and p is prime, the p-component of $n!$ is*

$$\left[\frac{n}{p}\right] + \left[\frac{n}{p^2}\right] + \left[\frac{n}{p^3}\right] + \cdots.$$

Remark. The sum has, of course, only finitely many nonzero terms.

Proof. The multiples of p from among the numbers $1, 2, \ldots, n$ are counted once each in $[n/p]$, those which are also multiples of p^2 are counted again in $[n/p^2]$, etc. Thus if $p^r \,\|\, m$, the total contribution to the sum

$$\left[\frac{n}{p}\right] + \left[\frac{n}{p^2}\right] + \cdots$$

from the number m is exactly r, as it should be. ∎

PROBLEMS

1. Carry out the proofs of the properties of the brackets function mentioned in the text.
2. Prove that $[2x] + [2y] \geq [x] + [y] + [x + y]$, where x and y are arbitrary real numbers. [*Hint:* Consider separately the cases that neither, one, or both of $((x))$ and $((y))$ are smaller than $\frac{1}{2}$.]
3. Let $f(x, n)$ be the number of integers less than or equal to x and prime to n. Show that

   a) $\displaystyle\sum_{d \mid n} f\left(\frac{x}{d}, \frac{n}{d}\right) = [x]$. [Parallel the proof of Theorem 3.12.]

   b) $\displaystyle f(x, n) = \sum_{d \mid n} \mu(d) \cdot \left[\frac{x}{d}\right]$.

   Do you see a way to obtain (b) from (a) by the inversion formula?
4. a) Let x be a number between 0 and 1. Let a_1 be the smallest positive integer such that $x_1 = x - a_1^{-1} \geq 0$, let a_2 be the smallest positive integer such that $x_2 = x_1 - a_2^{-1} \geq 0$, etc. Show that this leads to a finite expansion

   $$x = \frac{1}{a_1} + \frac{1}{a_2} + \cdots + \frac{1}{a_n}$$

   (that is, that $x_{n+1} = 0$ for some n) if and only if x is rational. (The resulting sum is sometimes called the *Egyptian* or *Ahmes expansion* of x, because the ancient Egyptians wrote rational numbers as sums of reciprocals of integers. The allusion is misleading, however, since the Egyptians did not use this "smallest integer" algorithm.)

   b) Show that if the integers $1 < b_1 < b_2 < \cdots$ increase so rapidly that

   $$\frac{1}{b_{k+1}} + \frac{1}{b_{k+2}} + \cdots < \frac{1}{b_k - 1} - \frac{1}{b_k} \qquad \text{for } k \geq 1,$$

   then the number $\sum b_k^{-1}$ is irrational. Prove that $\sum_0^\infty (2^{3^k} + 1)^{-1}$ is irrational.
5. Show that if $(n, k) = 1$, then

   $$\sum_{j=1}^{n} \left(\left(\frac{jk}{n}\right)\right) = \frac{n - 1}{2}.$$

6. a) Show that $(ab)!/a!\,(b!)^a$ is an integer if a and b are positive integers. [*Hint:* Use induction on a.]

   b) Show that $(2a)!\,(2b)!/a!\,b!\,(a + b)!$ is an integer. [*Hint:* Use Problem 2.]

6.4 THE SYMBOLS "O", "o", "$\ll$", AND "$\sim$"

If we construct tables of values of the common number-theoretic functions, we are immediately struck by how erratically they behave. Thus $\tau(n)$ can be arbitrarily large, since for example $\tau(2^m) = m + 1$, and yet $\tau(n) = 2$ whenever n is prime.

Neither φ nor σ varies quite so wildly, in the sense that each of them definitely grows with n, but they are still far from being monotonic. It is one of the objects of this chapter to see what can be said about the size of these and other function values simply in terms of the size of their arguments.

Some convenient notations have been introduced for use in this connection. Let $g(x)$ be defined and positive for all x in some unbounded set S of positive numbers (which will usually be either the set of positive integers or the set of positive real numbers but might, for example, be the set of primes.) Then if $f(x)$ is defined on S, and if there is a constant M such that

$$\frac{|f(x)|}{g(x)} < M$$

for all sufficiently large $x \in S$, then we write either $f(x) = O(g(x))$ or $f(x) \ll g(x)$. If

$$\lim_{\substack{x\to\infty \\ x \in S}} \frac{f(x)}{g(x)} = 0,$$

we write $f(x) = o(g(x))$, and if

$$\lim_{\substack{x\to\infty \\ x \in S}} \frac{f(x)}{g(x)} = 1,$$

we write $f(x) \sim g(x)$, and say that $f(x)$ is *asymptotically equal* to $g(x)$. For example, for $x \in \mathbf{R}$ and $n \in \mathbf{Z}^+$,

$$\begin{aligned}
\sin x &\ll x, \\
\sin x &= o(x), \\
\sin x &= O(1), \\
\varphi(n) &= O(n), \\
\sqrt{x} &= o(x), \\
x^k &= o(e^x) && \text{for every constant } k, \\
\log^k x &= o(x^\alpha) && \text{for every pair of constants } \alpha > 0 \text{ and } k, \\
[x] &\sim x.
\end{aligned}$$

Here each of the second and third relations gives more information than the one preceding it; the first says that $\sin x$ does not grow any faster than x itself, the second that it does not grow as fast, and the third that $\sin x$ remains bounded as x increases. In the fourth equation, $O(n)$ could not be replaced by $o(n)$, since $\varphi(p) = p - 1 \sim p$.

The purpose of introducing the Landau symbols "big oh" and "little oh" is that, by their use, a complicated expression can be replaced by its principal or largest term, plus a remainder or error term whose possible size is indicated. Retaining an estimate for the error term is necessary because if several such expressions are combined, one has eventually to show that the sum of the error terms is still of smaller order of magnitude than the principal term. This in turn makes it necessary to combine terms involving "O" and "o". The following abbreviated rules apply:

a) $O(O(g(x))) = O(g(x))$,
b) $O(o(g(x)))$, $o(O(g(x)))$ and $o(o(g(x))) = o(g(x))$,
c) $O(g(x)) \pm O(g(x)) = O(g(x)) \pm o(g(x)) = O(g(x))$,
d) $o(g(x)) \pm o(g(x)) = o(g(x))$,
e) $\{O(g(x))\}^2 = O(g^2(x))$,
f) $O(g(x)) \cdot o(g(x)) = \{o(g(x))\}^2 = o(g^2(x))$.

The meaning of the first statement, for example, is that if $f(x) = O(g(x))$ and $h(x) = O(f(x))$, then $h(x) = O(g(x))$; this follows from the fact that if $0 < f(x) < M_1 g(x)$ and $|h(x)| < M_2 f(x)$, then $|h(x)| < M_1 M_2 g(x)$. The other assertions are equally straightforward; they need not be remembered explicitly, but are listed here to help orient the student, who should analyze all of them. Notice that suitable combinations of these rules give more general ones; for example, rules (a) and (c) show that

$$O(f(x)) \pm O(g(x)) = O(\max(f(x), g(x))).$$

A useful fact to remember is that the implication

$$f(x) = O(g(x)) \qquad \text{implies} \qquad h(f(x)) = O(h(g(x)))$$

does not hold in general; a sufficient condition is that h be monotonic and that $h(kx) = O(h(x))$ for every positive constant k, if $h(x)$ and $f(x) \to \infty$ as $x \to \infty$. Thus if $f(x)$ is larger than some positive constant for every $x > 0$, then $f(x) = O(g(x))$ implies that $\log f(x) = O(\log g(x))$, but it does not imply that

$$e^{f(x)} = O(e^{g(x)}),$$

since, for example, $\log x = O(\log \sqrt{x})$ but $x \neq O(\sqrt{x})$.

The situation is quite different for the "o" symbol. If $f(x) = o(g(x))$, then

$$e^{f(x)} = o(e^{g(x)})$$

if $f(x)$ increases indefinitely with x, but the relation $\log f(x) = o(\log g(x))$ may be false, for example, if $f(x) = \sqrt{x}$, $g(x) = x$.

Another important point arises when we want to add together a set of error terms, the number $a(x)$ of such terms being an increasing function of x. It is not true without restriction that

$$\sum_{k=1}^{a(x)} O(g_k(x)) = O\left(\sum_{k=1}^{a(x)} g_k(x)\right),$$

since, for example,

$$x = O(x), \quad 2x = O(x), \quad \ldots,$$

but

$$\sum_{k=1}^{[x]} kx \neq O\left(\sum_{k=1}^{[x]} x\right).$$

What is needed here, of course, is that the constants implied in the symbols $O(g_k(x))$ all be bounded above by some number independent of k. The corresponding principle for the "o" symbol is this: if $f_k(x) = o(g_k(x))$, then we can write $f_k(x) = \varepsilon_k(x)g_k(x)$, where $\varepsilon_k(x) \to 0$ as $x \to \infty$, for fixed k, and if

$$\max(|\varepsilon_1(x)|, \ldots, |\varepsilon_{a(x)}(x)|) \to 0 \text{ as } x \to \infty,$$

then

$$\sum_{k=1}^{a(x)} f_k(x) = o\left(\sum_{k=1}^{a(x)} g_k(x)\right).$$

The reason for all these special caveats, of course, is that "$= O$ of" and "$= o$ of" are not equivalence relations; we are abusing the equality sign rather badly, and we cannot expect to retain all the simple rules of manipulation enjoyed by actual equations.

The oh-notation is especially useful when certain "principal" terms are given explicitly while others are merely estimated, for example $[x] = x + O(1)$ or $(x^2 + 1)/(x + 1)^2 = 1 + o(1)$. When the object is simply to get an upper bound for some expression, accurate to within a constant factor, the Vinogradov symbol "$\ll$" may be more convenient, since it is more reminiscent of the statement actually being made than is the equivalent "$= O$ of".

Turning now to the relation $f(x) \sim g(x)$, notice first that it is equivalent to the equation $f(x) = g(x) + o(g(x))$. Hence if $g(x) \to \infty$ as x increases indefinitely, the difference $f(x) - g(x)$ need not remain bounded; all that is asserted is that it is of smaller order of magnitude than $g(x)$ itself.

To give more precise information about $f(x)$, we must consider not $f(x)$ but $f(x) - g(x)$. As an example of this, consider the following theorem, which is not strictly a number-theoretic result but will be useful in what follows:

Theorem 6.10 *There is a constant* $\gamma = 0.57721\ldots$ *(called* **Euler's constant**) *such that*

$$\sum_{k=1}^{n} \frac{1}{k} = \log n + \gamma + O\left(\frac{1}{n}\right). \tag{1}$$

Remark. The relation

$$\sum_{k=1}^{n} \frac{1}{k} - \log n \sim \gamma, \qquad \text{or} \qquad \lim_{n\to\infty}\left(\sum_{k=1}^{n} \frac{1}{k} - \log n\right) = \gamma, \tag{2}$$

is weaker than (1), since it says nothing about the error except that it approaches zero. Note that (2) is not equivalent to

$$\sum_{k=1}^{n} \frac{1}{k} \sim \log n + \gamma$$

(that is, terms may not always be "transposed" in an asymptotic relation), for this last relation has no more content than the simpler relation

$$\sum_{k=1}^{n} \frac{1}{k} \sim \log n.$$

Proof. Put

$$\alpha_k = \log k - \log(k-1) - \frac{1}{k}, \qquad k = 2, 3, \ldots,$$

and put

$$\gamma_n = \sum_{k=1}^{n} \frac{1}{k} - \log n, \qquad n = 1, 2, \ldots,$$

so that

$$1 - \gamma_n = \sum_{k=2}^{n} \alpha_k, \qquad n = 2, 3, \ldots.$$

Geometrically, the number α_k represents the difference between the area of the region between the x-axis and the curve $y = 1/x$ in the interval $k - 1 \leq x \leq k$, and the area of the rectangle inscribed in this region; it is therefore positive. The regions having areas α_2, α_3, and α_4 are shaded in Fig. 6.1. If the regions having areas $\alpha_2, \ldots, \alpha_n$ are translated parallel to the x-axis into the interval $0 \leq x \leq 1$, it becomes obvious that $0 < 1 - \gamma_n < 1/2$ and that $1 - \gamma_{n+1} > 1 - \gamma_n$, for $n = 1, 2, \ldots$. Since every bounded increasing sequence is convergent, we have

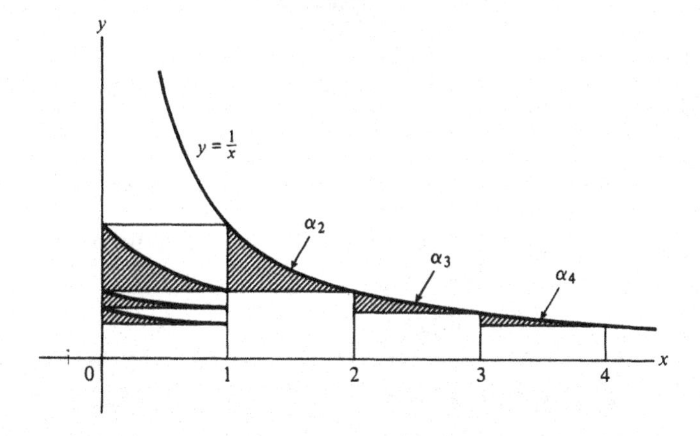

Figure 6.1

that $\lim_{n\to\infty}(1 - \gamma_n)$ exists; we call the limit $1 - \gamma$. Referring again to the square $0 \le x \le 1, 0 \le y \le 1$, we see that the region whose area is

$$\gamma_n - \gamma = (1 - \gamma) - (1 - \gamma_n) = \sum_{k=n+1}^{\infty} \alpha_k$$

is contained in the rectangle $0 \le x \le 1, 0 \le y \le 1/n$, of area $1/n$, so that

$$\gamma_n - \gamma = O\left(\frac{1}{n}\right). \quad \blacksquare$$

If $f(x) \sim g(x)$ and $g(x) \to \infty$ as $x \to \infty$, then $\log f(x) \sim \log g(x)$. The relation $e^{f(x)} \sim e^{g(x)}$ is usually false, however; it is true only when $f(x) - g(x) = o(1)$. Finally, under the above suppositions, together with that of the continuity of f and g, one may deduce that

$$\int_a^x f(t)\,dt \sim \int_a^x g(t)\,dt$$

for sufficiently large fixed a, by applying l'Hôpital's rule, but the corresponding relation $f'(x) \sim g'(x)$ is not always valid; two curves need not be going in the same direction merely because they are close together.

PROBLEMS

1. Carry out the proofs of all the unproved statements in this section.
2. Show that

$$\sum_{\substack{i,j=1\\ p_i<p_j}}^{\infty} \left[\frac{x}{p_ip_j}\right] = x \sum_{\substack{p_ip_j \le x\\ p_i<p_j}} \frac{1}{p_ip_j} + O(x),$$

where p_i is the ith prime.

†3. Prove the Euler-Maclaurin summation formula: *If $m, n \in \mathbf{Z}$ and f has a continuous derivative for $m \le x \le n$, then*

$$\sum_{k=m}^{n} f(k) = \int_m^n f(x)\,dx + \tfrac{1}{2}f(m) + \tfrac{1}{2}f(n) + \int_m^n f'(x)\eta(x)\,dx,$$

where $\eta(x) = x - [x] - \frac{1}{2}$. [*Hint:* For $m \le k < n$, integrate by parts with $\int dx = \eta(x)$ to obtain

$$\int_k^{k+1/2} f(x)\,dx = \tfrac{1}{2}f(k) - \int_k^{k+1/2} f'(x)\eta(x)\,dx,$$

$$\int_{k+1/2}^{k+1} f(x)\,dx = \tfrac{1}{2}f(k+1) - \int_{k+1/2}^{k+1} f'(x)\eta(x)\,dx,$$

and sum over k.] From it, obtain the following results:

a) Theorem 6.10.

b) $\sum_{k=1}^{n} k^{-\alpha} = \frac{n^{1-\alpha}}{1-\alpha} + c_\alpha + O(n^{-\alpha})$ if $0 < \alpha < 1$, for suitable constant c_α.

c) $\log(n!) = (n + \frac{1}{2}) \log n - n + O(\log n)$, for $n > 1$. (The error term could be improved to $C + O(n^{-1})$ by integrating $\int_1^n (\eta(x)/x)\,dx$ by parts and using the fact that $\int_a^b \eta(x)\,dx \ll 1$ for all $a, b \in \mathbf{R}$.)

4. Prove that
$$\operatorname{li}(x) = \int_2^x \frac{dt}{\log t} = \frac{x}{\log x} + O\left(\frac{x}{\log^2 x}\right).$$

5. Show that if $f(x)$ tends to zero monotonically as x increases without limit, and is continuous for $x > 0$, and if the series
$$\sum_{k=1}^{\infty} f(k)$$
diverges, then
$$\sum_{k=1}^{n} f(k) \sim \int_1^n f(x)\,dx.$$
If $g(x)$ is a second function satisfying the same hypotheses as $f(x)$, and if $g(x) = o(f(x))$, show that
$$\sum_{k=1}^{n} g(k) = o\left(\sum_{k=1}^{n} f(k)\right).$$

6. Find a function $f(x)$ such that (a) $f'(x) > 0$ for large x, (b) for every constant $\delta > 0$, $f(x) = o(x^\delta)$, and (c) for every $a > 0$, $f(x) \neq O(\log^a x)$.

7. Show that if $h(x) = o(1)$, then $\int_a^x h(t)\,dt = o(x)$.

8. In the notation of Problem 11 of Section 6.1, show that
$$\left(\frac{\varphi(n)}{n}\right)^2 \ll \frac{\theta(n)}{n} \ll \left(\frac{\varphi(n)}{n}\right)^2.$$

9. Let $\log_k x$ be the kth iterate of the logarithm:
$$\log_1 x = \log x, \qquad \log_k x = \log(\log_{k-1} x).$$
Is there a continuous increasing function $f(x)$ such that $\lim_{x\to\infty} f(x) = \infty$ but $f(x) = o(\log_k x)$ for each fixed k? If so, exhibit one.

6.5 THE SIEVE OF ERATOSTHENES

We now turn to the study of $\pi(x)$, and shall obtain many of the classical elementary results concerning the distribution of primes. None of these estimates is the best of its kind that is known, but to obtain more accurate results would require either too long a discussion to be worth while or the use of tools not available here, as, for example, the theory of functions of a complex variable. For many purposes our results are quite as useful as the better estimates.

One method of estimating $\pi(x)$ is based upon the observation that if n is less than or equal to x and is not divisible by any prime less than or equal to $\sqrt{x}$, then it is prime. Thus if we eliminate from the integers between 1 and x first all multiples of 2, then all multiples of 3, then all multiples of 5, etc., until all multiples of all primes less than or equal to $\sqrt{x}$ have been eliminated, then the numbers remaining are prime. This method of eliminating the composite numbers is known as the **sieve of Eratosthenes**; it has been adapted by V. Brun and others into a powerful method of estimating the number of integers in a certain interval having specified divisibility properties with respect to a certain set of primes. (See Section 6.12.)

We can modify the process just described by striking out the multiples of the first r primes $p_1, \ldots, p_r$, retaining r as an independent variable until the best choice for it can be clearly seen. If p_r is not the largest prime less than or equal to $\sqrt{x}$, but is some smaller prime, then of course it is no longer the case that all the integers remaining are primes, but certainly none of the primes except $p_1, \ldots, p_r$ have been removed. Thus if $A(x, r)$ is the number of integers remaining after all multiples of $p_1, \ldots, p_r$ (including $p_1, \ldots, p_r$ themselves, of course) have been removed from the integers less than or equal to x, then

$$\pi(x) \leq r + A(x, r).$$

In order to estimate $A(x, r)$ we use Theorem 6.4. If we take the $N = [x]$ objects to be the positive integers $\leq x$ and take S_k, for $1 \leq k \leq r$, to be the set of elements of S divisible by p_k, then

$$N_1 = \left[\frac{x}{2}\right], \quad N_2 = \left[\frac{x}{3}\right], \quad \ldots, \quad N_{12} = \left[\frac{x}{2 \cdot 3}\right], \quad \ldots,$$

and so

$$A(x, r) = [x] - \sum_{i=1}^{r} \left[\frac{x}{p_i}\right] + \sum_{1 \leq i < j \leq r} \left[\frac{x}{p_i p_j}\right] - \sum_{1 \leq i < j < k \leq r} \left[\frac{x}{p_i p_j p_k}\right] + \cdots$$
$$+ (-1)^r \left[\frac{x}{p_1 p_2 \cdots p_r}\right].$$

The difference between this and the expression resulting from it by omitting the brackets, namely

$$x - \sum_{1 \leq i \leq r} \frac{x}{p_i} + \sum_{1 \leq i < j \leq r} \frac{x}{p_i p_j} - \cdots + (-1)^r \frac{x}{p_1 p_2 \cdots p_r},$$

does not exceed

$$1 + \binom{r}{1} + \binom{r}{2} + \cdots + \binom{r}{r} = 2^r,$$

and consequently

$$\pi(x) \leq r + x\prod_{i=1}^{r}\left(1 - \frac{1}{p_i}\right) + 2^r.$$

This result would be utterly useless if we chose r so that the rth prime is the largest one $< \sqrt{x}$, for then (as will turn out later) we should have $r > x^{(1/2)-\varepsilon}$ and $2^r > x$, whereas obviously $\pi(x) < x$. The trouble is that with this r, the expression for $A(x, r)$ has many terms with value 0, and we are making an error of almost 1 in dropping the brackets from each.

To obtain something useful from the above bound, we need an estimate for the product occurring there.

Theorem 6.11 *If* $x \geq 2$, *then*

$$\prod_{p \leq x}\left(1 - \frac{1}{p}\right) < \frac{1}{\log x}.$$

Proof. We have

$$\prod_{p \leq x}\frac{1}{1 - 1/p} = \prod_{p \leq x}\left(1 + \frac{1}{p} + \frac{1}{p^2} + \cdots\right),$$

and, by the unique factorization theorem, when the product on the right is multiplied out it gives the sum of the reciprocals of all integers having only primes not exceeding x as prime divisors. In particular, all integers less than or equal to x are of this form, and so

$$\prod_{p \leq x}\frac{1}{1 - 1/p} > \sum_{k=1}^{x}\frac{1}{k} > \int_{1}^{[x]+1}\frac{du}{u} > \log x. \quad ■$$

We can now prove

Theorem 6.12

$$\pi(x) \ll \frac{x}{\log\log x}.$$

Proof. As above,

$$\pi(x) \leq r + 2^r + x\prod_{i=1}^{r}\left(1 - \frac{1}{p_i}\right) \leq 2^{r+1} + x\prod_{i=1}^{r}\left(1 - \frac{1}{p_i}\right),$$

and by Theorem 6.11,

$$\pi(x) \leq 2^{r+1} + \frac{x}{\log p_r}.$$

But $p_r \geq r$, and so

$$\pi(x) < \frac{x}{\log r} + 2^{r+1}.$$

Taking $r = [\log x] + 1$, this becomes

$$\pi(x) < \frac{x}{\log\log x} + 4 \cdot 2^{\log x} = \frac{x}{\log\log x} + O(x^{\log 2}).$$

Since $\log 2 < 1$, the last term is $o(x/(\log\log x))$. Hence

$$\pi(x) < \frac{x}{\log\log x} + o\left(\frac{x}{\log\log x}\right) = O\left(\frac{x}{\log\log x}\right). \quad ■$$

We have proved a little more than the theorem claims, in the final line, and we shall soon get even closer to the truth.

Theorems 6.11 and 6.12 bear a rather peculiar relation to each other: Theorem 6.11 was used in the proof of Theorem 6.12, yet the import of Theorem 6.11 is that the primes are not too infrequent, while that of Theorem 6.12 is that they are not too frequent. For if, for example, the primes were so scarce that $p_n > n^2$, say, for all $n > n_o > 1$, we would have

$$\begin{aligned}\prod_{n=1}^{N}\left(1 - \frac{1}{p_n}\right) &\gg \prod_{n=n_o}^{N}\left(1 - \frac{1}{n^2}\right) \\ &= \prod_{n=n_o}^{N}\frac{n-1}{n} \cdot \prod_{n=n_o}^{N}\frac{n+1}{n} \\ &= \frac{n_o - 1}{N} \cdot \frac{N+1}{n_o} > \frac{1}{2}\end{aligned}$$

for all $N > n_o$, and this is false by Theorem 6.11. It follows from Theorem 6.12, however, that there is no constant c such that $p_n < cn$ for all n, or all large n.

PROBLEMS

1. Show by a sieve argument that the number of square-free integers not exceeding x is less than

$$x \prod_p \left(1 - \frac{1}{p^2}\right) + o(x),$$

where the product extends over all primes.

2. Let $\omega(n)$ be the number of distinct primes dividing n, as in Problems 8 and 9 of Section 6.1.

   a) For $x \in (0, 1)$, let $\varphi(x, n)$ be the number of positive integers $m \leq xn$ which are prime to n. Show that $\varphi(x, n) = x\varphi(n) + O(\tau(n))$.

   b) Deduce that as $n \to \infty$, $\varphi(x, n) \sim x\varphi(n)$.

3. Let $p_1, \ldots, p_m$ be the first m odd primes, and let $P(x, m)$ be the number of odd integers $\leq x$ and not divisible by any of these primes. Let $\{u\} = [u + \frac{1}{2}]$ be the integer nearest to u. Show that $P(x, m) = \sum_a \{x/2a\} - \sum_b \{x/2b\}$, where a and b

run over all products of an even and an odd number of primes among $p_1, \ldots, p_m$, respectively. Show also that $\pi(x) = \pi(\sqrt{x}) + P(x, \pi(\sqrt{x}) - 1) - 1$, and use this to evaluate $\pi(200)$.

6.6 SUMS INVOLVING PRIMES

Theorem 6.11 is one of several theorems which have elementary proofs and which give, indirectly, information about the frequency of the primes. We devote this section to two others, namely, asymptotic estimates for $\sum_{p\le x} 1/p$ and $\sum_{p\le x}(\log p)/p$, and indicate in Problem 1 how these can be converted to direct statements about $\pi(x)$. We begin with a crude lower bound which is the additive equivalent of Theorem 6.11.

Theorem 6.13 *The series $\sum_p 1/p$ diverges. In fact,*

$$\sum_{p\le x} p^{-1} > \tfrac{1}{2}\log\log x.$$

Proof. By Theorem 6.11,

$$\log \prod_{p\le x}\left(1 - \frac{1}{p}\right) = \sum_{p\le x}\log\left(1 - \frac{1}{p}\right) < -\log\log x.$$

On the other hand, we have the well-known inequality

$$\frac{t}{1+t} < \log(1+t) < t \qquad \text{for } t > -1, \quad t \ne 0. \tag{3}$$

(This results almost immediately from the law of the mean, which guarantees that for some T between 0 and t,

$$\log(1+t) = \{(1+t) - 1\}\frac{d}{du}\log(1+u)\Big]_{u=T}.$$

(3) is useful only for $|t|$ not too large, of course.) Since $t/(1+t) \ge 2t$ for $0 > t \ge -\frac{1}{2}$, and since $p \ge 2$, we obtain, using the left-hand side of (3),

$$-\frac{2}{p} < \log\left(1 - \frac{1}{p}\right)$$

for all primes p (as is also evident from Fig. 6.2), and so

$$\sum_{p\le x}\frac{2}{p} > \log\log x. \quad \blacksquare$$

In order to get more precise information about the behavior of the sum

$$\sum_{p\le x}\frac{1}{p},$$

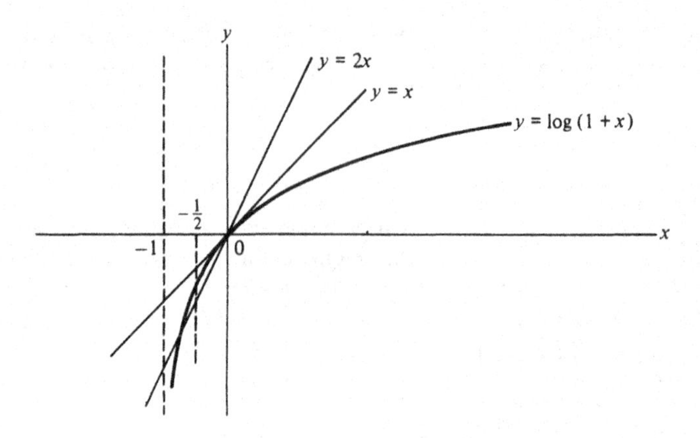

Figure 6.2

we proceed in a rather roundabout way, making use of the connection established in Theorem 6.9 between $n!$ and the primes not exceeding n.

Theorem 6.14

$$\sum_{p \le x} \frac{\log p}{p} = \log x + O\left(\frac{\log x}{\log \log x}\right).$$

Proof. By Theorem 6.9,

$$n! = \prod_{p \le n} p^{[n/p]+[n/p^2]+\cdots},$$

and so

$$\log n! = \sum_{p \le n} \left[\frac{n}{p}\right] \log p + \sum_{p \le n} \left(\left[\frac{n}{p^2}\right] + \left[\frac{n}{p^3}\right] + \cdots\right) \log p.$$

Now

$$\sum_{p \le n} \left[\frac{n}{p}\right] \log p \le \sum_{p \le n} \frac{n}{p} \log p,$$

and

$$\sum_{p \le n} \left[\frac{n}{p}\right] \log p \ge \sum_{p \le n} \left(\frac{n}{p} - 1\right) \log p = \sum_{p \le n} \frac{n}{p} \log p - \sum_{p \le n} \log p$$

$$\ge \sum_{p \le n} \frac{n}{p} \log p - \log n \sum_{p \le n} 1.$$

Moreover,

$$0 \le \sum_{p \le n} \left(\left[\frac{n}{p^2}\right] + \left[\frac{n}{p^3}\right] + \cdots\right) \log p \le \sum_{p \le n} \left(\frac{n}{p^2} + \frac{n}{p^3} + \cdots\right) \log p.$$

Thus

$$\log n! = \sum_{p \le n} \frac{n}{p} \log p + O(\pi(n) \log n)$$

$$+ O\left(n \sum_{p \le n} \left(\frac{1}{p^2} + \frac{1}{p^3} + \cdots\right) \log p\right)$$

$$= n \sum_{p \le n} \frac{\log p}{p} + O(\pi(n) \log n) + O\left(n \sum_{p \le n} \frac{\log p}{p(p-1)}\right),$$

and since the series

$$\sum_{k=2}^{\infty} \frac{\log k}{k(k-1)}$$

converges, this gives

$$\log n! = n \sum_{p \le n} \frac{\log p}{p} + O(\pi(n) \log n) + O(n).$$

On the other hand, by Problem 3(c) of Section 6.4 we have

$$\log n! = n \log n + O(n).$$

Combining these two estimates for $\log n!$, we have

$$n \sum_{p \le n} \frac{\log p}{p} = n \log n + O(n) + O(\pi(n) \log n), \tag{4}$$

so that by Theorem 6.12,

$$\sum_{p \le n} \frac{\log p}{p} = \log n + O\left(\frac{\log n}{\log \log n}\right).$$

This proves the theorem when x is an integer n. Since $\log x = \log [x] + O(1)$, the result also holds for $x \in \mathbf{R}$. ∎

To go from this estimate for $\sum_{p \le x} (\log p)/p$ to one for $\sum_{p \le x} 1/p$, we invoke the following theorem, in which the displayed equation is called the formula for **partial summation**. It is, in fact, a generalization (to the so-called Stieltjes integral) of the usual formula for integration by parts, and it has enormously wide applicability.

Theorem 6.15 *Suppose that $\lambda_1, \lambda_2, \ldots$ is a nondecreasing sequence of real numbers with limit infinity, that $c_1, c_2, \ldots$ is an arbitrary sequence of real or complex numbers, and that $f(x)$ has a continuous derivative for $x \ge \lambda_1$. Put*

$$C(x) = \sum_{\lambda_n \le x} c_n,$$

where the summation is over all n for which $\lambda_n \le x$. Then for $x \ge \lambda_1$,

$$\sum_{\substack{n \\ \lambda_n \le x}} c_n f(\lambda_n) = C(x) f(x) - \int_{\lambda_1}^{x} C(t) f'(t)\, dt.$$

Proof. We have

$$\sum_{\lambda_n \le x} c_n f(\lambda_n) = C(\lambda_1)f(\lambda_1) + (C(\lambda_2) - C(\lambda_1))f(\lambda_2) + \cdots + (C(\lambda_\nu) - C(\lambda_{\nu-1}))f(\lambda_\nu),$$

where λ_ν is the greatest λ_n which does not exceed x. Regrouping the terms, we have

$$\begin{aligned}\sum_{\lambda_n \le x} c_n f(\lambda_n) &= C(\lambda_1)(f(\lambda_1) - f(\lambda_2)) + \cdots \\ &\quad + C(\lambda_{\nu-1})(f(\lambda_{\nu-1}) - f(\lambda_\nu)) \\ &\quad + C(\lambda_\nu)(f(\lambda_\nu) - f(x)) + C(\lambda_\nu)f(x) \\ &= -\int_{\lambda_1}^{x} C(t)f'(t)\,dt + C(x)f(x),\end{aligned}$$

since $C(t)$ is a step function, constant over each of the intervals $(\lambda_{i-1}, \lambda_i)$ and over the interval (λ_ν, x). ■

Theorem 6.16

$$\sum_{p \le x} \frac{1}{p} \sim \log\log x.$$

Proof. Take $\lambda_n = p_n$, $c_n = (\log p_n)/p_n$, $f(t) = 1/\log t$ in Theorem 6.15. Using Theorem 6.14 in the third line below, we obtain

$$\begin{aligned}\sum_{p \le x} \frac{1}{p} &= \sum_{2<p\le x}\left(\frac{\log p}{p}\cdot\frac{1}{\log p}\right) + \frac{1}{2} \\ &= \frac{1}{\log x}\sum_{2<p\le x}\frac{\log p}{p} - \int_3^x\left(\sum_{p\le t}\frac{\log p}{p}\right)\frac{-dt}{t\log^2 t} + O(1) \\ &= \frac{1}{\log x}\left\{\log x + O\left(\frac{\log x}{\log\log x}\right)\right\} \\ &\quad + \int_3^x\left\{\log t + O\left(\frac{\log t}{\log\log t}\right)\right\}\frac{dt}{t\log^2 t} + O(1) \\ &= \int_3^x \frac{dt}{t\log t} + \int_3^x O\left(\frac{1}{t\log t\log\log t}\right)dt + O(1) \\ &= \log\log x + \int_3^x O\left(\frac{1}{t\log t\log\log t}\right)dt + O(1).\end{aligned}$$

Now for some constant M,

$$\begin{aligned}\left|\int_3^x O\left(\frac{1}{t\log t\log\log t}\right)dt\right| &< M\int_3^x \frac{dt}{t\log t\log\log t} \\ &= M\log\log\log x + O(1),\end{aligned}$$

so that $\sum_{p\le x} 1/p = \log\log x + O(\log\log\log x)$. ■

Notice that in all the sums for which we have obtained asymptotic estimates, the nth term tends to 0 as $n \to \infty$. The difficulty with $\pi(x) = \sum_{p \le x} 1$ is that it gives equal weight to all primes, and this is a more difficult kind of sum to estimate. Still, some information about $\pi(x)$ can be elicited from Theorem 6.16 by another partial summation, as is indicated in Problem 1 below. Chebyshev (Tchebychef, Tschebyschew, etc.) published this result in 1851, but with a different proof.

Chebyshev's name is well known in many branches of mathematics—probability theory, approximation theory and numerical analysis, and elementary integrability among them. In number theory, besides his work on primes, he initiated the study of inhomogeneous Diophantine approximation. He also did important work in applied mechanics (linkages and hinge mechanisms) and cartography. He was a master of the mathematics of inequalities, insisting always on strict error bounds. A popular and influential teacher, he was the prime force in establishing the "St. Petersburg school" in Russian mathematics.

Pavnuty L. Chebyshev (1821–1894)

PROBLEMS

†1. a) Use Theorem 6.16 and partial summation to show that

$$\int_2^x \frac{\pi(t)}{t^2}\,dt = \sum_{p \le x} \frac{1}{p} + o(1) \sim \log\log x.$$

b) Let $\rho(x)$ be the ratio of the two functions involved in the prime number theorem:

$$\rho(x) = \frac{\pi(x)}{x/\log x}.$$

Show that for no $\delta > 0$ is there a $T = T(\delta)$ such that $\rho(x) > 1 + \delta$ for all $x > T$, nor is there a T such that $\rho(x) < 1 - \delta$ for all $x > T$. This means that

$$\liminf \rho(x) \le 1 \le \limsup \rho(x),$$

so that if $\lim \rho(x)$ exists, it must have the value 1. (The existence of the limit remained in doubt until 1896.)

c) Deduce the result in (b) from Theorem 6.14 rather than Theorem 6.16.

2. What can you deduce from Theorem 6.13 about $\prod_{p \le x} (1 - p^{-1})$?
3. Derive the prime number theorem, $\pi(x) \sim x/\log x$, from the hypothetical relation $\sum_{p \le x} \log p \sim x$. [*Hint:* l'Hôpital's rule may prove useful, eventually.]
4. Show that $\sum (p \log p)^{-1}$ converges.
5. Use (3) to show that

$$\sum_{p \le x} \frac{1}{p - 1} > \log \log x,$$

and conclude that

$$\sum_{p \le x} \frac{1}{p} > \log \log x - 1.$$

†6. a) In Theorem 6.15, it is supposed that one factor in the summand is the value of a real differentiable function. More generally, show that if $\{a_k\}$ and $\{b_k\}$ are arbitrary complex sequences, and $A_k = a_1 + \cdots + a_k$, then for $n > m \ge 1$,

$$\sum_{k=m}^{n} a_k b_k = \sum_{k=m}^{n} A_k(b_k - b_{k+1}) + A_n b_{n+1} - A_{m-1} b_m.$$

This is called **Abel's (partial) summation formula.**

b) Show that an infinite series $\sum_1^\infty a_k b_k$ ($b_k \in \mathbf{R}$) is convergent if either $\sum a_k$ converges and $\{b_k\}$ is monotonic and bounded, or if $\sum a_k$ has bounded partial sums and $\{b_k\}$ tends monotonically to 0.

c) Show that $\sum_{k=1}^\infty (-1)^k k^{-s}$ converges for $s > 0$.

d) Show that for each prime $p > 2$, the series $\sum_{k=1}^\infty (k/p) k^{-s}$ converges for $s > 0$. Here (k/p) is the Legendre symbol.

e) Show that if the series $\sum_1^\infty a_k k^{-s}$ converges for $s = s_0$, then it converges for all $s \ge s_0$. Also, the series $\sum a_k (\log k) k^{-s}$ then converges for all $s > s_0$.

7. Show that for each fixed positive integer n,

$$\mathrm{li}(x) = \frac{x}{\log x} + \frac{1!\,x}{\log^2 x} + \frac{2!\,x}{\log^3 x} + \cdots + \frac{(n-1)!\,x}{\log^n x} + O\left(\frac{x}{\log^{n+1} x}\right).$$

6.7 THE TRUE ORDER OF $\pi(x)$

In a companion paper [1852] to the one mentioned in the preceding section, Chebyshev succeeded in showing that the actual order of magnitude of $\pi(x)$ is $x/\log x$:

Theorem 6.17 *There are positive finite constants c_1 and c_2 such that for $x \ge 2$,*

$$c_1 \frac{x}{\log x} < \pi(x) < c_2 \frac{x}{\log x}.$$

Remark. Coupled with Problem 1 of the preceding section, this gives

$$0 < c_1 < \liminf \rho(x) \le 1 \le \limsup \rho(x) < c_2 < \infty.$$

In fact, Chebyshev proved that $c_1 > 0.92\ldots$ and $c_2 < 1.105\ldots$. (The argument given below yields constants farther from 1 and we do not evaluate them, although it would be easy to do.) These were the first theorems proved about $\pi(x)$ since Euclid's time.

Proof. Chebyshev's idea was to find a combination of factorials in which the exponents on the prime-power factors would be small, so that the combination would be nearly the product of all primes in an interval. Problem 3 of the preceding section indicates that an asymptotic estimate for the logarithm of such a product would yield the prime number theorem. We shall use a simpler combination than that used by Chebyshev, namely, the binomial coefficient

$$\binom{2n}{n}.$$

Take $n \ge 2$. Corresponding to each $p \le 2n$ there is a unique integer r_p such that $p^{r_p} \le 2n < p^{r_p+1}$. We first prove that

$$\prod_{n<p\le 2n} p \,\Big|\, \frac{(2n)!}{n!\,n!} \quad \text{and} \quad \frac{(2n)!}{n!\,n!} \,\Big|\, \prod_{p\le 2n} p^{r_p}. \tag{5}$$

The first part is obvious, since any prime between n and $2n$ occurs as a factor of $(2n)!$ but does not occur in the denominator $(n!)^2$. For the second part, we have by Theorem 6.9 that the highest power of p which divides the numerator $(2n)!$ has exponent

$$\sum_{m=1}^{r_p} \left[\frac{2n}{p^m}\right],$$

while the highest power of p which divides the denominator has exponent

$$2\sum_{m=1}^{r_p} \left[\frac{n}{p^m}\right],$$

so that the highest power of p dividing $\binom{2n}{n}$ has exponent

$$\sum_{m=1}^{r_p} \left\{\left[\frac{2n}{p^m}\right] - 2\left[\frac{n}{p^m}\right]\right\} \le \sum_{m=1}^{r_p} 1 = r_p.$$

Here we have used property (f) of the function $[x]$, from the list in Section 6.3. From (5) we get

$$n^{\pi(2n)-\pi(n)} \le \prod_{n<p\le 2n} p \le \binom{2n}{n} \le \prod_{p\le 2n} p^{r_p} \le (2n)^{\pi(2n)},$$

whence $(\pi(2n) - \pi(n)) \log n \leq \log \binom{2n}{n} \leq \pi(2n) \log 2n$. Clearly,

$$\binom{2n}{n} < \sum_{k=0}^{n} \binom{2n}{k} = 2^{2n},$$

and also

$$\binom{2n}{n} = \frac{(n+1)\cdots(2n)}{1\cdots n} = \prod_{a=1}^{n} \frac{n+a}{a} \geq \prod_{a=1}^{n} 2 = 2^n;$$

thus

$$(\pi(2n) - \pi(n))\log n \leq 2n \log 2,$$

or

$$\pi(2n) - \pi(n) \ll \frac{n}{\log n}, \tag{6}$$

and

$$\pi(2n) \log 2n \geq n \log 2,$$

or

$$\pi(2n) \gg \frac{n}{\log n}. \tag{7}$$

We get from (7) that

$$\pi(x) \geq \pi\left(2\left[\frac{x}{2}\right]\right) \gg \frac{[x/2]}{\log[x/2]} \gg \frac{x}{\log x},$$

and since $\pi(x) \geq 1$ for $x \geq 2$, we obtain $\pi(x) > c_1 x/\log x$ for $x \geq 2$.

If $y \geq 4$, we get from (6) that

$$\pi(y) - \pi\left(\frac{y}{2}\right) = \pi(y) - \pi\left(\left[\frac{y}{2}\right]\right) \leq 1 + \pi\left(2\left[\frac{y}{2}\right]\right) - \pi\left(\left[\frac{y}{2}\right]\right)$$

$$\ll \frac{[y/2]}{\log[y/2]} \ll \frac{y}{\log y},$$

and so $\pi(y) - \pi(y/2) < cy/\log y$ for $y \geq 2$, since this inequality also holds for $2 \leq y \leq 4$, for suitable c. Using the trivial bound $\pi(y/2) \leq y/2$, we get

$$\pi(y) \log y - \pi\left(\frac{y}{2}\right) \log \frac{y}{2} = \left\{\pi(y) - \pi\left(\frac{y}{2}\right)\right\} \log y + \pi\left(\frac{y}{2}\right) \log 2$$

$$< \frac{cy}{\log y} \cdot \log y + \frac{y}{2} < c'y.$$

If we put $y = x/2^m$ with $2^m \leq x/2$ and $m \geq 0$, this becomes

$$\pi\left(\frac{x}{2^m}\right) \log \frac{x}{2^m} - \pi\left(\frac{x}{2^{m+1}}\right) \log \frac{x}{2^{m+1}} < c' \frac{x}{2^m};$$

summing over all such m's, we have

$$\pi(x)\log x - \pi\left(\frac{x}{2^{\mu+1}}\right)\log\frac{x}{2^{\mu+1}} < 2c'x,$$

where $2^\mu \le x/2 < 2^{\mu+1}$. But $x/2^{\mu+1} < 2$, so that $\pi(x/2^{\mu+1}) = 0$, and we obtain $\pi(x)\log x < c_2 x$. ■

Theorem 6.18 *There are positive constants c_3, c_4 such that for $r > 1$,*

$$c_3 r\log r < p_r < c_4 r\log r.$$

Proof. Taking x to be p_r in Theorem 6.17, we get

$$c_1\frac{p_r}{\log p_r} < r < c_2\frac{p_r}{\log p_r}.$$

The right-hand inequality gives immediately

$$p_r > c_3 r\log p_r > c_3\, r\log r.$$

Using the other inequality and the fact that $\log u = o(\sqrt{u})$, we have that for r sufficiently large,

$$\frac{\log p_r}{\sqrt{p_r}} < c_1 < \frac{r\log p_r}{p_r},$$

$$p_r < r^2,$$

$$\log p_r < 2\log r,$$

and so for such r,

$$p_r < \frac{1}{c_1} r\cdot 2\log r,$$

whence finally $p_r < c_4 r\log r$ for all $r > 1$. ■

We can use Theorem 6.17 to improve Theorems 6.14 and 6.16. Examining the proof of Theorem 6.14, we see from (4) that the error term can now be reduced to $O(n)$. This gives the first sentence in the following theorem.

Theorem 6.19 *We have*

$$\sum_{p\le x}\frac{\log p}{p} = \log x + O(1).$$

Furthermore, there is a constant C such that

$$\sum_{p\le x}\frac{1}{p} = \log\log x + C + O\left(\frac{1}{\log x}\right).$$

Proof. Following the argument used for Theorem 6.14, we now have

$$\sum_{p \le x} \frac{1}{p} = \frac{1}{\log x}(\log x + O(1)) + \int_2^x \log t \cdot \frac{dt}{t \log^2 t}$$
$$+ \int_2^x \left(\sum_{p \le t} \frac{\log p}{p} - \log t\right) \frac{dt}{t \log^2 t}$$
$$= 1 + O\left(\frac{1}{\log x}\right) + \log\log x - \log\log 2$$
$$+ \int_2^\infty \left(\sum_{p \le t} \frac{\log p}{p} - \log t\right) \frac{dt}{t \log^2 t} - \int_x^\infty \frac{O(1)\, dt}{t \log^2 t}.$$

Here the first integral is convergent, since $\int_2^\infty (t \log^2 t)^{-1}\, dt$ is, and the second is clearly $O(1/\log x)$. ■

These two estimates were obtained by F. Mertens in 1874.

PROBLEMS

1. Apply Theorem 6.19 to show that for some constant B,

$$\sum_{p \le x} \log\left(1 - \frac{1}{p}\right) = -\log\log x - B + O\left(\frac{1}{\log x}\right).$$

Deduce that

$$\prod_{p \le x}\left(1 - \frac{1}{p}\right) = \frac{e^{-B}}{\log x} + O\left(\frac{1}{\log^2 x}\right).$$

By Theorem 6.11, $B \ge 0$; although we do not prove it, B is Euler's constant. [Use the law of the mean to show that if $f(x) \to 0$ as $x \to \infty$, then $e^{f(x)} = 1 + O(f(x))$.]

2. a) Show that for each $\varepsilon > 0$, there are infinitely many pairs p_n and p_{n+1} of consecutive primes such that $p_{n+1} < (1 + \varepsilon)p_n$. (This does not quite follow from the inequality preceding Theorem 6.4.)

   b) Show that for some $c > 0$, there is a prime between x and cx for all $x > 2$. [*Hint:* Suppose that a is such that

$$\left|\sum_{p \le x} \frac{\log p}{p} - \log x\right| < a \qquad \text{for } x > 2,$$

   and obtain a contradiction to the hypothesis that there is no prime between x and $e^{2a}x$. Alternatively, use Theorem 6.17 directly.]

3. Attempt to prove the prime number theorem from Theorem 6.19 by partial summation.

4. Adopting a commonly used notation, we define

$$\vartheta(x) = \sum_{p \le x} \log p.$$

a) Show from Theorem 6.17 that $\vartheta(x) < c_2 x$, and hence that $\prod_{p \le x} p < A_1^x$ for some constant A_1.

b) Show by partial summation that

$$\pi(x) = \frac{\vartheta(x)}{\log x} + O\left(\frac{x}{\log^2 x}\right),$$

and hence in particular that $x \ll \vartheta(x) \ll x$. [*Hint:* To estimate $\int_2^x (\log^2 t)^{-1}\,dt$, split the interval $[2, x]$ at $\sqrt{x}$.]

5. Another commonly used function is the quantity

$$\psi(x) = \sum_{\substack{p,m \\ p^m \le x}} \log p;$$

for example, $\psi(6) = \log 2 + \log 3 + \log 2 + \log 5$.

a) Show that for $n \in \mathbf{Z}^+$,

$$\psi(n) = \log\,[1, 2, \ldots, n].$$

b) Show that $\psi(x) = \sum_{m=1}^{\infty} \vartheta(x^{1/m})$, where all the terms for which $x^{1/m} < 2$ are zero. Conclude that $\psi(x) = \vartheta(x) + O(x^{1/2}) = O(x)$ and that $[1, 2, \ldots, n] < A_2^n$ for some constant A_2.

6. Show that if $F(n)$ is defined and positive for $n \ge 2$, and is such that $F(n)/\log n$ is nonincreasing, then the two series $\sum_p F(p)$ and $\sum_n F(n)/\log n$ both converge or both diverge. [*Hint:* Write

$$\sum_{p \le N} F(p) = \sum_{n=2}^{N} (\vartheta(n) - \vartheta(n-1))\frac{F(n)}{\log n},$$

use Abel's summation formula, then the monotonicity and the known upper (lower) bound for $\vartheta(n)$, and then reverse the partial summation.]

7. Let p_n be the nth prime ($n \ge 1$) and put $d_n = p_n - p_{n-1}$ for $n > 1$ and $d_1 = 2$. Let $N\{S\}$ denote the number of elements in a finite set S.

a) Show that for every $c > 0$,

$$N\{n\colon p_n \le x,\ d_n > c \log x\} \cdot c \log x < \sum_{\substack{p_n \le x \\ d_n > c \log x}} d_n \le x.$$

b) Deduce that

$$N\{n\colon p_n \le x,\ d_n > c \log p_n\} < N\{n\colon \sqrt{x} < p_n < x,\ d_n > c \log p_n\} + O(\sqrt{x})$$

$$< \frac{2x}{c \log x} + O(\sqrt{x}).$$

Conclude that for suitable c,

$$N\{n\colon p_n \le x,\ d_n < c \log p_n\} \gg \pi(x).$$

6.8 PRIMES IN ARITHMETIC PROGRESSIONS

The 1852 work of Chebyshev, beautiful as it was, was perhaps a wrong turning; at least no one ever has been able to push it on to a complete proof of the prime number theorem. In his 1851 paper, however, Chebyshev had started from a relationship between the set of primes and the set of all integers which was of an entirely different sort from that in Theorem 6.9. It has already been adumbrated in the proof of Theorem 6.11, where we find the germ of another and, it has turned out, much more fecund approach to problems concerning the primes. The loose association developed there between the finite product $\prod_{p \le x} (1 - p^{-1})^{-1}$ and the finite sum $\sum_{k \le x} k^{-1}$ can be refined into an identity, with the help of an auxiliary real variable $s > 1$, as follows, and it was this identity from which Chebyshev proceeded.

Theorem 6.20 *For $s > 1$,*

$$\prod_p \frac{1}{1 - p^{-s}} = \sum_{k=1}^{\infty} \frac{1}{k^s}.$$

Proof. As in the proof of Theorem 6.11, we have

$$\prod_{p \le x} \frac{1}{1 - p^{-s}} = \prod_{p \le x} \left(1 + \frac{1}{p^s} + \frac{1}{p^{2s}} + \cdots\right),$$

and each of the finitely many geometric series on the right converges absolutely for $s > 0$, so that the (finite) product can be formed by adding products of terms in any convenient order. Hence, by the unique factorization theorem,

$$\begin{aligned} \prod_{p \le x} \frac{1}{1 - p^{-s}} &= \sum_{\substack{k \\ p \mid k \Rightarrow p \le x}} \frac{1}{k^s} \\ &= \sum_{k \le x} \frac{1}{k^s} + \sum_{\substack{k \ge x \\ p \mid k \Rightarrow p \le x}} \frac{1}{k^s} \\ &= \Sigma_1(x) + \Sigma_2(x), \text{ say.} \end{aligned}$$

The convergence of the series $\sum_1^\infty k^{-s}$ for $s > 1$ implies that for each such s, $\Sigma_1(x)$ has a limit as $x \to \infty$, and also that $\Sigma_2(x) \to 0$, and the theorem follows. ∎

The series $\sum_1^\infty k^{-s}$ defines what is called the **Riemann zeta-function**, $\zeta(s)$, for $s > 1$. The honor goes to Riemann because he was the first (1859) to study its properties as a function of a *complex* variable s, a step which was to be crucial in the later proofs of the prime number theorem and a host of other results.

But in fact the equation in Theorem 6.20 had been applied by Euler (1737) in the case $s = 1$ (when it is not valid!) to give a new proof of Euclid's theorem that there are infinitely many primes: if there were not, the left side would be finite while the right side is infinite. This argument becomes correct if instead we consider what

Before his death from tuberculosis at age 39, Riemann had revolutionized substantial portions of analysis, geometry, and mathematical physics. Gauss, not given to hyperbole, characterized his doctoral dissertation as exhibiting a "gloriously fertile originality." Riemannian manifolds, Riemann surfaces, the Cauchy-Riemann equations, the Riemann hypothesis—all this and more is packed into his one-volume collected works. He wrote only one paper on number theory, eight pages in length.

Riemann succeeded Dirichlet, whom he had known when he was a student at Berlin, when Dirichlet died in 1859, four years after replacing Gauss in Göttingen.

Bernhard Riemann (1826–1866)

happens when $s \to 1^+$, and Dirichlet [1837] used the same principle in his proof that there are infinitely many primes of the form $ak + b$ if $(a, b) = 1$. Dirichlet's proof involved not only the zeta-function but also certain related series called L-functions. As a result, series of the form $\sum_{k=1}^{\infty} a_k k^{-s}$ are called **Dirichlet series.** We cannot pursue any of these ideas very far, but perhaps the following will give some of the flavor. In the remainder of this section, q and r vary over all primes such that $q \equiv 1 \pmod 4$ and $r \equiv -1 \pmod 4$.

Theorem 6.21 *There are infinitely many primes q of the form $4k + 1$. They are so frequent, in fact, that*

$$\lim_{s \to 1^+} \prod_q (1 - q^{-s})^{-1} = \infty.$$

Proof. First, we need to learn more about the behavior of $\zeta(s)$ as $s \to 1^+$. To do so, we apply partial summation to obtain (for $s > 1$)

$$\begin{aligned}\sum_{k \le x} \frac{1}{k^s} &= \frac{[x]}{x^s} + s\int_1^x \frac{[t]}{t^{s+1}}\,dt \\ &= \frac{[x]}{x^s} + s\int_1^x \frac{t - ((t))}{t^{s+1}}\,dt \\ &= \frac{[x]}{x^s} + \frac{s}{s-1}(1 - x^{1-s}) - s\int_1^x \frac{((t))}{t^{s+1}}\,dt,\end{aligned}$$

where $((t))$ is the fractional part of t. Taking the limit as $x \to \infty$, this gives

$$\zeta(s) = \frac{s}{s-1} - s\int_1^\infty \frac{((t))}{t^{s+1}}\,dt,$$

the integral being convergent in fact for $s > 0$. It follows that

$$\lim_{s\to 1^+} (s-1)\zeta(s) = 1, \quad \text{and hence} \lim_{s\to 1^+} \zeta(s) = \infty.$$

Second, we introduce the relevant L-function, namely

$$L(s) = \sum_{k=1}^{\infty} \frac{(-1/k)}{k^s} = 1 - \frac{1}{3^s} + \frac{1}{5^s} - \frac{1}{7^s} + \frac{1}{9^s} - \cdots,$$

where $(-1/k)$ is almost the same as the Jacobi symbol introduced in the preceding chapter, except that it is also defined for even second entry:

$$(-1/k) = \begin{cases} 0 & \text{if } 2 \mid k, \\ (-1)^{(k-1)/2} & \text{if } 2 \nmid k. \end{cases}$$

It should not be surprising that the symbol occurs in the present context, since we have seen that the odd prime solutions of the equation $(-1/p) = 1$ are exactly the primes $p = q \equiv 1 \pmod 4$. The symbol $(-1/k)$ is a **completely multiplicative** function of k, in the sense that

$$(-1/k)(-1/l) = (-1/kl) \qquad \text{for all } k, l > 0.$$

Hence

$$(-1/p_1^{e_1} \cdots p_r^{e_r}) = (-1/p_1)^{e_1} \cdots (-1/p_r)^{e_r},$$

where

$$(-1/p) = \begin{cases} 0 & \text{for } p = 2, \\ 1 & \text{for } p = q \equiv 1 \pmod 4, \\ -1 & \text{for } p = r \equiv -1 \pmod 4. \end{cases}$$

Now Theorem 6.20 has an easy generalization to Dirichlet series of a type which includes $L(s)$: *if a_k is a completely multiplicative function of k, and the series $\sum a_k k^{-s}$ converges absolutely for all $s > s_0$, then for such s,*

$$\sum_{k=1}^{\infty} \frac{a_k}{k^s} = \prod_p \left(1 - \frac{a_p}{p^s}\right)^{-1}.$$

The proof is just the same as that of Theorem 6.20. Since the series for $L(s)$ is obviously absolutely convergent for $s > 1$ we have, for such s,

$$\begin{aligned} L(s) &= \prod_p \left(1 - \frac{(-1/p)}{p^s}\right)^{-1} \\ &= \prod_q \left(1 - \frac{1}{q^s}\right)^{-1} \cdot \prod_r \left(1 + \frac{1}{r^s}\right)^{-1}. \end{aligned}$$

Next, we examine the product $\zeta(s)L(s)$. By what we have proved, it has the factorization

$$\zeta(s)L(s) = \left(1 - \frac{1}{2^s}\right)^{-1} \prod_q \left(1 - \frac{1}{q^s}\right)^{-2} \prod_r \left(1 - \frac{1}{r^{2s}}\right)^{-1},$$

valid for $s > 1$. Now as $s \to 1^+$, $(1 - 2^{-s})^{-1} \to 2$, and for all $s \geq 1$,

$$0 < \prod_r \left(1 - \frac{1}{r^{2s}}\right)^{-1} < \prod_r (1 - r^{-2})^{-1} < \prod_p (1 - p^{-2})^{-1} = \zeta(2).$$

If it were true that also $\prod_q (1 - q^{-s})^{-1}$ remains bounded as $s \to 1^+$, then the same would be true for $\zeta(s)L(s)$. But $\zeta(s) \to \infty$ as $s \to 1^+$, so if we can show that

$$\lim_{s \to 1^+} L(s) \neq 0, \tag{8}$$

then $\zeta(s)L(s)$ cannot remain bounded, and the proof will be complete.

Next we prove that $L(s)$ is continuous at $s = 1$, so that (8) is equivalent to

$$L(1) \neq 0. \tag{9}$$

To prove continuity, we invoke the standard result that any series $\sum_1^\infty u_k(s)$ of functions all continuous on a closed interval I, which converges uniformly on I, converges to a function which is itself continuous on I. In the present case, since the terms $(-1/k)k^{-s}$ are continuous on the whole s-axis, it suffices to show that the series for $L(s)$ converges uniformly in each interval I: $0 < s_1 \leq s \leq s_2$. That is, we wish to show that for each $\varepsilon > 0$ there is an $m_0(\varepsilon)$ such that

$$\left|\sum_{k=m}^{n} \frac{(-1/k)}{k^s}\right| < \varepsilon \quad \text{for all } m \text{ and } n \text{ with } n > m > m_0(\varepsilon), \quad \text{and all } s \in I.$$

Now clearly

$$\left|\sum_{k_1}^{k_2} (-1/k)\right| \leq 1 \quad \text{for all } k_1, k_2,$$

so if we put $A_{m-1} = 0$ and $A_k = \sum_m^k (-1/l)$ for $k \geq m$, then for $s \in I$, by Problem 6 of Section 6.6,

$$\begin{aligned}
\left|\sum_{k=m}^{n} \frac{(-1/k)}{k^s}\right| &= \left|\sum_{k=m}^{n} (A_k - A_{k-1})k^{-s}\right| \\
&= \left|\sum_{k=m}^{n} A_k(k^{-s} - (k+1)^{-s}) + A_n(n+1)^{-s} - A_{m-1}m^{-s}\right| \\
&\leq \sum_{k=m}^{n} (k^{-s} - (k+1)^{-s}) + (n+1)^{-s} \\
&= m^{-s} \leq m^{-s_1},
\end{aligned}$$

and clearly $m^{-s_1} < \varepsilon$ for all sufficiently large m.

The last step is the verification of (9), and this is trivial (in this case):

$$L(1) = 1 - \tfrac{1}{3} + \tfrac{1}{5} - \cdots = (1 - \tfrac{1}{3}) + (\tfrac{1}{5} - \tfrac{1}{7}) + \cdots > \tfrac{2}{3}. \quad \blacksquare$$

We have, in fact, just outlined in a rough way the proof of the general Dirichlet theorem on primes in a progression $ak + b$. Two Dirichlet series enter when $a = 4$; in general, $\varphi(a)$ different series are involved, corresponding to the various residue classes in the group U_a. One of them behaves like the ζ-function as $s \to 1^+$, while the others are continuous at $s = 1$; both of these facts are proved just as above. A slightly different combination of these functions has to be used for each residue class (mod a), and usually the proof is arranged so as to obtain a contradiction with the assumption that $\sum_{p \equiv b \pmod a} p^{-1}$ is finite, but these changes are minor. The unexpectedly difficult point is the proof of the analogue of (9) for the L-functions continuous at $s = 1$.

It must be clear that this new approach to problems concerning primes, in which the entire machinery of real and complex analysis can be brought to bear, can be expected to yield results not obtainable by more direct ad hoc arguments involving clever combinations of factorials. This expectation has been amply fulfilled during the twentieth century, as ways have been found to apply analysis to more and more kinds of questions in number theory. There is a strong parallel with algebraic number theory; in each case another branch of mathematics has provided powerful tools for attacking arithmetical problems—and also, in each case the resolution of the problems has necessitated or stimulated the development of whole new areas in the other field.

PROBLEMS

1. Show that the sum of the reciprocals of the primes $\equiv 1 \pmod 4$ diverges.
2. Show that
$$\lim_{s \to 1^+} (s - 1) \prod_{q \equiv 1(4)} (1 - q^{-s})^{-1} = A$$
exists and is a finite nonzero real number. Conclude that
$$\lim_{s \to 1^+} \prod_{r \equiv -1(4)} (1 - r^{-s})^{-1} = \infty,$$
and hence that the sum of the reciprocals of the primes $\equiv -1 \pmod 4$ diverges.
3. Suppose that $f(n)$ is multiplicative, and that the series $\sum_1^\infty f(n)$ converges absolutely. Show that
$$\sum_1^\infty f(n) = \prod_p \sum_{k=0}^\infty f(p^k).$$
Show that if in addition $f(n)$ is completely multiplicative, then $f(p) \neq 1$ for every p, and
$$\sum_1^\infty f(n) = \prod_p (1 - f(p))^{-1}.$$

4. A **character (mod k)** is a completely multiplicative function $\chi(a)$ defined on the residue classes (mod k), such that $\chi(a) = 0$ if and only if $(a, k) > 1$. (The function $(-1/a)$ used in this section is a character (mod 4).)

   a) Show that if $(a, k) = 1$, then $\chi(a)$ is an hth root of unity, where $h = \varphi(k)$.

   b) There are four characters (mod 8), defined as follows:

a	1	3	5	7 (mod 8)
$\chi_0(a)$	1	1	1	1
$\chi_1(a)$	1	-1	1	-1
$\chi_2(a)$	1	1	-1	-1
$\chi_3(a)$	1	-1	-1	1

   If multiplication of characters is defined by $(\chi_i\chi_j)(a) = \chi_i(a)\chi_j(a)$, verify that the above characters form a multiplicative group isomorphic to U_8.

   c) Define the **Dirichlet L-functions** (for $k = 8$) for $s > 1$ by

$$L(s, \chi_i) = \sum_{n=1}^{\infty} \frac{\chi_i(n)}{n^s}, \qquad i = 0, 1, 2, 3.$$

   Show that

$$L(s, \chi_0) = (1 - 2^{-s})\zeta(s),$$

   and hence that $L(s, \chi_0) \sim \frac{1}{2}(s - 1)^{-1}$ as $s \to 1^+$. On the other hand, show that $L(s, \chi_i)$ is continuous and nonzero at $s = 1$, and find the factorization of $L(s, \chi_i)$, for $i = 1, 2, 3$.

   d) Show that for $s > 1$,

$$\prod_{i=0}^{3} L(s, \chi_i) = \prod_{p\equiv 1(8)} (1 - p^{-s})^{-4} \prod_{p\equiv 3,5,\text{ or } 7(8)} (1 - p^{-2s})^{-2}.$$

   e) Show that for $s > 1$,

$$\frac{L(s, \chi_2)L(s, \chi_3)}{L(s, \chi_0)L(s, \chi_1)} = \prod_{p\equiv 5(8)} \frac{(1 - p^{-s})^4}{(1 - p^{-2s})^2}$$

   and obtain similar identities by interchanging subscripts 1 and 2, or 1 and 3.

   f) Show that $\sum_{p\equiv a(8)} p^{-1}$ diverges, for $a = 1, 3, 5,$ and 7.

6.9 BERTRAND'S HYPOTHESIS

In 1845 J. Bertrand showed empirically that there is a prime between n and $2n$ for all n greater than 1 and less than six million, and predicted that this is true for all positive integers n. Chebyshev proved this in 1852, and indeed that for every $\varepsilon > \frac{1}{5}$ there is a ξ such that for every $x > \xi$ there is a prime between x and $(1 + \varepsilon)x$. (The prime number theorem immediately implies that this last theorem is true for every fixed $\varepsilon > 0$.) We shall content ourselves here with a proof of Bertrand's original conjecture.

It is worth noting that Theorem 6.19 implies a weak form of the theorem: *there exists a positive constant c such that there is a prime between n and cn for all n.* For by Theorem 6.19, there is a constant A such that

$$\log n - A < \sum_{p \le n} \frac{\log p}{p} < \log n + A$$

for all n. Suppose that there is no prime between n and ne^{2A}. Then

$$\sum_{p \le n} \frac{\log p}{p} = \sum_{p \le ne^{2A}} \frac{\log p}{p},$$

and so,

$$\sum_{p \le n} \frac{\log p}{p} > \log(ne^{2A}) - A = \log n + A.$$

With this contradiction, the result is proved with $c = e^{2A}$.

For the proof of the more exact theorem we need two lemmas.

Theorem 6.22 *For every positive integer* n, $\prod_{p \le n} p < 4^n$.

Proof. We use induction on n. The inequality is obvious if $n = 1$ or 2. Suppose it is true for $1, 2, \ldots, n - 1$, where $n \ge 3$. Then we can restrict attention to odd n, since otherwise

$$\prod_{p \le n} p = \prod_{p \le n-1} p < 4^{n-1} < 4^n,$$

so we can put $n = 2m + 1$. From its definition, the binomial coefficient

$$\binom{2m+1}{m} = \frac{(2m+1)!}{m!\,(m+1)!}$$

is divisible by every prime p with $m + 2 \le p \le 2m + 1$. Hence

$$\prod_{p \le 2m+1} p \le \binom{2m+1}{m} \cdot \prod_{p \le m+1} p < \binom{2m+1}{m} 4^{m+1}.$$

But the numbers

$$\binom{2m+1}{m} \quad \text{and} \quad \binom{2m+1}{m+1}$$

are equal, and both occur in the expansion of $(1 + 1)^{2m+1}$, so that

$$\binom{2m+1}{m} \le \frac{1}{2} \cdot 2^{2m+1} = 4^m,$$

and so

$$\prod_{p \le 2m+1} p < 4^m \cdot 4^{m+1} = 4^{2m+1}.$$

The theorem follows by induction on n. ∎

Theorem 6.23 *If* $n \ge 3$ *and* $\frac{2}{3}n < p \le n$, *then* $p \nmid \binom{2n}{n}$.

Proof. The restrictions on n and p are such that

a) p is greater than 2,

b) p and $2p$ are the only multiples of p which are less than or equal to $2n$, since $3p$ is greater than $2n$,

c) p itself is the only multiple of p which is less than or equal to n.

From (a) and (b), $p^2 \parallel (2n)!$, and from (c), $p^2 \parallel (n!)^2$, so that

$$p \nmid (2n)!/(n!)^2. \quad \blacksquare$$

Theorem 6.24 *For every positive integer n, there is a prime p such that $n < p \leq 2n$.*

Proof. There is such a prime for $n = 1$ or 2. Assume there is none for a certain integer $n \geq 3$. Then by Theorem 6.23, every prime which divides $\binom{2n}{n}$ must be less than or equal to $2n/3$. Let p be such a prime, and suppose that

$$p^e \,\Big\|\, \binom{2n}{n}.$$

Then by the proof of Theorem 6.17, since

$$\binom{2n}{n} \,\Big|\, \prod_{p \leq 2n} p^{r_p}, \quad \text{where} \quad p^{r_p} \leq 2n < p^{r_p+1},$$

it follows that $p^e \leq 2n$. Thus if $e \geq 2$, then $p \leq \sqrt{2n}$, and so there are at most $[\sqrt{2n}]$ primes appearing in the prime-power factorization of $\binom{2n}{n}$ with exponent larger than 1, and in each case, $p^e < 2n$. Hence

$$\binom{2n}{n} \leq (2n)^{[\sqrt{2n}]} \cdot \prod_{p \leq 2n/3} p.$$

But $\binom{2n}{n}$ is the largest of the $2n + 1$ terms in the expansion of $(1 + 1)^{2n}$, so that

$$4^n < (2n + 1)\binom{2n}{n},$$

and so

$$\frac{4^n}{2n + 1} < (2n)^{\sqrt{2n}} \cdot \prod_{p \leq 2n/3} p.$$

By Theorem 6.22, this implies that

$$\frac{4^n}{2n + 1} < (2n)^{\sqrt{2n}} \cdot 4^{2n/3},$$

and since $2n + 1 < 4n^2$, this gives

$$4^n < (2n)^{\sqrt{2n}+2} \cdot 4^{2n/3}, \qquad \text{or} \qquad 4^{n/3} < (2n)^{\sqrt{2n}+2}.$$

Taking logarithms, we have

$$\frac{n \log 4}{3} < (\sqrt{2n} + 2) \log 2n.$$

This inequality is false for $n > 512$, so there is a prime between n and $2n$ for $n > 512$. But in the sequence of primes

$$2, 3, 5, 7, 13, 23, 43, 83, 163, 317, 557,$$

each number is smaller than twice the one preceding it, so there is also such a prime for all $n \leq 512$, and hence for all $n \geq 1$. ∎

PROBLEMS

1. It follows from Problem 1 of Section 6.6 that in Theorem 6.17, $c_1 < 1 < c_2$. If estimates had been made of c_1 and c_2 in the proof of Theorem 6.17 (which would be simple to do), we would know, as a consequence, two particular constants c_1 and c_2 for which the inequality of Theorem 6.17 holds. Suppose that this is the case, and that $c_2/c_1 = \beta > 1$. Show that if $\alpha > 0$, then
$$\frac{\pi((1 + \alpha)x) - \pi(x)}{x/\log x} > c_1(1 + \alpha - \beta) + O\left(\frac{1}{\log x}\right).$$
Deduce that if $1 + \alpha > \beta$, the number of primes between x and $(1 + \alpha)x$ tends to infinity with x. (For the values of c_1 and c_2 obtained by Chebyshev, $\beta = 6/5$, and this is how Chebyshev gave the first proof of (a strengthened version of) Theorem 6.24.)
2. By taking logarithms and applying partial summation, deduce the second inequality of Theorem 6.17 from the inequality of Theorem 6.22.
3. Show that every integer $n > 6$ is the sum of distinct primes. [*Hint:* This is true for $6 < n \leq 19$, with all $p < 13$; hence it is true for $6 < n \leq 26$, with all $p \leq 13$.]
4. It is known that $\pi(x) = x/\log x + O(x/\log^2 x)$.
   a) Use this estimate to show that $\vartheta(x) = x + O(x/\log x)$. (See Problem 4, Section 6.7, for $\vartheta(x)$.)
   b) Deduce that for every $\varepsilon > 0$, each of the inequalities
   $$\prod_{p \leq x} p > e^{(1+\varepsilon)x} \qquad \text{and} \qquad \prod_{p \leq x} p < e^{(1-\varepsilon)x}$$
   is false for all sufficiently large x.
   c) Decide whether or not the product of the primes $\leq x$ is asymptotic to e^x.
5. a) Show that if, among a finite set of consecutive integers, one is prime, then one is relatively prime to all the others.

b) Conversely, by concentrating on the sets $2, 3, \ldots, 2n$, show that the implication of part (a) implies Bertrand's hypothesis.

6. Show that if $n, k \in \mathbf{Z}$, $n > 1$, $k > 0$, then the number

$$\frac{1}{n} + \frac{1}{n+1} + \cdots + \frac{1}{n+k}$$

is not an integer. [*Hint:* First show that $k \geq n$.]

7. Show that for $n > 1$, there is a prime p such that $p \parallel n!$.

6.10 THE ORDER OF MAGNITUDE OF φ, σ, AND τ

The quantity $\pi(x)$ is reasonably well-behaved, and so one can make fairly precise statements about its size as a function of x. This is not true of the other functions we have considered, which vary much too wildly to permit asymptotic approximations. There are, however, various weaker statements which can be made about their size and which still yield considerable information.

Consider, for example, the quantity $\tau(n)$. A moment's thought shows that the number of divisors of n is much smaller than n itself, for large n; it is to be expected that $\tau(n) = o(n)$. And while $\tau(n) = 2$ infinitely many times, it is also possible to make $\tau(n)$ arbitrarily large for suitable n. Thus if the points $(n, \tau(n))$ are plotted in a coordinate system, as in Fig. 6.3, there is a unique "lowest" polygonal path extending upward and to the right from $(1, 1)$ which is concave downward and is such that every point $(n, \tau(n))$ lies on or below it. Suppose that this path is described by the equation $y = T(x)$. While we shall not obtain an asymptotic estimate for $T(x)$, the following theorem shows that it increases more rapidly than any power of $\log x$, and less rapidly than any positive power of x.

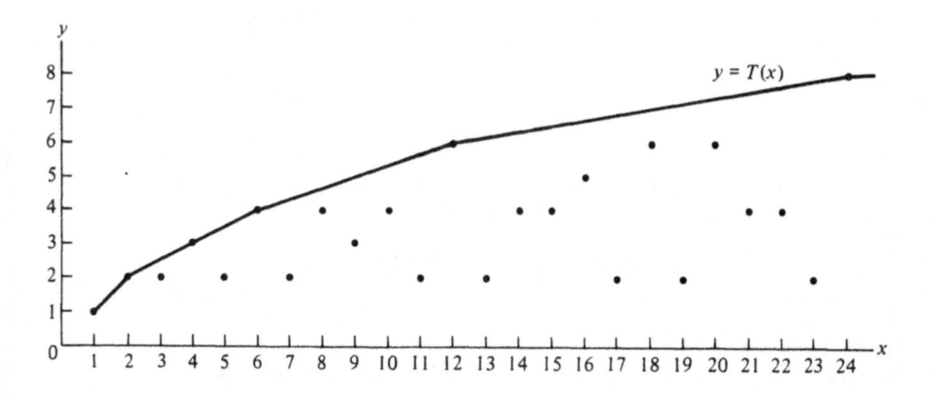

Figure 6.3

Theorem 6.25

a) *The relation* $\tau(n) \ll \log^h n$ *is false for every constant* h.

b) *The relation* $\tau(n) \ll n^\delta$ *is true for every fixed* $\delta > 0$.

Proof.

a) Let n be any of the numbers $(2 \cdot 3 \cdots p_r)^m$, $m = 1, 2, \ldots$; here r is arbitrary but fixed. Then

$$\tau(n) = \prod_1^r (m + 1) = (m + 1)^r > m^r.$$

But $m = \log n / \log (2 \cdot 3 \cdots p_r)$, so that

$$\tau(n) > \frac{\log^r n}{(\log (2 \cdot 3 \cdots p_r))^r} \gg \log^r n,$$

where the implied constant depends only on r, and not on n.

b) Let

$$f(n) = \frac{\tau(n)}{n^\delta};$$

then f is multiplicative. But $f(p^m) = (m + 1)/p^{m\delta}$, so that $f(p^m) \to 0$ as $p^m \to \infty$, that is, as either p or m, or both, increases. This clearly implies that $f(n) \to 0$ as $n \to \infty$, which is the assertion.

The argument can be pushed a little further. Let δ be positive, and let

$$n = \prod_1^r p_i^{e_i}.$$

Then

$$\frac{\tau(n)}{n^\delta} = \frac{e_1 + 1}{p_1^{e_1\delta}} \cdots \frac{e_r + 1}{p_r^{e_r\delta}} \le \prod_{p_i \mid n} \max_{x \ge 0} \left(\frac{x + 1}{p_i^{\delta x}} \right).$$

For fixed δ, the quantity

$$\max_{x \ge 0} \frac{x + 1}{p^{\delta x}}$$

is equal to 1 for sufficiently large p, and is never smaller than 1, as one sees by sketching the curve $y = (x + 1)a^{-x}$. Hence

$$\frac{\tau(n)}{n^\delta} \le \prod_p \max_{x \ge 0} \left(\frac{x + 1}{p^{\delta x}} \right) = c_\delta,$$

which gives the inequality in (b) with an explicit constant for each δ. ■

As regards the φ-function, we have the trivial upper bound $\varphi(n) \le n - 1$ for $n > 1$, equality being attained whenever n is prime. A lower bound was indicated in Problem 9 of Section 6.1, but a larger one can be established; as is indicated in Problem 1 below, this new lower bound is the "correct" one.

Theorem 6.26

$$\varphi(n) \gg \frac{n}{\log\log n}.$$

Proof. We have

$$\frac{\varphi(n)}{n} = \prod_{p\mid n}\left(1 - \frac{1}{p}\right),$$

so that

$$\begin{aligned}\log\frac{\varphi(n)}{n} &= \sum_{p\mid n}\log\left(1 - \frac{1}{p}\right)\\ &= -\sum_{p\mid n}\frac{1}{p} + \sum_{p\mid n}\left\{\log\left(1 - \frac{1}{p}\right) + \frac{1}{p}\right\}\\ &> -\sum_{p\mid n}\frac{1}{p} + O(1),\end{aligned}$$

since by (3),

$$\begin{aligned}\sum_{p}\left\{\log\left(1 - \frac{1}{p}\right) + \frac{1}{p}\right\} &> \sum_{p}\left(\frac{1}{p} - \frac{1}{p-1}\right)\\ &> -\sum_{n=2}^{\infty}\left(\frac{1}{n-1} - \frac{1}{n}\right) = -1.\end{aligned}$$

Now let $p_1, \ldots, p_{r-\rho}$ be the distinct primes less than $\log n$ which divide n, and let $p_{r-\rho+1}, \ldots, p_r$ be the larger prime divisors of n, so that

$$\sum_{p\mid n}\frac{1}{p} = \sum_{k=1}^{r-\rho}\frac{1}{p_k} + \sum_{k=r-\rho+1}^{r}\frac{1}{p_k} = S_1 + S_2,$$

say. Then

$$\log^{\rho} n \le p_{r-\rho+1}^{\rho} \le \prod_{k=r-\rho+1}^{r} p_k \le n,$$

so that

$$\rho \le \frac{\log n}{\log\log n},$$

and

$$S_2 \le \frac{1}{\log n}\cdot\frac{\log n}{\log\log n} = o(1).$$

By Theorem 6.19,

$$S_1 < \log\log p_{r-\rho} + O(1) < \log\log\log n + O(1).$$

Combining these results, we get

$$\log \frac{\varphi(n)}{n} > -\log \log \log n + O(1).$$

and so

$$\frac{\varphi(n)}{n} \gg \frac{1}{\log \log n}. \quad \blacksquare$$

We can use Theorem 6.26 to obtain a corresponding upper bound for $\sigma(n)$, with the help of the following theorem.

Theorem 6.27

$$\frac{1}{2} < \frac{\sigma(n)\varphi(n)}{n^2} < 1.$$

Proof. If $n = \prod p^e$, then

$$\begin{aligned}\sigma(n)\varphi(n) &= \prod_{p \mid n} \left(\frac{p^{e+1} - 1}{p - 1}\right) n \prod_{p \mid n} \left(1 - \frac{1}{p}\right) \\ &= n \prod_{p \mid n} \frac{1 - p^{-(e+1)}}{1 - 1/p} \cdot n \prod_{p \mid n} \left(1 - \frac{1}{p}\right) \\ &= n^2 \prod_{p \mid n} (1 - p^{-(e+1)}).\end{aligned}$$

Here the coefficient of n^2 is clearly less than 1 and greater than or equal to

$$\prod_{p \mid n} \left(1 - \frac{1}{p^2}\right) > \prod_{k=2}^{n} \left(1 - \frac{1}{k^2}\right) > \frac{1}{2},$$

the final inequality having already been proved at the end of Section 6.5. $\blacksquare$

Theorem 6.28 $\sigma(n) \ll n \log \log n$.

Proof. By Theorem 6.26,

$$\frac{\varphi(n)}{n} \gg \frac{1}{\log \log n},$$

so by Theorem 6.27,

$$\frac{\sigma(n)}{n} < \frac{n}{\varphi(n)} \ll \log \log n. \quad \blacksquare$$

PROBLEMS

1. Show that Theorem 6.26 is best-possible, in the sense that there is an increasing sequence of positive integers $n_1, n_2, \ldots$ such that

$$\varphi(n_k) \ll \frac{n_k}{\log \log n_k}.$$

2. Show that only finitely many positive integers n have the property that for all m for which $1 < m < n$ and $(m, n) = 1$, m is prime. (In fact, 30 is the largest such n.)
3. If Theorem 6.25 had been available, what conclusion could have been drawn in Problem 9 of Section 6.1?
4. Let $\theta(n)$ be the number of positive integers $m \le n$ such that $(m, n) = (m + 1, n) = 1$. (See Problem 11 of Section 6.1.) Show that $\theta(n) = 0$ if $2 \mid n$, while
$$\theta(n) \gg n (\log \log n)^{-2}$$
for odd n.
5. Let $\tau_k(n)$ be the number of (ordered) factorizations of n as the product of exactly k positive integers, so that $\tau_1(n) = 1$ and $\tau_2(n) = \tau(n)$. Show that for fixed k,
   a) $\tau_k(n)$ is multiplicative;
   b) $\tau_k(p^e) < (e + 1)^k$ (more difficult, and not needed for (c):
   $$\tau_k(p^e) = \binom{e + k - 1}{k - 1}\Bigg);$$
   c) $\tau_k(n) = O(n^\delta)$ for every constant $\delta > 0$.

6.11 AVERAGE ORDER OF MAGNITUDE

Another way of describing the behavior of a number-theoretic function is in terms of its average order, that is, in terms of the quantity
$$\frac{1}{n} \sum_{m=1}^{n} f(m).$$
Summing the values of a function has the effect of smoothing out the irregularities, so that it is frequently possible to make quite precise statements about the size of the sum.

The following theorem seldom gives very accurate results, but it provides a simple example of change of order of summation in a double arithmetic sum.

Theorem 6.29 *If*
$$F(m) = \sum_{d \mid m} f(d),$$
then
$$\sum_{m=1}^{n} F(m) = \sum_{m=1}^{n} \left[\frac{n}{m}\right] f(m).$$

Proof. By the definition of $F(m)$,
$$\sum_{m=1}^{n} F(m) = \sum_{m=1}^{n} \sum_{d \mid m} f(d).$$
This order of summation associates with each m $(1 \le m \le n)$ all its divisors d.

Instead, one could associate with each integer d $(1 \le d \le n)$ all its multiples kd $(1 \le kd \le n)$; clearly k can assume any of the values $1, 2, \ldots, [n/d]$. Hence

$$\sum_{m=1}^{n} F(m) = \sum_{d=1}^{n} \sum_{k=1}^{[n/d]} f(d) = \sum_{d=1}^{n} f(d) \sum_{k=1}^{[n/d]} 1,$$

and the theorem follows. ■

Applying this to $\tau(n)$ and $\sigma(n)$ gives $\sum_1^n \tau(m) = n \log n + O(n)$ and $\sum_1^n \sigma(n) = O(n^2)$, if one simply approximates $[n/m]$ by $n/m + O(1)$. Both of these are very weak estimates, and we seek better approaches.

The strongest result that can be obtained quite easily about the average of the τ-function is this.

Theorem 6.30 $\sum_{m=1}^{n} \tau(m) = n \log n + (2\gamma - 1)n + O(n^{1/2})$, *where γ is Euler's constant.*

Proof. By Theorem 6.29,

$$\sum_{m=1}^{n} \tau(m) = \sum_{m=1}^{n} \left[\frac{n}{m}\right].$$

Geometrically, the sum on the right represents the number of *lattice points* (x, y) (that is, points such that x and y are integers) with positive coordinates, on or below the hyperbola $xy = n$, since for fixed x the number of integers y such that $1 \le y \le n/x$ is exactly $[n/x]$.

By symmetry, the number of lattice points (x, y) with $0 < xy \le n$, $y > x$, is equal to the number with $0 < xy \le n$, $y < x$ (see Fig. 6.4). Hence the number of

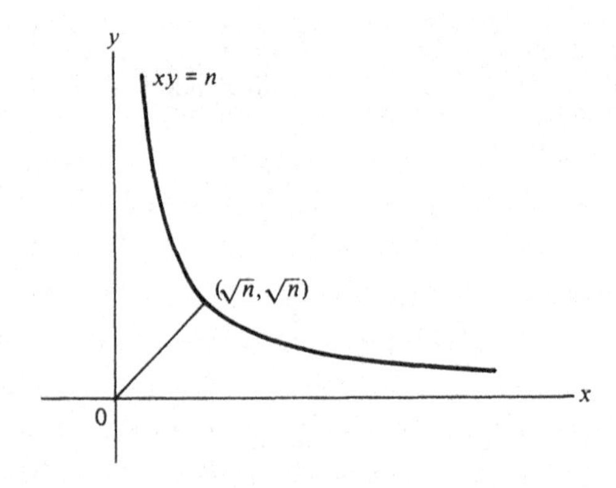

Figure 6.4

points (x, y) with $0 < xy \le n$ is twice the number of those with $y > x$, plus the number with $y = x$; with the help of Theorem 6.10 we obtain

$$\begin{aligned}\sum_{m=1}^{n} \tau(m) &= 2\sum_{x=1}^{\sqrt{n}}\left(\left[\frac{n}{x}\right] - x\right) + [\sqrt{n}]\\ &= 2n\sum_{1}^{\sqrt{n}}\frac{1}{x} + O(\sqrt{n}) - 2\frac{[\sqrt{n}]([\sqrt{n}]+1)}{2} + O(\sqrt{n})\\ &= 2n(\log\sqrt{n} + \gamma + O(1/\sqrt{n})) - n + O(\sqrt{n})\\ &= n\log n + (2\gamma - 1)n + O(\sqrt{n}). \quad \blacksquare\end{aligned}$$

The term $O(\sqrt{n})$ in Theorem 6.30 is not the best possible estimate of this error. The problem of increasing the accuracy of the estimate, usually called *Dirichlet's divisor problem*, has received a large amount of study. It is known that $O(n^{1/2})$ can be replaced by $O(n^{1/3})$, but not by $O(n^{1/4})$. The exact exponent, if such exists, is still unknown.

There is another rather similar geometric problem, that of estimating the number of lattice points in the circular disk $x^2 + y^2 \le n$. It is easy to see that this number is $\pi n + O(n^{1/2})$, and the same remarks apply to the error term in this *circle problem* as in the Dirichlet divisor problem.

It is interesting to compare Theorems 6.25 and 6.30: even though the τ-function occasionally becomes larger than any fixed power of the logarithm, the average of its first n values is very nearly $\log n$.

To get an asymptotic estimate for the sum of the first n values of the φ-function, we need a preliminary result concerning the ζ-function.

Theorem 6.31 *For* $s > 1$,

$$\frac{1}{\zeta(s)} = \sum_{n=1}^{\infty}\frac{\mu(n)}{n^s}.$$

Proof. The series of the theorem and that for $\zeta(s)$ converge absolutely for $s > 1$, so that they may be multiplied together by adding all possible products of a term from one series and a term from the other, and the resulting terms may be arranged in any convenient order. Hence by Theorem 6.5,

$$\sum_{m=1}^{\infty}\frac{1}{m^s}\sum_{n=1}^{\infty}\frac{\mu(n)}{n^s} = \sum_{m,n=1}^{\infty}\frac{\mu(n)}{(mn)^s} = \sum_{t=1}^{\infty}\frac{1}{t^s}\sum_{d\mid t}\mu(d) = 1. \quad \blacksquare$$

In the next theorem we have replaced $\zeta(2)$ by its value, which is known to be $\pi^2/6$. All that is important here is that it is a finite number. In Problems 4 and 11 we use the fact that $\zeta(2) < 2$, but this is easily verified by taking one term and bounding the remainder from above, as in the integral test for convergence:

$$\zeta(2) < 1 + \int_1^{\infty}\frac{dt}{t^2} = 2.$$

Theorem 6.32

$$\sum_{m=1}^{n} \varphi(m) = \frac{3n^2}{\pi^2} + O(n \log n).$$

Proof. Since

$$\varphi(m) = m \sum_{d \mid m} \frac{\mu(d)}{d},$$

and since $[x]^2 = x^2 + O(x)$ and $\mu(m) = O(1)$, we have

$$\begin{aligned}
\sum_{m=1}^{n} \varphi(m) &= \sum_{m=1}^{n} m \sum_{d \mid m} \frac{\mu(d)}{d} = \sum_{dd' \le n} d'\mu(d) = \sum_{d=1}^{n} \mu(d) \sum_{d'=1}^{n/d} d' \\
&= \sum_{d=1}^{n} \mu(d) \frac{[n/d]^2 + [n/d]}{2} \\
&= \frac{1}{2} \sum_{d=1}^{n} \mu(d) \frac{n^2}{d^2} + O\left(\sum_{d=1}^{n} \frac{n}{d}\right) \\
&= \frac{n^2}{2}\left(\sum_{d=1}^{\infty} \frac{\mu(d)}{d^2} - \sum_{d=n+1}^{\infty} \frac{\mu(d)}{d^2}\right) + O(n \log n) \\
&= \frac{n^2}{2} \frac{1}{\zeta(2)} + O\left(n^2 \sum_{n+1}^{\infty} \frac{1}{d^2}\right) + O(n \log n) \\
&= \frac{3n^2}{\pi^2} + O(n) + O(n \log n) \\
&= \frac{3n^2}{\pi^2} + O(n \log n). \quad \blacksquare
\end{aligned}$$

Since $\sum_1^n m \sim \frac{1}{2}n^2$, it might be said that the "average value" of $\varphi(n)$ is $6n/\pi^2 \approx 0.608n$.

We leave the corresponding estimate for $\sigma(n)$ as an exercise.

PROBLEMS

1. Show that

$$\sum_{m=1}^{n} \sigma(m) = \frac{\pi^2 n^2}{12} + O(n \log n).$$

2. Show, using partial summation or otherwise, that

a) $\displaystyle\sum_{n \le x} \frac{\tau(n)}{n} = \frac{1}{2} \log^2 x + 2\gamma \log x + O(1).$

b) $\displaystyle\sum_{n \le x} \frac{\tau(n)}{\log n} = x + 2\gamma \frac{x}{\log x} + O\left(\frac{x}{\log^2 x}\right).$

3. Let $\delta(n)$ be the largest odd divisor of n. Show that

$$\sum_{n\le x}\delta(n) = \frac{x^2}{3} + O(x) \qquad \text{and} \qquad \sum_{n\le x}\frac{\delta(n)}{n} = \frac{2x}{3} + O(\log x).$$

[*Hint:* Classify the numbers less than or equal to x according to the exponents of the powers of 2 dividing them, and show that

$$\sum_{n\le x}\delta(n) = \sum_{n=0}^{(x-1)/2}(2n+1) + \sum_{n=0}^{(x-2)/4}\frac{4n+2}{2} + \sum_{n=0}^{(x-4)/8}\frac{8n+4}{4} + \cdots.]$$

4. Show that

$$\sum_{n\le x}\frac{\varphi(n)}{n} \sim \frac{x}{\zeta(2)} > \frac{x}{2}.$$

Deduce that the numbers $\varphi(n)/n$ are not uniformly distributed in the interval $[0, 1]$. (A sequence $\{a_n\}$ of numbers in $[0, 1]$ is said to be *uniformly distributed* if, for every α and β with $0 \le \alpha < \beta \le 1$,

$$\lim_{N\to\infty}\frac{1}{N}\sum_{\substack{n\le N\\ \alpha<a_n<\beta}} 1 = \beta - \alpha.]$$

5. a) Show that

$$\frac{1}{\varphi(n)} = \frac{1}{n}\sum_{d\mid n}\frac{\mu^2(d)}{\varphi(d)}.$$

[*Hint:* Note that $(1 - p^{-1})^{-1} = 1 + (p - 1)^{-1}$.]

b) Deduce that

$$\sum_{n\le x}\frac{1}{\varphi(n)} = A\log x + 0(1), \qquad A = \sum{}'\frac{1}{n\varphi(n)} \approx 1.93,$$

where the accent designates summation over all square-free integers.

6. Suppose that $f(n)$ is a number-theoretic function such that $\sum_1^x f(n) \sim xg(x)$, where g is a positive differentiable function such that $xg'(x) = o(g(x))$. Show that $\sum_1^x f(n)/g(n) \sim x$. (E.g., 2(b) above.)

7. Show that $\tau(mn) \le \tau(m)\tau(n)$ for all $m, n > 0$. Then show that

$$\sum_{n\le x}\tau^2(n) \ll x\log^3 x$$

by writing

$$\sum_{n\le x}\tau^2(n) = \sum_{n\le x}\tau(n)\sum_{d\mid n}1$$

and interchanging summations.

8. Show by summing the relation in Theorem 6.5 that

$$\sum_{k=1}^{n} \mu(k) \left[\frac{n}{k}\right] = 1,$$

and conclude that

$$\left|\sum_{k=1}^{n} \frac{\mu(k)}{k}\right| \le 1.$$

9. Let $\omega(n)$ be the number of distinct prime divisors of n, and let $\Omega(n)$ be the total number of prime divisors, so that $\omega(12) = 2$, $\Omega(12) = 3$. Show, perhaps with the help of Theorem 6.29, that for suitable constants c, c',

$$\sum_{n \le x} \omega(n) = x \log \log x + cx + o(x),$$

$$\sum_{n \le x} \Omega(n) = x \log \log x + c'x + o(x).$$

10. Show that the number of lattice points in the disk $x^2 + y^2 \le n$ is $\pi n + O(n^{1/2})$.

11. Let (k, l) be the GCD, as usual, and let ε be positive.
    a) Show that for every ε, $\sum_{l=1}^{k} (k, l) = o(k^{1+\varepsilon})$.
    b) Show that if $f(n) = O(n^{1-\delta})$ for some $\delta > 0$, then

$$\sum_{k,l=1}^{n} f((k, l)) \sim \frac{6n^2}{\pi^2} \sum_{m=1}^{\infty} \frac{f(m)}{m^2}.$$

    [*Hint:* First estimate the sum over $1 \le l \le k$, $1 \le k \le n$.]

12. a) Show that the congruence $x \equiv yz \pmod p$ establishes a 1–1 correspondence between pairs of integers x, y with $(x, y) = 1$, $0 < x < \sqrt{p}$ and $0 < y < \sqrt{p}$ on the one hand and certain residue classes $z \pmod p$ on the other.
    b) Show that the number of such pairs x, y is

$$1 + 2 \sum_{m=2}^{\sqrt{p}} \varphi(m),$$

    and that this is larger than $(p - 1)/2$ for all large p.
    c) Conclude that some z is a quadratic nonresidue of p, for such p, and hence that some x or y is also a nonresidue. This shows that there is a quadratic nonresidue of p between 1 and $\sqrt{p}$, for all large p. (With careful treatment of the error term, it can be shown that $p > 23$ suffices.)

13. The fact that $\sum_{d \mid n} \mu(d) \ne 0$ if and only if $n = 1$ can be used to estimate sums over reduced residue systems.
    a) Show that if $r \in \mathbf{R}^+$, then

$$F_r(m) = \sum_{\substack{1 \le a \le m \\ (a,m)=1}} a^r = \sum_{a=1}^{m} a^r \sum_{\substack{d \mid a \\ d \mid m}} \mu(d).$$

b) By reversing the order of summation, conclude that

$$F_r(m) = \frac{m^r \varphi(m)}{r+1} + O(m^r \tau(m)).$$

[*Hint:* Use Problem 3 of Section 6.4 to estimate the sum of the rth powers of the first n integers, for large n.]

6.12 BRUN'S THEOREM ON TWIN PRIMES

Many of the threads in this chapter come together in the proof of the following theorem. Unfortunately, the theorem destroys any hope of proving that there are infinitely many twin primes, at least by proceeding as Dirichlet did for primes in a progression.

Theorem 6.33 *The series of reciprocals of the twin primes either is a finite sum or forms a convergent infinite series:*

$$(\tfrac{1}{3} + \tfrac{1}{5}) + (\tfrac{1}{5} + \tfrac{1}{7}) + (\tfrac{1}{11} + \tfrac{1}{13}) + \cdots < \infty.$$

In fact, if $\pi_2(x)$ is the number of such pairs not exceeding x, then

$$\pi_2(x) \ll P(x), \qquad \text{where } P(x) = \frac{x(\log\log x)^2}{\log^2 x}. \tag{10}$$

The first assertion follows from the second by an easy application of Theorem 6.15.

Proof. We remove from among the integers $a_n = n(n+2)$, for $1 \le n \le x$, those which are divisible by at least one prime $\le y$, where y is some integer $\le \sqrt{x}$. If the number of a_n's remaining ("sieved out," as one says) is $A(x, y)$, then just as in Section 6.5,

$$\pi_2(x) \le \pi(y) + A(x, y), \tag{11}$$

and the problem reduces to that of showing that $A(x, y) \ll P(x)$ for some $y = y(x)$ such that $\pi(y) \ll P(x)$. Put

$$R = \prod_{p \le y} p;$$

then

$$\begin{aligned} A(x, y) &= [x] - \sum_{p \mid R} \sum_{\substack{n \le x \\ p \mid a_n}} 1 + \sum_{\substack{p_1 < p_2 \\ p_1 p_2 \mid R}} \sum_{\substack{n \le x \\ p_1 p_2 \mid a_n}} 1 - \cdots \\ &= S_0 - S_1 + S_2 - \cdots, \text{ say.} \end{aligned}$$

If we proceeded as in Section 6.5, we should obtain an inequality so weak as to be useless. A modification which is central to "Brun's method" is to replace the above equation by the inequality

$$A(x, y) \le S_0 - S_1 + S_2 - \cdots + S_{2k}, \tag{12}$$

which is valid for every *even* subscript, $0 \le 2k \le \pi(y)$. For suppose that a_n is divisible by exactly m of the primes $\le y$. If $m = 0$, then a_n is counted once on the right-hand side of (12), as it should be. If $m > 0$, the total contribution from a_n is

$$C(a_n) = 1 - \binom{m}{1} + \binom{m}{2} - \cdots + \binom{m}{2k},$$

and we are asserting in (12) that this number is nonnegative. This follows from the fact that the binomial coefficients $\binom{m}{l}$, $0 \le l \le m$, increase to the middle term (or pair of equal middle terms) and then decrease. Thus clearly if $0 \le 2k \le \frac{1}{2}(m + 1)$, then

$$C(a_n) = 1 + \left\{\binom{m}{2} - \binom{m}{1}\right\} + \left\{\binom{m}{4} - \binom{m}{3}\right\} + \cdots > 0.$$

If $2k \ge m$, then $C(a_n) = (1 - 1)^m = 0$, and if $\frac{1}{2}(m + 1) < 2k < m$, then again

$$C(a_n) = (1 - 1)^m - \left\{-\binom{m}{2k + 1} + \binom{m}{2k + 2} - \cdots\right\}$$

$$= \left\{\binom{m}{2k + 1} - \binom{m}{2k + 2}\right\} + \cdots \ge 0.$$

Now we must estimate the inner sums in the terms S_l, $0 \le l \le 2k$, a typical one being

$$T_d = \sum_{\substack{n \le x \\ d \mid a_n}} 1, \quad d = p_1 p_2 \cdots p_l, \quad d \mid R, \quad \mu(d) \ne 0.$$

Suppose first that d is odd. Then the congruence $n(n + 2) \equiv 0 \pmod{d}$ holds if and only if for some factorization $d = d_1 d_2$,

$$n \equiv 0 \pmod{d_1}, \qquad n + 2 \equiv 0 \pmod{d_2},$$

and for each factorization, n is unique (mod d). Different factorizations yield different n's (mod d), since if also $n \equiv 0 \pmod{d_1'}$ and $n + 2 \equiv 0 \pmod{d_2'}$, then by Theorem 3.16, $(d_1, d_2') \mid 2$ and $(d_1', d_2) \mid 2$, and this is impossible if $d_1 \ne d_1'$. (Why?) Hence there are exactly $\tau(d)$ solutions (mod d) of $n(n + 2) \equiv 0 \pmod{d}$; so the number of $n \le x$ satisfying this congruence is

$$T_d = \left[\frac{x}{d}\right]\tau(d) + \theta_1\tau(d) = \frac{x}{d}\tau(d) + \theta\tau(d), \quad 0 \le \theta_1 \le 1, \quad |\theta| \le 1.$$

If, on the other hand, d is even then n must be even, say $n = 2m$, and we are led to count the $m \le \frac{1}{2}x$ for which $m(m + 1) \equiv 0 \pmod{\frac{1}{2}d}$; by the same argument,

$$T_d = \frac{x/2}{d/2}\tau(d/2) + \theta\tau(d/2), \qquad |\theta| \le 1.$$

Hence if we put $\tau'(d) = \tau(d)$ or $\tau(d/2)$ according as $2 \nmid d$ or $2 \mid d$, then always

$$T_d = \frac{x}{d}\,\tau'(d) + \theta\tau'(d), \qquad |\theta| \le 1. \tag{13}$$

Returning to (12), we now have

$$\begin{aligned} A(x, y) \le x\Bigg(1 - \sum_{p \mid R} \frac{\tau'(p)}{p} + \sum_{p_1 p_2 \mid R} \frac{\tau'(p_1 p_2)}{p_1 p_2} - \cdots \\ + \sum_{p_1 \cdots p_{2k} \mid R} \frac{\tau'(p_1 \cdots p_{2k})}{p_1 \cdots p_{2k}}\Bigg) \\ + \Bigg(\sum_{p \mid R} \tau'(p) + \sum_{p_1 p_2 \mid R} \tau'(p_1 p_2) + \cdots \\ + \sum_{p_1 \cdots p_{2k} \mid R} \tau'(p_1 \cdots p_{2k})\Bigg); \end{aligned} \tag{14}$$

here and in the remainder of the proof we adopt the convention that $p_1 < p_2 < \cdots$ in these sums. Since, for each l,

$$\begin{aligned} \sum_{p_1 \cdots p_l \mid R} \tau'(p_1 \cdots p_l) &\le 2^l \binom{\pi(y)}{l} \\ &= 2^l \frac{\pi(y)(\pi(y) - 1)\cdots(\pi(y) - l + 1)}{l!} \\ &< \frac{2^l \pi^l(y)}{l!}, \end{aligned}$$

we have a bound for the total error term in (14):

$$\sum_{p \mid R} \tau'(p) + \cdots + \sum_{p_1 \cdots p_{2k} \mid R} \tau'(p_1 \cdots p_{2k}) \le \pi^{2k}(y) \sum_{1}^{2k} \frac{2^l}{l!} < e^2 \pi^{2k}(y). \tag{15}$$

To estimate the main term in (14)—the term involving x—we note that since $\tau'(n)$ is multiplicative, the coefficient of x would be

$$\prod_{p \mid R} \left(1 - \frac{\tau'(p)}{p}\right) = \frac{1}{2} \prod_{\substack{p \mid R \\ p > 2}} \left(1 - \frac{2}{p}\right)$$

if the (possibly) missing terms

$$-\sum_{p_1 \cdots p_{2k+1} \mid R} \frac{\tau'(p_1 \cdots p_{2k+1})}{p_1 \cdots p_{2k+1}} + \sum_{p_1 \cdots p_{2k+2} \mid R} \frac{\tau'(p_1 \cdots p_{2k+2})}{p_1 \cdots p_{2k+2}} - \cdots = V_k, \text{ say},$$

were added in. Including these terms, we have

$$A(x, y) \ll x \prod_{2<p\le y}\left(1 - \frac{2}{p}\right) + \pi^{2k}(y) + x|V_k|. \tag{16}$$

To bound the new error term, $|V_k|$, note that for $l \le \pi(y)$,

$$\sum_{p_1\cdots p_l | R} \frac{\tau'(p_1\cdots p_l)}{p_1\cdots p_l} \le \sum_{p_1\cdots p_l | R} \frac{2^l}{p_1\cdots p_l} \le \frac{1}{l!}\left(\sum_{p\le y}\frac{2}{p}\right)^l.$$

The last inequality holds because in the multinomial expansion of $(t_1 + \cdots + t_s)^l$, each of the $\binom{s}{l}$ products $t_{h_1}\cdots t_{h_l}$, where $1 \le h_1 < \cdots < h_l \le s$, occurs with coefficient $l!$. (Alternatively, the inequality can be proved by induction.) Hence, by Theorem 6.19,

$$|V_k| < \sum_{2k<l\le\pi(y)} \frac{1}{l!}\left(\sum_{p\le y}\frac{2}{p}\right)^l < \sum_{l>2k}\frac{1}{l!}(2\log\log y + c)^l,$$

for some $c \ge 0$. Now

$$e^l = \sum_{u=0}^{\infty}\frac{l^u}{u!} > \frac{l^l}{l!},$$

so $l! > (l/e)^l$, and hence

$$|V_k| < \sum_{l>2k}\left(\frac{2e\log\log y + ec}{l}\right)^l.$$

If we choose $k = [6\log\log y]$, then for y large, $k > 2e\log\log y + ec$, so for such y,

$$|V_k| < \sum_{l>2k} 2^{-l} = 2^{-2k} < 2^{-12\log\log y} < (\log y)^{-8}. \tag{17}$$

Furthermore, we see by Theorem 6.11 that

$$\prod_{2<p\le y}\left(1-\frac{2}{p}\right) = \prod_{2<p\le y}\left\{\left(1-\frac{1}{p}\right)^2 - \frac{1}{p^2}\right\} < \prod_{2<p\le y}\left(1-\frac{1}{p}\right)^2 \ll \frac{1}{\log^2 y}. \tag{18}$$

Combining (11), (16), (17), and (18), we obtain, for large y,

$$\pi_2(x) \ll \frac{x}{\log^2 y} + \pi^{2k}(y) + \frac{x}{\log^8 y} \ll \frac{x}{\log^2 y} + \left(\frac{y}{\log y}\right)^{2k}.$$

Finally, choosing $y = x^{1/(12\log\log x)}$, we have (by Theorem 6.17)

$$\pi_2(x) \ll \frac{x(\log\log x)^2}{\log^2 x} + \left\{\frac{x^{1/12\log\log x}}{(\log x)/12\log\log x}\right\}^{12\log\log x}$$

$$\ll \frac{x(\log\log x)^2}{\log^2 x}. \quad \blacksquare$$

PROBLEMS

1. Show that for fixed $a > 0$, $\prod_{a<p\le x}(1 - a/p) \ll \log^{-a} x$.
2. Parallel the proof of Theorem 6.33 to show that

$$\pi(x) \ll \frac{x \log\log x}{\log x}.$$

   (Note! The inequality $\pi(y) \ll y/\log y$ which was used in the final step is no longer available, of course; y must be chosen slightly differently.)
3. In equation (13), $\tau'(p)$ is the number of solutions (mod p) of $n(n + 2) \equiv 0 \pmod{p}$, whether $p = 2$ or $p > 2$. Rederive (13) with the help of Theorem 3.21.
4. Using the technique suggested in the preceding problem, show that the number of n $(1 \le n \le x)$ such that $n(n + 2)(n + 6) \equiv 0 \pmod{d}$, where $d = p_1 \cdots p_l$ and $\mu(d) \ne 0$, is $(x/d + \theta)\tau^*(d)$, where τ^* is multiplicative, $\tau^*(2) = 1$, $\tau^*(3) = 2$, $\tau^*(p) = 3$ for $p > 3$, and $|\theta| \le 1$. Conclude that the number $\pi_3(x)$ of "prime triples" $p, p + 2, p + 6$, all prime and less than x, is $\ll x(\log\log x)^3/\log^3 x$. What goes wrong if one considers instead the triples $p, p + 2, p + 4$?

NOTES AND REFERENCES

Section 6.1

It is known that an odd perfect number must have at least six distinct prime factors (U. Kühnel, 1949) and must be larger than 10^{400} (Buxton and Elmore [1976]).

Fermat came to his theorem that $a^{p-1} \equiv 1 \pmod{p}$ while studying perfect numbers.

Section 6.2

A. F. Möbius defined the μ-function in 1832, using it in the "reversion" of a Dirichlet series. The inversion formula as given in Theorem 6.6 is due to Dedekind (1857). Many variants are known; see Dickson [1919] and Rota [1964].

Section 6.5

For the definitive account of sieve methods, see Halberstam and Richert [1974].

The result in Problem 3, which is due to Legendre, was modified by E. Meissel (1870) to a formula by which he actually computed $\pi(10^8)$. See Uspensky and Heaslet [1939].

Section 6.6

Chebyshev [1851] started from the identity in Theorem 6.20, rather than that in Theorem 6.9. He obtained the much stronger result that for every n, each of the inequalities

$$\pi(x) - \mathrm{li}(x) < x/\log^n x \quad \text{and} \quad \pi(x) - \mathrm{li}(x) > -x/\log^n x$$

holds for arbitrarily large values of x. This implies in particular that if $\pi(x)$ can be approximated by a rational function of x and $\log x$ with an error $O(x/\log^{n+1} x)$ for some $n > 0$, then the function must be of the form

$$\frac{x}{\log x} + \frac{1!\,x}{\log^2 x} + \frac{2!\,x}{\log^3 x} + \cdots + \frac{(n-1)!\,x}{\log^n x} + O\left(\frac{x}{\log^{n+1} x}\right).$$

(The latter can be deduced with the aid of the relation in Problem 7.) It follows that $\mathrm{li}(x)$ is a better approximation to $\pi(x)$ for large x than is $x/\log x$, if either is consistently good.

For an extensive discussion of both elementary and nonelementary theory of primes, see Landau [1909] or Ingham [1932].

Section 6.7

Instead of the binomial coefficient $\binom{2n}{n}$, Chebyshev [1852] used the more complicated function

$$\frac{[x]!\left[\dfrac{x}{30}\right]!}{\left[\dfrac{x}{2}\right]!\left[\dfrac{x}{3}\right]!\left[\dfrac{x}{5}\right]!}.$$

This yielded $c_1 = \log(2^{1/2}3^{1/3}5^{1/5}/30^{1/30})$ and $c_2 = 6c_1/5$, the values mentioned in the text.

Chebyshev [1851] obtained Theorem 6.19 under the assumption that the prime number theorem is valid; Mertens removed the hypothesis and evaluated C (as well as B in Problem 1.) Chebyshev and Bertrand thus originated what has become an honored tradition in prime number theory, of proving theorems under unproved hypotheses. After the prime number theorem succumbed in 1896, its place was taken by the so-called "Riemann hypothesis" (still unproved), which is concerned with the complex numbers s for which the function $\zeta(s)$ of Section 6.8 vanishes.

Section 6.8

For a complete proof of Dirichlet's theorem (and of the prime number theorem), see for example LeVeque [1955].

Section 6.9

Theorem 6.24 is called Bertrand's "hypothesis" rather than "conjecture" because he took it as a working tool in his study of a problem in group theory. This must have seemed entirely safe, considering the actual density of primes in the tables. There is not merely one prime between 500,000 and 1,000,000, say, there are 36,960 of them! The proof given here is a modification of that given by P. Erdös in 1932. The proof of Theorem 6.22 is simpler than that originally given by Erdös; it was found independently by Erdös and L. Kalmár in 1939, but was not published.

Section 6.10

For more on extreme and average orders of magnitude of number-theoretic functions, see Hardy and Wright [1960].

Section 6.12

It is conjectured that the series in Theorem 6.33 converges to a number in the interval $1.9021604 \pm 5 \cdot 10^{-7}$. For this and other conjectures and empirical data on twin primes, see Brent [1975]. For a portion of the graph of $y = \pi_2(x)$ and a conjectured approximation, see graph 2 at the end of this book.

7
Sums of Squares

7.1 PRELIMINARIES

The principal objectives of this chapter are to find the number of representations of a positive integer as a sum of two squares, and to show that every integer is a sum of four squares. We shall make both investigations hinge on the following theorem, which, as we shall see, also has other applications.

Theorem 7.1 (Brauer and Reynolds [1951]) *Let r, s, and m be positive integers with $m > 1$ and $r < s$, and let $\lambda_1, \ldots, \lambda_s$ be positive real numbers such that*

$$\lambda_1 < m, \ldots, \lambda_s < m, \qquad \lambda_1 \cdots \lambda_s > m^r.$$

Then the $r \times s$ system of homogeneous linear congruences

$$\sum_{j=1}^{s} a_{ij}x_j \equiv 0 \pmod{m}, \qquad i = 1, 2, \ldots, r,$$

where the $a_{ij} \in \mathbf{Z}$, has a solution in integers $X_1, \ldots, X_s$, not all zero, such that $|X_j| < \lambda_j$ for $j = 1, \ldots, s$.

Proof. It was pointed out in Section 1.3 that Dirichlet's pigeon-hole principle is useful for proving existence theorems; we shall now see it in action. For $i = 1, \ldots, r$, put

$$y_i = y_i(x_1, \ldots, x_s) = \sum_{j=1}^{s} a_{ij}x_j.$$

For each j, let x_j range over the integers

$$(1) \qquad 0 \leq x_j \leq [\lambda_j]^*,$$

where $[t]^*$ is the largest integer strictly smaller than t. This gives $1 + [\lambda_j]^*$ values of x_j and these are distinct (mod m), since $\lambda_j < m$. Hence there are

$$l = \prod_{j=1}^{s} (1 + [\lambda_j]^*)$$

different s-tuples $\{x_1, \ldots, x_s\}$. Corresponding to each such s-tuple there is an r-tuple $\{y_1, \ldots, y_r\}$, and so we have found

$$l \geq \lambda_1 \cdots \lambda_s > m^r$$

r-tuples $\{y_1, \ldots, y_r\}$. But there are only m^r r-tuples which are distinct (mod m), so there must be two s-tuples of x's such that

$$\text{(2)} \qquad \{x_1, \ldots, x_s\} \not\equiv \{x_1', \ldots, x_s'\} \pmod{m}$$

while

$$y_i(x_1, \ldots, x_s) \equiv y_i(x_1', \ldots, x_s') \pmod{m} \qquad \text{for } i = 1, \ldots, r.$$

Using the linearity of the y's, this gives

$$y_i(x_1 - x_1', \ldots, x_s - x_s') \equiv 0 \pmod{m} \qquad \text{for } i = 1, \ldots, r.$$

We can take $X_j = x_j - x_j'$ for $j = 1, \ldots, s$, since not all the X_j are 0, by (2), and since

$$|X_j| = |x_j - x_j'| < \lambda_j \qquad \text{for } j = 1, \ldots, s,$$

by (1). ■

The following consequence can be ascribed, in various versions, to Aubry (1913), Thue (1917), and Vinogradov (1927).

Corollary *If* $\varepsilon, \lambda \in \mathbf{R}^+$, *and* $\lambda < m$, *then for each* $a \not\equiv 0 \pmod{m}$ *there are* x, y *such that* $ax \equiv y \pmod{m}$, *with* $1 \leq x < \lambda$ *and* $1 \leq |y| \leq m/\lambda$.

Proof. This is very nearly the case $r = 1$, $s = 2$ of the theorem, with $\lambda_1 = \lambda$, $\lambda_2 = (m + \varepsilon)/\lambda$. It differs, first, in the prescription $x > 0$, which can be enforced, since $x \neq 0$, by introducing a factor of -1 into the congruence if necessary. Second, we have replaced the inequality $|y| < (m + \varepsilon)/\lambda$ by $|y| \leq m/\lambda$; this is easily seen to be permissible, since y is an integer and ε is arbitrarily small, by considering the cases $m/\lambda \in \mathbf{Z}$ and $m/\lambda \notin \mathbf{Z}$. ■

To illustrate the power of these theorems, we note a consequence which is otherwise unrelated to this chapter.

Theorem 7.2 *Suppose that either* (a) k *is odd and* $(k, p - 1) = d > 1$, *or* (b) $k = 2$ *and* $p \equiv 1 \pmod{4}$. *Then there is a kth power nonresidue* n_k *of* p *with* $0 < n_k < \sqrt{p}$.

Proof. According to Theorem 4.13, the hypotheses of the present theorem guarantee that p has (kth power) nonresidues and that -1 is a residue. Let a be any nonresidue, and suppose $p \nmid z$. Then az is a nonresidue whenever z is a residue, and vice versa, again by Theorem 4.13. So if (for this a) x and y are chosen in accordance with the above corollary so that $ax \equiv y \pmod{p}$, $1 \leq x < \sqrt{p}$, $1 \leq |y| < \sqrt{p}$, then either x or y must be a nonresidue, and if y is a nonresidue, so also is $-y$. ■

PROBLEMS

1. Suppose that p is an odd prime, $1 \le g \le p$, $h = [p/g]$ and $(r/p) = 1$. Then one of the numbers $1^2, 2^2, \ldots, h^2$ is congruent (mod p) to one of $r, 2^2r, \ldots, (g-1)^2r$.
2. Prove the following analogue of Theorem 7.1 for systems of homogeneous equations in place of congruences: Suppose that the coefficients in the system

$$\sum_{j=1}^{s} a_{ij}x_j = 0 \qquad i = 1, 2, \ldots, r$$

are integers, with $\max_{i,j}(|a_{ij}|) = A$, and suppose that $s > r$. Then the system has a solution in integers $x_1, \ldots, x_s$, not all zero, such that

$$|x_j| < 1 + (sA)^{r/(s-r)} \qquad j = 1, \ldots, s.$$

[*Hint:* Let the x_j range independently over the $2H + 1$ integers $0, \pm 1, \ldots, \pm H$. Then $y_i = \sum a_{ij}x_j$ is easily bounded, and Dirichlet's principle shows that there is a nontrivial solution if $(2sAH + 1)^r < (2H + 1)^s$. Since $sA > 1$, $2sAH + 1 < sA(2H + 1)$, so it suffices to choose H so that $2H + 1 \ge (sA)^{r/(s-r)}$.]

7.2 PRIMITIVE REPRESENTATIONS AS A SUM OF TWO SQUARES

The goal in this and the next section is to find the exact conditions under which the equation

$$n = x^2 + y^2 \tag{3}$$

is solvable, and to find the number of solutions for each n for which it is solvable, in terms of the multiplicative structure of n. It is a measure of the subtlety of number theory that there is any connection of this sort, between additive and multiplicative properties of integers. This comes about in part, in the present instance, because of the ancient identities

$$(a^2 + b^2)(c^2 + d^2) = (ac \pm bd)^2 + (ad \mp bc)^2, \tag{4}$$

which can be deduced from the product of complex numbers $(a + bi)(c \mp di)$ in an obvious way, and which show that the set of integers representable as a sum of two squares is closed under multiplication.

A representation of n in the form (3) is said to be **primitive** if $(x, y) = 1$.

Theorem 7.3 *If n has a primitive representation* (3), *then* -1 *must be a quadratic residue of n. In particular, if $p \equiv 3$* (mod 4) *and $p \mid n$, then n has no representation* (3) *at all unless the p-component of n is a square, p^{2k}, and in that case, if* (3) *is solvable, p^k appears explicitly throughout, in every representation:*

$$n = p^{2k}n' = (p^kx')^2 + (p^ky')^2, \qquad n', x', y' \in \mathbf{Z}. \tag{5}$$

Proof. Suppose first that n has a primitive representation (3), and that p is *any* prime divisor of n. Then $p \nmid x$, so x^{-1} exists in U_p; thus (3), which implies $x^2 +$

$y^2 \equiv 0 \pmod{p}$, also implies $1 + (x^{-1}y)^2 \equiv 0 \pmod{p}$, so -1 is a quadratic residue of p (and therefore also of n). Hence $p \not\equiv 3 \pmod{4}$.

Next suppose that $p \equiv 3 \pmod{4}$, that $p^h \parallel n$, and that (3) holds with $(x, y) = d$. Put $x = dx_1$, $y = dy_1$. Then from (3), $n/d^2 = N \in \mathbf{Z}$, and we have

$$N = x_1^2 + y_1^2, \qquad (x_1, y_1) = 1, \qquad p^{h-2j} \mid N.$$

This contradicts what has already been proved unless $h = 2j$, and in that case we obtain (5) with $k = h$. ■

Theorem 7.4 *Suppose that $n > 1$ is an integer of which -1 is a quadratic residue. Then to each solution u of*

$$u^2 \equiv -1 \pmod{n} \tag{6}$$

there corresponds a unique pair of integers x, y such that

$$n = x^2 + y^2, \quad x > 0, \quad y > 0, \quad (x, y) = 1, \quad y \equiv ux \pmod{n}. \tag{7}$$

Conversely, every pair x, y satisfying the first four conditions in (7) *determines, via the fifth condition, a unique solution u of* (6).

Proof. Suppose first that u satisfies (6), and determine r and s in accordance with the Corollary to Theorem 7.1, with $\lambda = \sqrt{n}$ and $a = u$:

$$us \equiv r \pmod{n}, \quad 0 < s < \sqrt{n}, \quad |r| \le \sqrt{n}.$$

If $r > 0$, put $x = s$, $y = r$. If $r < 0$, note that $-ur \equiv s \pmod{n}$, by (6), and put $x = -r$, $y = s$. In either case,

$$x^2 + y^2 \equiv 0 \pmod{n}, \quad 0 < x \le \sqrt{n}, \quad 0 < y \le \sqrt{n}, \quad y \equiv ux \pmod{n},$$

and at least one of x and y is $< \sqrt{n}$. Hence

$$0 < x^2 + y^2 = tn < 2n, \qquad \text{so } x^2 + y^2 = n.$$

Moreover, if $u^2 + 1 = kn$ and $y = ux + ln$, then

$$\begin{aligned} n &= x^2 + y^2 \\ &= x^2 + (ux + ln)^2 \\ &= x^2(1 + u^2) + uxln + ln(ux + ln) \\ &= xn(kx + ul) + lny, \end{aligned}$$

so that $(k + ul)x + ly = 1$, and therefore $(x, y) = 1$. Hence (7) holds for this pair x, y.

If (7) holds also for a pair X, Y, then by (4),

$$n^2 = (x^2 + y^2)(X^2 + Y^2) = (xX + yY)^2 + (xY - Xy)^2.$$

But this implies that $0 < xX + yY \le n$; moreover,

$$xX + yY \equiv xX + u^2xX \equiv 0 \pmod{n}.$$

Hence $xX + yY = n$, so $xY - yX = 0$. Since $(x, y) = (X, Y) = 1$, clearly $x = X$, $y = Y$. Thus x and y are uniquely determined by u.

Conversely, suppose x and y are any integers satisfying (7), u being uniquely defined (mod n) by (7) since $(x, n) = 1$. Then $x^2 + y^2 \equiv x^2(1 + u^2) \equiv 0 \pmod{n}$, so u satisfies (6). ∎

We have already counted solutions of (6), when it is solvable, in Theorem 5.2, and by combining that information with the two preceding theorems we obtain the following.

Theorem 7.5 *The number $p_2(n)$ of primitive representations of $n > 1$ as a sum of two squares is four times the number of solutions of the congruence $u^2 \equiv -1 \pmod{n}$:*

$$p_2(n) = \begin{cases} 0 & \text{if } 4 \mid n \text{ or if some } p \equiv 3 \pmod{4} \text{ divides } n; \\ 4 \cdot 2^s & \text{if } 4 \nmid n, \text{ no } p \equiv 3 \pmod{4} \text{ divides } n, \text{ and } s \text{ is the number of} \\ & \text{distinct odd prime divisors of } n. \end{cases}$$

Proof. If $n > 1$, then $xy \neq 0$ in a primitive representation (3), so each such representation with $x > 0$, $y > 0$ gives rise to three others, $\{x, -y\}$, $\{-x, y\}$, $\{-x, -y\}$. This accounts for the factor of 4 in the value of $p_2(n)$. ∎

For $n = 2$, we have the four representations $2 = (\pm 1)^2 + (\pm 1)^2$. For $n = p \equiv 1 \pmod{4}$, there are exactly eight distinct primitive representations, $p = (\pm a)^2 + (\pm b)^2 = (\pm b)^2 + (\pm a)^2$, and no imprimitive ones. (Note that $a \neq b$ since $(a, b) = 1$ and $n > 1$.)

Corollary *A prime $p \not\equiv 3 \pmod{4}$ can be represented uniquely (to within order and sign) as a sum of two squares. Conversely, if $N \equiv 1 \pmod{2}$ has a unique representation (ignoring order and sign) and this is primitive, then N is prime. If $N \equiv 1 \pmod{2}$ has only one primitive representation, then $N = p^k$ for some $p \equiv 1 \pmod{4}$, $k \geq 1$.*

This result was asserted by Fermat; the first published proof is due to Euler.

PROBLEMS

1. Formulate and prove a result similar to Theorem 7.3 concerning the 2-component of n.
2. a) A simplified version of Theorem 7.4 is that every positive prime of which -1 is a quadratic residue can be represented in the form $x^2 + y^2$. Write a correspondingly simplified proof.

   b) Show that every positive prime of which -2 is a quadratic residue can be represented in the form $x^2 + 2y^2$.

   c) Show that every positive prime of which -3 is a quadratic residue can be represented in the form $x^2 + 3y^2$.

   d) Show that the obvious analogue concerning -5 and $x^2 + 5y^2$ is false, and in fact for all the primes in certain progressions.

e) Using an appropriate generalization of (4), prove the analogues of (a), (b), and (c) in which "prime" is replaced by "integer".

3. By the Corollary to Theorem 7.5, an odd integer N with two distinct representations $N = a^2 + b^2 = c^2 + d^2$ (a and c odd; b and d even; $a, b, c, d > 0$; $a \neq c$) cannot be prime. Show how to use these representations to obtain a nontrivial factorization of N. [*Hint:* Show that if $(a - c, d - b) = u$ and $(a + c, d + b) = v$, then $a - c = lu$, $d - b = mu$, $a + c = mv$, $d + b = lv$ for some m, l. Then verify that

$$N = \left\{\left(\frac{u}{2}\right)^2 + \left(\frac{v}{2}\right)^2\right\}(m^2 + l^2).\Bigg]$$

(The converse in the corollary is the method devised and used by Euler for testing for primality. More recently, it has been used, for example, to show that $2^{39} - 7 = 64045^2 + 738684^2$ is prime, this being the smallest positive prime value of $2^n - 7$.)

4. Factor $1{,}000{,}009 = 972^2 + 235^2$.

7.3 THE TOTAL NUMBER OF REPRESENTATIONS

We shall make the study of $r_2(n)$, the number of all (primitive or imprimitive) solutions of (3), depend upon the arithmetic of the Gaussian ring $\mathbf{Z}[i]$, partly because it is efficient and partly because it illustrates the utility of sometimes venturing beyond $\mathbf{Z}$ even for purely rational questions.

Recall that it was shown in Section 2.2 that $\mathbf{Z}[i]$ is a Euclidean domain and that the only units are the powers of i, namely ± 1 and $\pm i$. If $\alpha = a + bi \in \mathbf{Z}[i]$, then $\bar{\alpha} = a - bi$ is called the **conjugate** of α, and $\alpha\bar{\alpha} = (a + bi)(a - bi) = a^2 + b^2$ is called the **norm** of α and designated by $\mathcal{N}(\alpha)$, or simply $\mathcal{N}\alpha$. It is easily verified that $\mathcal{N}\alpha\beta = \mathcal{N}\alpha \cdot \mathcal{N}\beta$, and hence that if $\alpha \mid \gamma$ then $\mathcal{N}\alpha \mid \mathcal{N}\gamma$. Since the units are exactly the elements of $\mathbf{Z}[i]$ of norm ± 1, it follows that *if $\mathcal{N}\gamma$ is a rational prime* (i.e., a prime in $\mathbf{Z}$), *then γ is prime in* $\mathbf{Z}[i]$; for a factorization $\gamma = \alpha\beta$ in which neither α nor β is a unit implies the nontrivial factorization $\mathcal{N}\gamma = \mathcal{N}\alpha\mathcal{N}\beta$ in $\mathbf{Z}$. Thus $1 + i$ and $1 - i$ are both prime, since each has norm 2. In fact, $1 + i = i(1 - i)$, so these are associated primes. Similarly, if $p \equiv 1 \pmod 4$ then $p = a^2 + b^2 = \mathcal{N}(a + bi) = \mathcal{N}(a - bi)$, so $a + bi$ and $a - bi$ are also prime. But they are not associates: since $|a| \neq |b|$, the equation $a + bi = i^n(a - bi)$ is clearly false for all n.

The rational primes $p \equiv 3 \pmod 4$ do not split further in $\mathbf{Z}[i]$; that is, they remain prime in the larger set. For if

$$p = (c + di)(e + fi),$$

then

$$p^2 = (c^2 + d^2)(e^2 + f^2).$$

But the only factorizations of p^2 are $1 \cdot p^2$ and $p \cdot p$, and it is impossible that $p = c^2 + d^2 = e^2 + f^2$, by Theorem 7.3; hence one of the factors $c + di$, $e + fi$ is a unit.

Digressing for a moment, with the deliberate intention of being tantalizing, we point out the striking duality (reciprocity?) we have demonstrated:

In the ring $\mathbf{Z}[i]$ determined by the equation $x^2 + 1 = 0$:	In the ring of polynomials (mod p):
$p = 2$	
$2 = i(1 - i)^2$ is (the associate of) a square	$x^2 + 1 = (x + 1)^2$ in $\mathbf{Z}_2[x]$
$p \equiv 1 \pmod 4$	
p is the product of two distinct primes	$x^2 + 1 \equiv 0 \pmod p$ has two distinct solutions, so $x^2 + 1 = (x - a)(x - b)$ in $\mathbf{Z}_p[x]$
$p \equiv 3 \pmod 4$	
p is prime	$x^2 + 1 \equiv 0 \pmod p$ has no solutions, so $x^2 + 1$ is irreducible in $\mathbf{Z}_p[x]$

It would require remarkable skepticism to doubt that there is a theorem—no, a whole theory—lurking behind this table. Unfortunately, it is outside the scope of the present book.

It can be shown that the primes we have found in $\mathbf{Z}[i]$ are all there are, but we do not need this to evaluate $r_2(n)$.

Suppose that n has the prime-power decomposition

$$n = 2^u \cdot \prod_{p_j \equiv 1 \pmod 4} p_j^{t_j} \cdot \prod_{q_j \equiv 3 \pmod 4} q_j^{s_j},$$

in $\mathbf{Z}$, and designate by n' the first of the two products here, and by m the second product, so that $n = 2^u n' m$.

Theorem 7.6 *If $n \geq 1$, then the number $r_2(n)$ of representations of n as a sum of two squares is zero if m is not a square, and is $4\tau(n')$ if m is a square.*

Proof. The case in which m is not a square is covered by Theorem 7.3. If m is a square, each s_j is even, and we can put $s_j = 2r_j$. In this case we shall prove the theorem by establishing, by means of the identity $x^2 + y^2 = (x + iy)(x - iy)$, a one-to-one correspondence between the various representations of n on the one hand, and the factorizations of n as a product of two conjugate Gaussian integers, on the other. We must count these factorizations. Since $2 = i(1 - i)^2$, we can write the prime decomposition of n in $\mathbf{Z}[i]$ in the form

$$n = i^u(1 - i)^{2u} \prod ((a + bi)(a - bi))^t \prod q^{2r},$$

where the subscripts in the products have been omitted for clarity, and where

$$a > 0, \quad b > 0, \quad p = a^2 + b^2.$$

Then every divisor of n in $\mathbf{Z}[i]$ is of the form

$$x + iy = i^v(1 - i)^{u_1} \prod ((a + bi)^{t_1}(a - bi)^{t_2}) \prod q^{r_1},$$

where

$$0 \le v \le 3, \quad 0 \le u_1 \le 2u, \quad 0 \le t_1 \le t, \quad 0 \le t_2 \le t, \quad 0 \le r_1 \le 2r.$$

Not every such divisor leads to a representation; for this to be the case, it is necessary and sufficient that the complex conjugate,

$$\begin{aligned} x - iy &= (-i)^v(1 + i)^{u_1} \prod ((a - bi)^{t_1}(a + bi)^{t_2}) \prod q^{r_1} \\ &= i^{u_1 - v}(1 - i)^{u_1} \prod ((a + bi)^{t_2}(a - bi)^{t_1}) \prod q^{r_1}, \end{aligned}$$

be such that $(x + iy)(x - iy) = n$. It is clear that this is true if and only if $u_1 = u$, $t_1 + t_2 = t$, $r_1 = r$. Since the powers of i are periodic, with period 4, we obtain all the distinct factorizations of n into conjugate factors by listing the numbers

$$i^v(1 - i)^u \prod ((a + bi)^{t_1}(a - bi)^{t - t_1}) \prod q^r,$$

where u, t, and r are fixed, v is one of the integers 0, 1, 2, 3, and t_1 is one of $0, 1, \ldots, t$. Their total number is $4 \prod (t + 1) = 4\tau(n')$. ■

PROBLEMS

1. Use an argument involving $\mathbf{Z}[i]$ to prove that any prime in $\mathbf{Z}$ has at most one representation as a sum of two squares.
2. Show that every prime $\rho \in \mathbf{Z}[i]$ divides some rational prime. [Consider $\mathcal{N}\rho$.] Conclude that there are no other primes in $\mathbf{Z}[i]$ than those listed in the table in the text.
3. Show that $(1 + i) \mid (a + bi)$ if and only if $a \equiv b \pmod 2$.
4. Show from Theorem 7.3 that if n is positive and square-free, then $p_2(n) = r_2(n)$. Show that Theorems 7.5 and 7.6 are consistent with this equation.
5. Show that $\sum_{m=1}^{n} r_2(m) = \pi n + O(\sqrt{n})$. [*Hint:* The sum on the left is the number of lattice points inside or on the circle $x^2 + y^2 = n$. Associate each such point with the unit square of which it is the lower left corner. The resulting region has a polygonal boundary, no point of which is at distance greater than $\sqrt{2}$ from the circle.]
6. Let $\tau_1(n)$ and $\tau_3(n)$ be the numbers of divisors of n which are congruent to 1 and to 3 (mod 4), respectively. Show that $r_2(n) = 4(\tau_1(n) - \tau_3(n))$. [*Hint:* Put $f(m) = 0$, 1 or -1 according as m is even, or $m \equiv 1 \pmod 4$, or $m \equiv 3 \pmod 4$. Show that $f(ab) = f(a)f(b)$ for all a, b, deduce a factored form of $\sum_{d \mid n} f(d)$, and evaluate the factors.]
7. Working within $\mathbf{Z}$, show that the Diophantine equation $x^2 + 1 = y^n$ $(x > 0, n > 1)$ has no solution with x odd or n even. Conclude that if x, y is a solution, then $(x - i, x + i) = 1$, so $x + i = (a + bi)^n$ for some $a, b \in \mathbf{Z}$. By splitting the binomial expansion into its real and imaginary parts, conclude that $b = \pm 1$ always, and that the equation has no solution for $n = 3, 5, 7$.

7.4 SUMS OF THREE SQUARES

The problem of the solvability of the equation

$$n = x^2 + y^2 + z^2 \tag{8}$$

is much more difficult than the corresponding question for the sum of either two or four squares. The result is simple, however: (8) is solvable if and only if n is not of the form $4^t(8k + 7)$. We prove here only the trivial half of this theorem, that if n is of the specified form then (8) has no integral solutions.

Since a square can have only the values 0, 1, or 4 (mod 8), the sum of three squares is congruent to 0, 1, 2, 3, 4, 5, or 6 (mod 8), so that no $n \equiv 7 \pmod 8$ is so representable. If $4 \mid n$ and (8) holds, then x, y, and z must all be even, so that $n/4$ must also be a sum of three squares. Therefore n cannot be a power of 4 times a nonrepresentable number.

It might be mentioned that one reason that problems concerning three squares are more difficult than those concerning either two or four is that there is no composition identity in this case analogous to (4), or to that given below for four squares. Indeed, the fact that 3 and 5 are sums of three squares, while 15 is not, shows that no such identity is possible.

7.5 SUMS OF FOUR SQUARES

Theorem 7.7 *Every positive integer can be represented as a sum of four squares of nonnegative integers.*

Proof. Since

$$\begin{aligned}(x_1^2 + x_2^2 + x_3^2 + x_4^2)(y_1^2 + y_2^2 + y_3^2 + y_4^2) \\ = (x_1y_1 + x_2y_2 + x_3y_3 + x_4y_4)^2 + (x_1y_2 - x_2y_1 + x_3y_4 - x_4y_3)^2 \\ + (x_1y_3 - x_3y_1 + x_4y_2 - x_2y_4)^2 + (x_1y_4 - x_4y_1 + x_2y_3 - x_3y_2)^2,\end{aligned}$$

the product of representable numbers is representable. Since 1 is also representable, it suffices to prove that every prime p is representable.

The idea of the proof is again to replace the equation $p = x^2 + y^2 + z^2 + t^2$ by the congruence $x^2 + y^2 + z^2 + t^2 \equiv 0 \pmod p$, together with bounds on x, y, z, t. To solve the congruence, we look back at the identity (4), and see that it suffices to find a and b such that $a^2 + b^2 \equiv -1 \pmod p$. So what we need first is the following fact: *If p is a prime, then the congruence*

$$x^2 + y^2 + 1 \equiv 0 \pmod p$$

has a solution.

The assertion is clearly correct for $p = 2$. For odd p, let x and y range independently over the numbers $0, 1, \ldots, (p - 1)/2$. Then all the numbers x^2 are distinct (mod p), and the same is true of the numbers $-(1 + y^2)$. (For if

$x_i^2 \equiv x_j^2 \pmod{p}$, then $p \mid (x_i - x_j)(x_i + x_j)$. But $0 < x_i + x_j < p$, unless $x_i = x_j = 0$, so $p \mid (x_i - x_j)$, $x_k \equiv x_j \pmod{p}$, and so $x_i = x_j$.) But now we have altogether

$$\frac{p+1}{2} + \frac{p+1}{2} = p + 1$$

numbers x^2 and $-1 - y^2$, so some x^2 is congruent to some $-1 - y^2$, modulo p, which is the assertion.

Suppose that $a^2 + b^2 + 1 \equiv 0 \pmod{p}$. By Theorem 7.1, the congruences

$$x \equiv az + bt \pmod{p},$$
$$y \equiv bz - at \pmod{p}$$

have a nontrivial solution x, y, z, t with

$$\max(|x|, |y|, |z|, |t|) \le \sqrt{p} + \varepsilon;$$

here $r = 2$, $s = 4$, $m = p$, and we have chosen all $\lambda_j = \sqrt{p} + \varepsilon$, where $\varepsilon > 0$ is so small that $\sqrt{p} + \varepsilon < p$. Now x, y, z, and t are integers, while $\sqrt{p}$ is not; if ε is chosen so small that $\sqrt{p} + \varepsilon < 1 + [\sqrt{p}]$, it follows that

$$\max(|x|, |y|, |z|, |t|) < \sqrt{p}.$$

We now have

$$x^2 + y^2 \equiv (a^2 + b^2)(z^2 + t^2) \equiv -(z^2 + t^2) \pmod{p},$$

while

$$0 < x^2 + y^2 + z^2 + t^2 < p + p + p + p = 4p,$$

so that

$$x^2 + y^2 + z^2 + t^2 = Ap,$$

where $A = 1, 2$, or 3.

If $A = 1$, we are finished. If $A = 2$, then x is congruent to y, z, or t (mod 2). If $x \equiv y \pmod{2}$, then $z \equiv t \pmod{2}$, and

$$p = \left(\frac{x+y}{2}\right)^2 + \left(\frac{x-y}{2}\right)^2 + \left(\frac{z+t}{2}\right)^2 + \left(\frac{z-t}{2}\right)^2,$$

where the quantities in parentheses are integers.

In the case $A = 3$, we note first that $p = 3$ has a representation,

$$3 = 1^2 + 1^2 + 1^2,$$

so that we need only consider $p \neq 3$. The square of an integer is congruent to 0 or 1 (mod 3), and the equation

$$x^2 + y^2 + z^2 + t^2 = 3p$$

implies that

$$x^2 + y^2 + z^2 + t^2 \equiv 0 \pmod{3},$$

while

$$x^2 + y^2 + z^2 + t^2 \not\equiv 0 \pmod{9}.$$

By the congruence, one of the quantities—say x—is divisible by 3, and either all the others are, or all are not, divisible by 3. Because of the incongruence, $3 \nmid yzt$, so that y, z, and t are all congruent to $\pm 1 \pmod 3$. Let z' be that one of $\pm z$ such that $z' \equiv y \pmod 3$, and let t' be that one of $\pm t$ such that $t' \equiv y \pmod 3$. Then

$$p = \left(\frac{y + z' + t'}{3}\right)^2 + \left(\frac{x + z' - t'}{3}\right)^2 + \left(\frac{x - y + t'}{3}\right)^2 + \left(\frac{x + y - z'}{3}\right)^2,$$

where the quantities in parentheses are again integers. ■

It is possible to determine the number $r_4(n)$ of representations of $n > 0$ as a sum of four squares (differences in sign and order being taken as significant). The result, due to Jacobi, is that $r_4(n)$ is 8 times the sum of those divisors of n which are not multiples of 4. Elementary proofs are known, but they are too long for inclusion here.

PROBLEMS

1. Show that for each $\varepsilon > 0$ and for each sufficiently large N, say $N > N_0(\varepsilon)$, at least $\frac{1}{6} - \varepsilon$ of the integers up to N actually require four squares—they cannot be represented as the sum of three or fewer. Assume the theorem stated in Section 7.4.

†2. Show that if $p \nmid abc$, then the congruence

$$ax^2 + by^2 + cz^2 \equiv 0 \pmod p$$

has a solution other than 0, 0, 0.

3. Use the identity $2n + 1 = (n + 1)^2 - n^2$ to show that every integer can be represented in the form $\pm x^2 \pm y^2 \pm z^2$, and show that 6 actually requires all three terms.

4. Use Problem 2 of Section 7.2 to show that every prime $\equiv 1$ or $3 \pmod 8$ is a sum of three squares.

7.6 WARING'S PROBLEM

In the same year (1770) that Lagrange proved the four-squares theorem, E. Waring conjectured a sweeping generalization of it, namely that for each positive exponent k, some fixed number of nonnegative kth powers is sufficient to represent all positive integers n. If the minimum number that suffices is $g(k)$, then $g(2) = 4$, and Waring conjectured that $g(3) \leq 9$, $g(4) \leq 19$, "and so on," whatever that means. This turned out to be a splendid problem, tractable enough to be attractive, but difficult enough to inspire the development of new techniques useful also for other problems.

The existence of $g(k)$ for every k was first proved by D. Hilbert in 1909. A better way of attacking the problem was developed by G. H. Hardy and J. E. Littlewood in the 1920s and improved by I. M. Vinogradov; their method led to an asymptotic formula for the number of representations of a large integer n as a sum of s kth powers, for s larger than an explicit function of k. Their work focused attention on the quantity $G(k)$, the minimal number of kth powers sufficient to represent all *sufficiently large* n; although, as follows from the preceding sections, $g(2) = G(2) = 4$, it is usually the case that $G(k) < g(k)$. For example, $g(3) = 9$

while $4 \leq G(3) \leq 7$. (Perhaps surprisingly, the value of $G(3)$ is still not known.)

There is a vast literature on Waring's problem and variants of it. We content ourselves with mentioning two tantalizing open problems: is every large integer the sum of 4 nonnegative cubes, and is every integer representable in the form $x^3 + y^3 - z^3 - t^3$ $(x, y, z, t > 0)$ in infinitely many ways?

NOTES AND REFERENCES

Concerning sums of squares as they pervade mathematics, see the wide-ranging survey article by Taussky [1970].

Section 7.2

The identity (4) was already known to Diophantus.

Section 7.3

The evaluation of $r_2(n)$ as in Theorem 7.6 is due to Legendre [1798]. The alternative form stated in Problem 6 was given by Jacobi; see below.

Section 7.4

Fermat correctly characterized the numbers representable as a sum of three squares, but did not claim to have a proof. Legendre [1798] attempted a demonstration, but needed Dirichlet's theorem on primes in progressions, as he had for the law of quadratic reciprocity. Gauss gave the first proof in the *Disquisitiones*, using his general theory of ternary quadratic forms. Somewhat simpler proofs have been found since; two of them are to be found in Landau [1958] and Mordell [1969]. A proof using p-adic fields is given in Serre [1973].

Section 7.5

Euler tried unsuccessfully over a period of some 25 years to prove the four-squares theorem. He recognized the importance of the congruence $x^2 + y^2 + 1 \equiv 0 \pmod{p}$ for the problem, and proved it solvable. Lagrange's first proof of the main theorem used this congruence, as have many later proofs.

The evaluation of $r_2(n)$ and $r_4(n)$ goes back to Jacobi's masterpiece of 1829 (which appeared when he was 25 years old), dealing with elliptic functions. In the same work he evaluated $r_6(n)$ and $r_8(n)$. By now, exact formulas are known for $r_s(n)$ for $s \leq 32$. But for each $s \geq 5$, $r_s(n)$ grows so regularly with n that an asymptotic estimate can be obtained, and besides being available for arbitrarily large s, it is actually more enlightening than the very complicated exact formula, even for $s \leq 32$. For a readable exposition of this approach, see Val'fish (= Walfisz) [1956].

For the elementary evaluation of $r_4(n)$, see Hardy and Wright [1960].

Section 7.6

For a discussion of elementary results concerning Waring's problem, see Hardy and Wright. For asymptotic estimates related to Waring's problem and various generalizations, especially involving cubic forms, see Davenport [1962].

8

Quadratic Equations and Quadratic Fields

8.1 LEGENDRE'S THEOREM

We commence the study of quadratic Diophantine equations with a beautiful theorem due to Legendre.

Theorem 8.1 *Suppose that $a, b, c \in \mathbf{Z}$ are nonzero, relatively prime in pairs, square-free, and not all of the same sign. Then the equation*

$$f(x, y, z) = ax^2 + by^2 + cz^2 = 0 \tag{1}$$

has a nontrivial solution in $\mathbf{Z}$ (i.e., with $\{x, y, z\} \neq \{0, 0, 0\}$) if and only if $-ab$, $-bc$, and $-ca$ are quadratic residues of $|c|$, $|a|$, and $|b|$, respectively, whenever the latter are $\neq 1$.

Remark. Every equation $Ax^2 + By^2 + Cz^2 = 0$, in which A, B, C are nonzero and not all of the same sign, can be transformed into one satisfying the hypotheses, so the theorem is more general than it may seem at first. For obviously any factor common to A, B, and C can be divided out, and any square factor in A, B, or C can be absorbed into x^2, y^2, or z^2. Suppose this has been done, and that still two coefficients have a common factor, say $(A, B) = d > 1$. Multiplying through by d gives coefficients d^2A', d^2B', dC', and the squares can again be absorbed into the variables. Repeat as necessary.

Proof. The necessity of the conditions is almost immediate. If a, b, c have the same sign, (1) clearly has only the trivial solution, even in $\mathbf{R}$. If there is a nontrivial solution, there is one in which $(x, y, z) = 1$. Then $(x, c) = (y, c) = 1$; for if $p \mid (x, c)$ then $p \mid by^2$, so $p \mid y$, so $p^2 \mid (ax^2 + by^2)$, so $p^2 \mid c$, contrary to hypothesis. Hence from

$$ax^2 + by^2 \equiv 0 \pmod{c}, \tag{2}$$

we obtain

$$(axy^{-1})^2 \equiv -ab \pmod{c},$$

so $-ab$ is a quadratic residue of c. (Of $|c|$, strictly speaking. We shall omit absolute-value bars in this proof when no confusion can ensue.) The proof of necessity is completed by exploiting the symmetric roles of a, b, and c.

Conversely, if $|c| > 1$ and $-ab$ is a quadratic residue of c, then the congruence (2) has a solution, say x_c, y_c, with $(c, x_c) = (c, y_c) = 1$. By Theorem 3.17, $at^2 + b$ factors in $\mathbf{Z}_c[t]$ as $a(t - x_c y_c^{-1})(t - t_0)$ for some t_0, so by putting $t = xy^{-1}$ we see that for certain coefficients whose exact values are of no interest,

$$ax^2 + by^2 = (a_1x + b_1y)(a_2x + b_2y) \qquad \text{in } \mathbf{Z}_c[x, y].$$

Hence there is a factorization (with $r_3 \equiv s_3 \equiv 0 \pmod{c}$, for example)

$$\begin{aligned} f(x, y, z) &\equiv (r_1x + r_2y + r_3z)(s_1x + s_2y + s_3z) \\ &\equiv R_c(x, y, z)S_c(x, y, z) \pmod{c}, \quad \text{say,} \end{aligned}$$

this being an identity in x, y, z. By symmetry, corresponding factorizations hold modulo $|a|$ and $|b|$, if these numbers are larger than 1:

$$\begin{aligned} f(x, y, z) &\equiv R_a(x, y, z)S_a(x, y, z) \pmod{a} \\ f(x, y, z) &\equiv R_b(x, y, z)S_b(x, y, z) \pmod{b}. \end{aligned}$$

By the Chinese remainder theorem, there are two homogeneous linear expressions $R(x, y, z)$ and $S(x, y, z)$, defined (mod abc), such that

$$R \equiv R_a, \quad S \equiv S_a \pmod{a}, \quad \text{etc.,}$$

and hence

$$f(x, y, z) \equiv R(x, y, z)S(x, y, z) \pmod{|abc|}.$$

This has been proved if all of $|a|$, $|b|$, $|c|$ are > 1; it obviously remains valid if at least one is > 1.

Now the theorem is clearly true if $|a| = |b| = |c| = 1$, so we exclude that case. Then since $|abc| > 1$ and abc is square-free, not all of

$$\lambda_1 = \sqrt{|bc|}, \quad \lambda_2 = \sqrt{|ca|}, \quad \lambda_3 = \sqrt{|ab|}$$

are integers; increase one which is not, very slightly, and apply Theorem 7.1: there are $x, y, z \in \mathbf{Z}$, not all zero, such that

$$R(x, y, z) \equiv 0 \pmod{|abc|}, \quad |x| < \lambda_1, \quad |y| < \lambda_2, \quad |z| < \lambda_3.$$

Since a, b, c are not of the same sign, we may suppose that $a > 0$, $b > 0$, and $c < 0$, in which case

$$f(x, y, z) < a|bc| + b|ca| + c \cdot 0 = 2|abc|,$$

and

$$f(x, y, z) > a \cdot 0 + b \cdot 0 + c|ab| = -|abc|.$$

Since also $f(x, y, z) \equiv 0 \pmod{|abc|}$, we see that

$$f(x, y, z) = 0 \text{ or } -abc.$$

In the first case, we are done. In the second case,

$$ax^2 + by^2 + c(z^2 + ab) = 0;$$

then

$$(ax^2 + by^2)(z^2 + ab) + c(z^2 + ab)^2 = 0,$$

or

$$a(xz + by)^2 + b(yz - ax)^2 + c(z^2 + ab)^2 = 0,$$

and $z^2 + ab$ is not 0 since it is positive. ∎

It is of some interest to note that the above proof yields an effective bound on the size of a solution of (1), when it is solvable: *If* (1) *has a solution, it has one with* $x, y, z > 0$ *and*

$$\max(x, y, z) < 2 \max(a^2, b^2, c^2). \tag{3}$$

(See Problem 2.) Better bounds are known, but what is significant is that it is a finite problem both to decide whether any instance of (1) is solvable, and to solve it when it is.

There are several reasons for being interested in equation (1). One is that with the help of Theorem 8.1 and certain elementary considerations, we can completely characterize all the *rational* solutions of the general binary (= two-variable) quadratic equation over **Q**,

$$ax^2 + bxy + cy^2 + dx + ey + f = 0. \tag{4}$$

Geometrically, this is a conic section—call it $\mathscr{C}$—and we are speaking about the rational points (= points with rational coordinates) on $\mathscr{C}$. If $b = 0$ and $ac = 0$, then (4) is linear in at least one of the variables (so $\mathscr{C}$ is a line or parabola) and the other variable can be given any rational value whatever. If $b = 0$ and $ac \neq 0$, a simple translation $x = x' + h$, $y = y' + k$ ($h, k \in \mathbf{Q}$) can easily be found which removes the linear terms in (4), resulting in an equation of the form

$$Ax^2 + By^2 + C = 0, \tag{5}$$

where $A, B, C \in \mathbf{Q}$. If $b \neq 0$ and $a = c = 0$, the substitution $x = x' - y'$, $y = x' + y'$ brings us back to the preceding case. If $b \neq 0$ and one of a and c is not 0, we may suppose (interchanging variables if necessary) that $a \neq 0$, and in that case the substitution $x = x' - by'/2a$, $y = y'$ again returns us to a case in which $b = 0$.

In short, the only interesting question is how to deal with (5), and this presents a problem only when $ABC \neq 0$. Multiplying through by a suitable factor, we may suppose $A, B, C \in \mathbf{Z}$. Finally, x and y are rational if and only if there are integers X, Y, and Z such that $x = X/Z$, $y = Y/Z$, and then (5) becomes (1) (in capital letters) upon multiplying through by Z^2. So as a result of what we have already learned, we can decide whether $\mathscr{C}$ has a rational point on it, and if so find one.

In fact, we can then do a great deal more: we can find *all* rational points on $\mathscr{C}$. For if $P_0 = (x_0, y_0)$ is a fixed rational point on $\mathscr{C}$, then the line $L(m)$ whose equation is

$$y - y_0 = m(x - x_0), \quad m \in \mathbf{Q}, \tag{6}$$

intersects $\mathscr{C}$ in P_0 and in a unique second point, say $P = P(m)$. Eliminating y between (4) and (6) gives a quadratic equation in x (with rational coefficients depending on m), of which $x - x_0$ must be a factor. Dividing out $x - x_0$ leaves a linear equation for the x-coordinate $x_1(m)$ of P, so $x_1(m)$ is rational, and by (6), P is a rational point. Thus every line $L(m)$ with rational slope m determines two rational points P_0 and $P(m)$ on $\mathscr{C}$, and two such points obviously determine such a line $L(m)$, and so we can express all rational points in terms of the parameter m. (Throughout, "∞", as the slope of the line $x = x_0$, or as a coordinate of $P(m)$ when $L(m)$ is parallel to an asymptote of $\mathscr{C}$, is to be regarded as rational.)

For an example, consider the circle $x^2 + y^2 = 1$ as an instance of (4). It passes through the point $(-1, 0)$, and the line $y = m(x + 1)$ intersects the circle at points where

$$x^2 - 1 + m^2(x + 1)^2 = 0,$$

so either $x = -1$ or

$$x - 1 + m^2(x + 1) = 0,$$

and then

$$x = \frac{1 - m^2}{1 + m^2}, \qquad y = \frac{2m}{1 + m^2}. \tag{7}$$

Putting $m = a/b$ with $(a, b) = 1$, we see that every solution in integers of the Pythagorean equation

$$x^2 + y^2 = z^2 \tag{8}$$

is of the form

$$x = c(a^2 - b^2), \quad y = 2abc, \quad z = c(a^2 + b^2). \tag{9}$$

Since $z + x$ and $z - x$ are integers, $2c \in \mathbf{Z}$. In a primitive solution—one for which $(x, y) = 1$—exactly one of x and y is even. Take it to be y. Then $c = \pm 1$ in (9), and a and b must be relatively prime and of opposite parity ($a \not\equiv b \pmod 2$). It is easily seen that these last conditions are also sufficient, so we have a very precise result:

Theorem 8.2 *Every primitive solution of* (8) *in integers such that* $2 \mid y$ *is given by* (9) *with* $c = \pm 1$, *for a suitably chosen and unique pair* $a, b \in \mathbf{Z}$ *such that* $(a, b) = 1$ *and* $a \not\equiv b \pmod 2$; *and every such pair gives a primitive solution. Every nonprimitive solution in which the 2-component of* y *is larger than that of* x *is of the form* (9) *with* $c \neq \pm 1$, $c \in \mathbf{Z}$, *and with* a, b *as before and still unique.*

(The uniqueness of a, b is evident from the geometry: changing the slope changes the point of intersection.)

As was remarked in Section 1.5, some rough equivalent of Theorem 8.2 seems to have been known for almost 4000 years.

Hermann Minkowski (1864–1909)

Minkowski entered mathematics dramatically at age 18 with a prize-winning 140-page paper on the number of representations of an integer as a sum of 5 squares. Throughout his short life he returned repeatedly to the theory of n-ary quadratic forms; many of his important results were obtained with the help of his highly original "geometry of numbers," still an important branch of number theory. With his generalized notion of distance in n-space, he paved the way for the theory of normed spaces and modern functional analysis. He also initiated the study of 4-dimensional "space–time," on which Einstein based his generalized relativity theory. Minkowski taught at Bonn and Zürich before Hilbert had a chair created for him at Göttingen.

Another reason for being interested in Theorem 8.1 is that it is, in disguised form, the special case $n = 3$ of the Hasse-Minkowski theorem stated at the end of Section 3.5, to the effect that a homogeneous quadratic equation $\sum_{i,j=1}^{n} a_{ij}x_i x_j = 0$, with the $a_{ij} \in \mathbf{Z}$, has a nontrivial solution in $\mathbf{Z}$ if and only if it has one in $\mathbf{R}$ and in every p-adic ring $\mathcal{O}_p$. The two theorems differ at the outset in that Theorem 8.1 is concerned only with the "diagonal" form $f(x, y, z) = ax^2 + by^2 + cz^2$, but the general ternary (3-variable) quadratic form can be reduced to this by a linear transformation with rational coefficients, in much the same way as we eliminated the xy-term in simplifying (4) to (5). What is left is to verify that the conditions imposed on $f(x, y, z)$ in Theorem 8.1, which concern the solvability of congruences (mod p) for p dividing abc, are sufficient to imply the solvability of $f(x, y, z) = 0$ (mod p^e) for all p and all e. For odd p this is straightforward: for $e = 1$ use the conditions of the theorem when $p \mid abc$, and use Problem 2 of Section 7.5 when $p \nmid abc$, to obtain a nonsingular solution of $f(x, y, z) \equiv 0 \pmod{p}$, and then apply Newton's method repeatedly to obtain a solution in $\mathcal{O}_p$. The prime $p = 2$ is more troublesome; one must first show that the conditions of Theorem 8.1 imply that $f(x, y, z) \equiv 0 \pmod{2^3}$ has a solution in which not all of x, y, z are even, and then apply Newton's method again to get a solution in $\mathcal{O}_2$. (Alternatively, the condition (mod 8) could be added at the end of Theorem 8.1; standing alone it obviously is necessary for the solvability of (1), but it happens to be implied by the other conditions, as a result of equation (31) of Chapter 3, so it may be included or not.) The interested student may pursue these matters in Problems 6, 7, and 8 following.

A third reason that number theorists are interested in rational points on conic sections is that there is a large family of algebraic curves (curves defined by polynomial equations $F(x, y) = 0$) of arbitrarily high degrees, with the property that the study of the rational points on any one of them can be reduced, through a sequence of rational substitutions with rational coefficients, to the corresponding problem for either a line or a conic section. These are the curves of genus 0. We cannot go into details, but it happens that with each algebraic curve is associated a nonnegative integer, its *genus*, and this quantity remains invariant when the curve is subjected to certain kinds of distortions called birational transformations. The curves of genus 0 are especially simple; they are the ones which allow parametrization by rational functions, such as is provided for the circle by equations (7). But these rational functions do not always have rational coefficients, even when the original equation has (see Problem 9 below), whereas the reduction to a line or conic section always yields rational coefficients. Also, it may not be easy to characterize the values of the parameter which correspond to rational values of x and y. It would lead too far afield to develop this subject in any detail, but here is an example.

Consider the curve

$$\mathscr{C}: \qquad 2(x^2 + y^2)^2 = x^2 - y^2.$$

This is called a lemniscate; its graph looks like an ∞-symbol, passing through the origin at angles of $\pm\pi/4$. It is easy to verify that it is of genus 0; for example, the circle $x^2 + y^2 = t(x - y)$ is tangent to $\mathscr{C}$ at the origin and otherwise intersects $\mathscr{C}$ in a unique finite point, whose coordinates are found to be

$$x = \frac{t(2t^2 + 1)}{4t^4 + 1}, \qquad y = \frac{t(2t^2 - 1)}{4t^4 + 1}.$$

This gives a parametrization of $\mathscr{C}$, and clearly rational values of t give rational points (x, y)—but are there other values of t with the same property? This is not obvious.

But consider instead the change of variables

$$u = \frac{x}{x^2 + y^2}, \qquad v = \frac{y}{x^2 + y^2}, \tag{10}$$

and restrict attention to points on $\mathscr{C}$. For them,

$$\mathscr{C}': \qquad u^2 - v^2 = 2,$$

and since $u^2 + v^2 = (x^2 + y^2)^{-1}$ we also have

$$x = \frac{u}{u^2 + v^2}, \qquad y = \frac{v}{u^2 + v^2}. \tag{11}$$

Transformation (10) is a birational transformation of $\mathscr{C}$, because it gives u and v as rational functions of x and y and on $\mathscr{C}$ the inverse transformation (11) gives x and y as rational functions of u and v. (Generally, the inverse of a rational function is not rational, for example, $w = z^2$.) And we see that, excluding the origin, equations (10) and (11) provide a 1–1 correspondence between the rational points on the quartic curve $\mathscr{C}$ and the conic section $\mathscr{C}'$. Thus we can find all the rational points on the lemniscate.

We have plucked equations (10) out of thin air. But given any algebraic curve with rational coefficients, there is a uniform finite procedure by which one can decide what its genus is and, when it is of genus 0, find a birational transformation with rational coefficients which carries it into an equation of degree ≤ 2. That procedure would yield transformation (10) (even though (10) happens to be a birational transformation of the entire xy-plane, not merely of $\mathscr{C}$.)

PROBLEMS

1. Decide whether the equation $18x^2 + 20y^2 - 35z^2 = 0$ has a nontrivial solution in integers.
2. Verify the italicized assertion involving inequality (3). [*Hint:* Use the inequality between the arithmetic and geometric means: if $u > 0$ and $v > 0$, then

   $$\sqrt{uv} \leq \tfrac{1}{2}(u + v).$$

   The truth of this becomes apparent upon squaring both sides.]
3. Show that if $c = \pm 1$, $(a, b) = 1$ and $a \not\equiv b \pmod 2$, then the x, y, z of (9) are relatively prime in pairs.
4. Why can c in (9) be chosen in $\mathbf{Z}$, as it is in Theorem 8.2?
5. Describe the positive primitive solutions of (8).
6. a) Use the method of Section 3.4 to prove the following result, which gives conditions under which a singular solution of a congruence can be developed to a p-adic solution of the corresponding equation:

   Let $f(x) \in \mathbf{Z}[x]$ and let p be prime. Suppose that for some $a \geq 0$ there is x_1 such that

   $$f(x_1) \equiv 0 \pmod{p^{2a+1}} \quad \text{and} \quad p^a \,\|\, f'(x_1).$$

   Then for every $n \geq 1$ there is an x_n such that

   $$f(x_n) \equiv 0 \pmod{p^{2a+n}}, \qquad x_{n+1} \equiv x_n \pmod{p^{a+n}}.$$

b) What is the connection between this result and the case $n = 2$ of Theorem 4.14?

c) Show that if abc is square-free, then the equation $ax^2 + by^2 + cz^2 = 0$ has a nontrivial 2-adic solution if the congruence $ax^2 + by^2 + cz^2 \equiv 0 \pmod 8$ has a solution in which not all of x, y, and z are even. [*Hint:* Show that there must be an odd term, say ax^2, and then hold y and z fixed.]

7. Reduce the ternary form $x^2 + 3y^2 + 5z^2 - 2xy + 60yz + 40xz$ to a diagonal form with rational coefficients, by "completing the square" on the terms involving x.

8. Use Theorem 5.10 to prove the Hasse-Minkowski theorem in the case $n = 2$. Then give a second proof without this device.

9. Prove that the circle $x^2 + y^2 = r$ $(r > 0)$ is of genus 0, but that for $r = 3$, for example, it has no parametrization $x = \phi(t)$, $y = \psi(t)$ in which ϕ and ψ are rational functions with rational coefficients.

10. Show that the equations $u = (x + 1)/y$, $v = 1/x$ define a birational transformation of the curve
$$(x + 2)y^2 = x^2(x + 1)^2,$$
and that the result is a conic section.

11. Show that the equations $u = x^2/(x^2 + y^2)$, $v = xy/(x^2 + y^2)$ define a birational transformation of the curve $y^2 = x^3 + x^2$ onto the circle $u^2 + v^2 - u = 0$, but that these equations are not invertible to give x and y as rational functions of independent variables u and v.

12. With reference to the parametric equations of the lemniscate, as obtained in the text, show that x and y are rational if and only if t is rational.

8.2 PELL'S EQUATION

The major objective now is to study the *integer* solutions of binary quadratic equations. This is a completely different kind of question from the problem of rational solutions, and we must forge new tools. As before, the general quadratic (4) can be reduced either to an equation which is linear in one of the variables and is thus uninteresting, or to the simplified form (5) of an ellipse or hyperbola. (The transformations effecting the simplification may have rational noninteger coefficients, and we are not asserting that there is a 1–1 correspondence between solutions in $\mathbf{Z}$ of (4) and of (5), but the difficulties are not serious.) It is convenient to modify (5) slightly: by multiplying through by A and changing names we can put the equation in the form called the **Pell equation,**

$$x^2 - dy^2 = k, \qquad d, k \in \mathbf{Z}, \quad d \text{ square-free.} \tag{12}$$

If $d < 0$ and $k > 0$, this is an ellipse and the problem is trivial: test every integer y such that $|y| \le \sqrt{-k/d}$ to see whether $dy^2 + k$ is a square. If $d < 0$ and $k < 0$, (12) has no real solutions, so no solutions in $\mathbf{Z}$. If $d = 1$, then $x \pm y$ are factors of k, and the problem is again finite. Only one interesting case is left:

for the remainder of the section, in (12) or otherwise, *d will be taken to be a fixed square-free integer larger than* 1. The left side of (12) then has a factorization involving the irrational number $\sqrt{d}$, namely $(x - y\sqrt{d})(x + y\sqrt{d})$, and the factors are elements of the domain $\mathbf{Z}[\sqrt{d}]$ consisting of all real numbers of the form $x + y\sqrt{d}$ with $x, y \in \mathbf{Z}$. We use Greek letters to designate elements of this ring. If $\alpha = x + y\sqrt{d}$, we call x and y the **components** of α, and we define the **conjugate** $\bar{\alpha} = x - y\sqrt{d}$ and the **norm** $\mathcal{N}\alpha = \alpha\bar{\alpha} = x^2 - dy^2$. Among the equations

$$\mathcal{N}\alpha\beta = (\alpha\beta)(\overline{\alpha\beta}) = \alpha\beta\bar{\alpha}\bar{\beta} = (\alpha\bar{\alpha})(\beta\bar{\beta}) = \mathcal{N}\alpha\mathcal{N}\beta,$$

only the second requires a moment's thought, and thus we see that the norm is completely multiplicative, just as it was in the case of the Gaussian integers. Solving (12) means finding all $\alpha \in \mathbf{Z}[\sqrt{d}]$ such that $\mathcal{N}\alpha = k$.

Since $\sqrt{d}$ is irrational, an equation $a_1 + b_1\sqrt{d} = a_2 + b_2\sqrt{d}$ $(a_i, b_i \in \mathbf{Z})$ implies that $a_1 = a_2$, $b_1 = b_2$, so each $\alpha \in \mathbf{Z}[\sqrt{d}]$ has unique components. Hence if $m > 1$ is in $\mathbf{Z}$, the elements of $\mathbf{Z}[\sqrt{d}]$ can be partitioned into m^2 residue classes (mod m), according to the pairs of residue classes in which their components lie. It is easy to deduce the analogues of Theorems 3.1 and 3.2: the relation $\alpha \equiv \beta$ (mod m) is an equivalence relation, and congruences can be added or multiplied together, and both sides of a congruence can be multiplied by a fixed factor from $\mathbf{Z}[\sqrt{d}]$.

We now return to equation (12), and restrict attention for the moment to positive values of x and y. If the equation has infinitely many solutions for some fixed k, then

$$x - y\sqrt{d} = \frac{k}{x + y\sqrt{d}} \to 0 \qquad \text{as } x, y \to \infty,$$

so the successive values of x/y must give better and better approximations to $\sqrt{d}$. In fact, if $x/y \approx \sqrt{d}$ then $x + y\sqrt{d} \approx 2y\sqrt{d}$, and if (12) holds, then

$$\left|\sqrt{d} - \frac{x}{y}\right| \approx \frac{|k|}{y^2(2\sqrt{d})},$$

and it is not at all obvious that such good approximations to $\sqrt{d}$ exist. We obtain the needed existence theorem with the help of the following general result.

Theorem 8.3 *If ξ is a real number and t is a positive integer, there are integers x and y such that*

$$|y\xi - x| \le \frac{1}{t+1}, \qquad 1 \le y \le t.$$

Proof. The $t + 1$ numbers

$$0\cdot\xi - [0\cdot\xi],\ 1\cdot\xi - [1\cdot\xi],\ \ldots,\ t\xi - [t\xi]$$

all lie in the interval $0 \le u < 1$. Call them, in increasing order of magnitude, $\xi_0, \xi_1, \ldots, \xi_t$. Mark the numbers $\xi_0, \ldots, \xi_t$ on a circle of unit circumference, that is, a unit interval on which 0 and 1 are identified. Then the $t + 1$ differences

$$\xi_1 - \xi_0, \quad \xi_2 - \xi_1, \quad \ldots, \quad \xi_t - \xi_{t-1}, \quad 1 - \xi_t$$

are the lengths of the arcs of the circle between successive ξ's, and so they are nonnegative and

$$(\xi_1 - 0) + (\xi_2 - \xi_1) + \cdots + (1 - \xi_t) = 1.$$

It follows that at least one of these $t + 1$ differences does not exceed $(t + 1)^{-1}$. But each difference is of the form

$$g_1\xi - g_2\xi - N,$$

where N is an integer, and we can take $y = |g_1 - g_2|$, $x = \pm N$. ■

Corollary *If ξ is real and irrational, the inequality*

$$|x - \xi y| < \frac{1}{y} \tag{13}$$

has infinitely many solutions.

Proof. According to Theorem 8.3, if ξ is irrational, the inequalities

$$0 < |x - \xi y| < \frac{1}{t}, \qquad 1 \le y \le t, \tag{14}$$

have a solution for each positive integer t. It is clear that each solution of (14) is also a solution of (13). Taking $t = 1$ in (14) gives a solution x_1, y_1 of (13). Then for suitable $t_1 > 1$,

$$|x_1 - \xi y_1| > \frac{1}{t_1},$$

and taking $t = t_1$ in (14) gives a solution x_2, y_2 of (13). Since

$$|x_2 - \xi y_2| < |x_1 - \xi y_1|,$$

the two solutions so far found are distinct. Now choose $t_2 > t_1$ so that

$$|x_2 - \xi y_2| > \frac{1}{t_2},$$

and for $t = t_2$ find x_3, y_3. Clearly this procedure can be continued indefinitely, yielding infinitely many solutions of (13). ■

Theorem 8.4 *There are infinitely many solutions of the equation* (12) *in positive integers x, y for some k with $|k| < 1 + 2\sqrt{d}$.*

Proof. If x, y is a solution of (13) with $\xi = \sqrt{d}$, then

$$|x + y\sqrt{d}| = |x - y\sqrt{d} + 2y\sqrt{d}| < \frac{1}{y} + 2y\sqrt{d} \le (1 + 2\sqrt{d})y,$$

and so

$$|x^2 - dy^2| < \frac{1}{y}(1 + 2\sqrt{d})y = 1 + 2\sqrt{d}.$$

Since there are infinitely many distinct pairs x, y available, but only finitely many integers numerically smaller than $1 + 2\sqrt{d}$, infinitely many of the numbers $x^2 - dy^2$ must have a common value, which is the theorem. ■

If we designate by $E(k)$ the set of $\alpha \in \mathbf{Z}[\sqrt{d}]$ such that $\mathcal{N}\alpha = k$, we have shown that $E(k)$ is infinite for some k; the bound on this k is of no importance here.

It is now a simple matter to analyze the equation

$$(15) \qquad x^2 - dy^2 = 1, \qquad d > 1 \text{ and square-free.}$$

Theorem 8.5 *Equation* (15) *has at least one solution with* $y \neq 0$.

Proof. According to Theorem 8.4, there is an integer k ($k > 0$) for which one of the two equations $\mathcal{N}\alpha = \pm k$ has infinitely many solutions α in $\mathbf{Z}[\sqrt{d}]$. Since there are only finitely many residue classes (mod k) in $\mathbf{Z}[\sqrt{d}]$, some residue class must contain at least three of these solutions (in fact, infinitely many!). Let us assume then that $\mathcal{N}\alpha_1 = \mathcal{N}\alpha_2 = \pm k$ and $\alpha_1 \equiv \alpha_2 \pmod{k}$, but that $\alpha_1 \neq \pm\alpha_2$. Then $\alpha_1\bar{\alpha}_2 \equiv \alpha_2\bar{\alpha}_2 \equiv 0 \pmod{k}$, so that $\beta = \alpha_1\bar{\alpha}_2/k$ is an element of $\mathbf{Z}[\sqrt{d}]$; that is, it has integral components. Since

$$\mathcal{N}\beta = \beta\bar{\beta} = \frac{\alpha_1\bar{\alpha}_2 \cdot \bar{\alpha}_1\alpha_2}{k^2} = \frac{\mathcal{N}\alpha_1\mathcal{N}\alpha_2}{k^2} = 1,$$

β yields a solution of (15). If the second component of β were 0, then $\mathcal{N}\beta = 1$ would imply that $\beta = \pm 1$, whence

$$\begin{aligned} \alpha_1\bar{\alpha}_2 &= \pm k = \pm\alpha_1\bar{\alpha}_1, \\ \bar{\alpha}_2 &= \pm\bar{\alpha}_1, \\ \alpha_2 &= \pm\alpha_1, \end{aligned}$$

contrary to hypothesis. ■

If the components of α are positive, then α, which is a real number as well as an element of $\mathbf{Z}[\sqrt{d}]$, is larger than 1. The four elements of $\mathbf{Z}[\sqrt{d}]$ which have the same components as α except for sign are α, $\bar{\alpha}$, $-\alpha$, and $-\bar{\alpha}$. If α has as components a positive solution of (15), then $\alpha\bar{\alpha} = 1$, and the four numbers just mentioned are α, $1/\alpha$, $-\alpha$, and $-1/\alpha$. Of these the first is larger than 1, the second is between 0 and 1, the third is smaller than -1, and the fourth lies between -1 and 0, so that the signs of x and y in (15) determine, and are determined by, the size of the associated α. To consider positive solutions of (15) is to consider elements $\alpha > 1$ of $\mathbf{Z}[\sqrt{d}]$.

These remarks also bring out the important fact that the set $E(1)$ of solutions $\alpha = x + y\sqrt{d}$ of (15) forms a multiplicative *group*, the group of units of norm 1 of $\mathbf{Z}[\sqrt{d}]$. For if $\alpha, \beta \in E(1)$, then $\mathcal{N}\alpha\beta = \mathcal{N}\alpha\mathcal{N}\beta = 1$, so $\alpha\beta \in E(1)$, and also $\alpha^{-1} = \bar{\alpha} \in E(1)$. Among all the elements of $E(1)$, there is a smallest one—call it δ—which is larger than 1. If $\delta = x_0 + y_0\sqrt{d}$, then each of x_0 and y_0 is minimal among all pairs $x, y > 0$ satisfying (15). The number δ is called the **fundamental solution** of (15), because of the following theorem.

Theorem 8.6 *If δ is the fundamental solution of* (15), *then every $\alpha \in E(1)$ has a unique representation in the form*

$$\alpha = \pm\delta^n, \qquad n \in \mathbf{Z}.$$

In other words, $E(1)$ is a multiplicative group on two generators, -1 and δ, and δ is of infinite order.

Proof. We have already seen that if $\alpha \in E(1)$, the four numbers α^n, $1/\alpha^n$, $-\alpha^n$, and $-1/\alpha^n$ give four solutions of (15) differing only in the signs of x and y, so we need only show that every $\alpha > 1$ such that $\mathcal{N}\alpha = 1$ is of the form $\alpha = \delta^n$ for suitable positive integer n.

Since $\alpha > 1$ and δ is minimal, we have $\alpha \geq \delta$. Hence there is a positive integer n such that $\delta^n \leq \alpha < \delta^{n+1}$. Now $\alpha/\delta^n = \alpha\bar{\delta}^n$ is in $\mathbf{Z}[\sqrt{d}]$, and $\mathcal{N}(\alpha/\delta^n) = 1$. In other words, the number $\alpha/\delta^n = \beta$ gives an integral solution of (15). From the definition of n it follows that $1 \leq \beta < \delta$, and by the definition of δ we cannot have $1 < \beta < \delta$. Hence $\beta = 1$, and $\alpha = \delta^n$. ∎

For reasons that will become clear in the next section, we wish also to consider the special cases $k = -1$ and $k = \pm 4$ in (12). As for the equation

$$(16) \qquad x^2 - dy^2 = -1, \qquad d > 1 \text{ and square-free},$$

it differs from (15) in one vital respect, in that it may have no solutions at all. For (16) implies, for example, that $x^2 \equiv -1 \pmod{d}$, which is not always possible. Moreover, the set $E(-1)$ of solutions of (16) does not form a group, since the product of two solutions is a solution of (15), not of (16).

Theorem 8.7 *If* (16) *is solvable, and if γ is the minimal solution with positive components, then $\delta = \gamma^2$ and all solutions of* (16) *are given by $\pm\gamma\delta^n$, $n \in \mathbf{Z}$. (In group-theoretic language, $E(\pm 1) = E(1) \cup E(-1)$ is a group with the two generators -1 and γ; $E(1)$ is a subgroup of index 2, and $E(-1)$ is the other coset.)*

Proof. $\mathcal{N}(\gamma^2) = (\mathcal{N}\gamma)^2 = 1$, so $1 < \delta \leq \gamma^2$ by the definition of δ. Since $1/\gamma = -\bar{\gamma}$, we have

$$\gamma^{-1} < -\delta\bar{\gamma} \leq \gamma,$$

and if we put $\beta = -\delta\bar{\gamma} \in \mathbf{Z}[\sqrt{d}]$, then $\mathcal{N}\beta = \mathcal{N}(-1)\mathcal{N}\delta\mathcal{N}\bar{\gamma} = -1$, so certainly $\beta \neq 1$. Hence either

$$\gamma^{-1} < \beta < 1 \quad \text{or} \quad 1 < \beta \leq \gamma,$$

and the first of these inequalities implies $1 < \beta^{-1} < \gamma$, which, like $1 < \beta < \gamma$, is impossible by the minimality of γ. Hence $\beta = \gamma$, and $\delta = \gamma^2$.

Now suppose that β' is any solution of (16); we can again restrict attention to the case $\beta > 1$. Then as in the proof of Theorem 8.6, there is a positive integer n such that

$$1 \leq \beta'\delta^{-n} < \delta = \gamma^2,$$

and dividing through by γ we obtain

$$\gamma^{-1} \leq \alpha < \gamma,$$

where $\alpha = \beta'\delta^{-n}\gamma^{-1}$ is a solution of (15). Since $1 < \gamma < \delta$, the last inequality implies that $\delta^{-1} < \alpha < \delta$, so that $\alpha = 1$ and $\beta' = \delta^n\gamma = \gamma^{2n+1}$. ■

Theorem 8.8 *Let $d > 1$ be square-free. If $\zeta \in \mathbf{Z}[\sqrt{d}]$ is the minimal solution, with positive components, of the equation*

$$x^2 - dy^2 = 4, \tag{17}$$

then every solution $\alpha \in \mathbf{Z}[\sqrt{d}]$ satisfies the equation

$$\frac{\alpha}{2} = \pm\left(\frac{\zeta}{2}\right)^n, \qquad n \in \mathbf{Z} \tag{18}$$

and every α of type (18) *lies in $\mathbf{Z}[\sqrt{d}]$ and satisfies* (17). *If the equation*

$$x^2 - dy^2 = -4, \qquad d > 1 \tag{19}$$

is solvable, and its minimal solution with positive components is η, then $\zeta/2 = (\eta/2)^2$, and a general solution of (19) *is given by*

$$\frac{\alpha}{2} = \pm\frac{\eta}{2}\left(\frac{\zeta}{2}\right)^n, \qquad n \in \mathbf{Z}.$$

Remark. If $\mathcal{N}\beta = 1$, then $\mathcal{N}(2\beta) = 4$, so (17) certainly has solutions. But not every solution need be of the form 2β with $\beta \in \mathbf{Z}[\sqrt{d}]$; for example, $3^2 - 5 \cdot 1^2 = 4$, but $(3 + \sqrt{5})/2 \notin \mathbf{Z}[\sqrt{d}]$.

Proof. Clearly, both or neither of x and y in (17) are odd, so each solution is $\equiv a(1 + \sqrt{d}) \pmod 2$, with $a = 0$ or 1. If odd solutions ($a = 1$) exist, then obviously d is odd. In any case, if $\alpha \equiv a(1 + \sqrt{d})$ and $\beta \equiv b(1 + \sqrt{d}) \pmod 2$ are two solutions of (17), with $ab = 0$ or 1, then

$$\alpha\beta \equiv ab(d + 1 + 2\sqrt{d}) \equiv 0 \pmod 2,$$

since if $ab = 1$, $d + 1$ is even. Hence for every pair of solutions of (17),

$$\frac{\alpha\beta}{2} = 2 \cdot \frac{\alpha}{2} \cdot \frac{\beta}{2} \in \mathbf{Z}[\sqrt{d}], \qquad \text{and} \qquad \mathcal{N}\frac{\alpha\beta}{2} = 4,$$

so in particular all the numbers (18) give integer solutions of (17). The proof that every solution of (17) is of type (18) follows the lines of the proof of Theorem 8.6, and the proof of the assertions regarding (19) is the same as that of Theorem 8.7. ∎

For general k, the set of solutions of (12) may not have such a simple structure as in the cases above. If $\mathcal{N}\alpha = k$, then obviously all the numbers $\beta\alpha$, where $\mathcal{N}\beta = 1$, also have norm k, and this gives us one (infinite) **class** of solutions. But there may be several classes; for example, the equation $x^2 - 2y^2 = 49$ has the solutions 7 and $9 + 4\sqrt{2}$, and neither can be obtained from the other by multiplying by an element of norm 1. The next theorem shows, for $k > 0$, that there are only finitely many classes, and that finding an element of each is a finite problem once the fundamental solution δ of (15) is known. There is a similar theorem for $k < 0$.

Theorem 8.9 *If the equation*

$$u^2 - dv^2 = k, \qquad d > 1 \textit{ and square-free}, \ k > 0,$$

is solvable, it has a solution with

$$\sqrt{k} < u \leq \sqrt{\Delta k}, \tag{20}$$

where the coefficient

$$\Delta = \frac{1}{2}\left(1 + \frac{\delta}{\delta - 1}x_1\right)$$

depends only on the fundamental solution $\delta = x_1 + y_1\sqrt{d}$ *of* (15). *If there are two or more classes of solutions of the equation, each contains an element for which* (20) *holds. There are only finitely many classes, for each* $k > 0$.

Proof. Since $u + v\sqrt{d}$ and $-u - v\sqrt{d}$ are always in the same class, we may assume that $u > 0$. We ask, given a solution $u_1 + v_1\sqrt{d}$ with $u_1 > 0$, when is it possible to find a smaller solution $u + v\sqrt{d}$, with $u > 0$, in the same class? That is, we want to find u and v such that

$$u + v\sqrt{d} = (x + y\sqrt{d})(u_1 + v_1\sqrt{d}), \quad 0 < u < u_1, \quad x^2 - dy^2 = 1.$$

Let $\delta = x_1 + y_1\sqrt{d}$ be the minimal positive solution of (15). If $v_1 > 0$, take $x + y\sqrt{d} = \delta^{-1} = x_1 - y_1\sqrt{d}$, while if $v_1 < 0$, take $x + y\sqrt{d} = \delta$; in either case, we get

$$\begin{aligned} u = u_1x_1 - y_1|v_1|d &= u_1\left(x_1 - y_1\sqrt{d}\,\frac{|v_1|\sqrt{d}}{u_1}\right) \\ &= u_1\left\{x_1 - y_1\sqrt{d} + y_1\sqrt{d}\left(1 - \sqrt{1 - \frac{k}{u_1^2}}\right)\right\}. \end{aligned}$$

Here $0 < k/u_1^2 < 1$. Since for $0 < t < 1$,

$$0 < 1 - \sqrt{1 - t} = \frac{t}{1 + \sqrt{1 - t}} < \frac{t}{2 - t},$$

we have

$$0 < u < u_1 \left(\delta^{-1} + \frac{y_1 \sqrt{d}\, k}{2u_1^2 - k} \right).$$

A little manipulation shows that the coefficient of u_1 is smaller than 1, so that $u < u_1$, if

$$u_1 > \sqrt{\frac{\delta' y_1 \sqrt{d} + 1}{2} k}, \qquad \text{where } \delta' = \frac{\delta}{\delta - 1}.$$

Since $y_1 \sqrt{d} = \sqrt{x_1^2 - 1} < x_1$, we have proved that as long as $u > \sqrt{\Delta k}$, a smaller solution can be found, so eventually (20) holds, since obviously $u^2 > k$ always.

Since there are only finitely many integers in the interval (20), there are only finitely many classes. It is easy to decide whether two solutions belong to the same class: they do if and only if their quotient belongs to $\mathbf{Z}[\sqrt{d}]$. ■

In the next section we shall consider the connection between Pell's equation and quadratic number fields. But it should be emphasized that the equation holds considerable interest in its own right. Some of the pre-European history of the equation was mentioned in Section 1.5. Fermat said he had proved the existence of a solution of $x^2 - dy^2 = 1$; the first published proof was that of Lagrange in 1766. John Pell had almost nothing to do with the equation (Euler got it wrong); probably the name sticks because it is unambiguous, since Pell also did little else of any great mathematical merit. Lagrange's proof, like that given here, was useless for actually computing the fundamental solution. But we already know where to look—namely among the good approximations to $\sqrt{d}$—and in the next chapter we shall develop an algorithm which leads straight to these approximations, and thence to the fundamental solution. A systematic procedure for finding the fundamental solution is important in practice, since the components of this solution can be quite unexpectedly large. For example, in the fundamental solution of $x^2 - 61y^2 = 1$, $x_1 = 1{,}766{,}319{,}049$.

The linear equation and Pell's equation play exceptional roles among binary Diophantine equations in being essentially the only such equations with infinitely many solutions in integers, or even in quasi-integers (a set of rational numbers with a fixed denominator). No curve of positive genus has infinitely many such solutions. If a curve of genus 0 has, it must have a parametrization

$$x = \frac{A(t)}{C^n(t)}, \quad y = \frac{B(t)}{C^n(t)}, \quad n \geq 0,$$

where A, B, C are polynomials with rational coefficients and $C(t)$ is either linear or is a quadratic polynomial assuming values of both signs, as does the "Pell polynomial" $t^2 - d$. In that case, the curve is reducible, by a very special birational transformation, either to a linear equation or to a Pell equation. (This exceedingly difficult theorem is due to C. L. Siegel [1929].) These last two cases are really very different from each other as regards their integral solutions: among the solutions x, y of $ax + by + c = 0$, those with $1 \leq x \leq X$ have cardinality $\gg X$ for large X, while the corresponding number for $x^2 - dy^2 = k$ is $\ll \log X$, since the solutions grow exponentially.

PROBLEMS

1. A quadratic equation $At^2 + Bt + C = 0$ $(A \neq 0)$ with coefficients in $\mathbf{Q}$ has solutions in $\mathbf{Q}$ if and only if $B^2 - 4AC$ is a square in $\mathbf{Q}$. Apply this principle twice, with $t = x$ and then $t = y$, to reduce the problem of finding all x, y such that

   (*) $$ax^2 + bxy + cy^2 + dx + ey + f = 0; \qquad 4apx \in \mathbf{Z}, \quad 2py \in \mathbf{Z},$$

   to the problem of finding all solutions of the Pell equation $w^2 - pz^2 = k$. It is supposed that $a, b, c, d, e, f \in \mathbf{Z}$, and that $a \neq 0$ and $p = b^2 - 4ac \neq 0$. (Here p is not necessarily prime.) We have put $k = q^2 - pr$, where $q = bd - 2ae$ and $r = d^2 - 4af$. (Here, in (*), is an illustration of the utility of the notion of a quasi-integral set of solutions, mentioned near the end of the section. Fixed denominators enter, in going from the general quadratic to the Pell equation.) Deduce that if (*) has one solution, it has infinitely many, provided that p is not a square and $k \neq 0$. (When $k = 0$, the polynomial in (*) factors.)
2. Find some integral solutions of the Diophantine equation
   $$x^2 + 6xy - 4y^2 - 4x - 12y - 19 = 0.$$
3. Find a general solution of the equation $x^2 - 2y^2 = 1$.
4. Describe all the integral solutions of the equation
   $$x^2 + 6xy + 7y^2 + 8x + 24y + 15 = 0.$$
5. Show that if $a_n = n(n\sqrt{2} - [n\sqrt{2}])$ and if $\varepsilon > 0$, then
   $$a_n < \frac{1 + \varepsilon}{2\sqrt{2}}$$
   for infinitely many n, while
   $$a_n > \frac{1 - \varepsilon}{2\sqrt{2}}$$
   for all sufficiently large n.
6. Show that a necessary condition that the equation $x^2 - dy^2 = -1$ be solvable is that d have a primitive representation as a sum of two squares.
7. Complete the proof of Theorem 8.8.

8. The statement obtained from Theorem 8.8 by replacing 2 and 4 by 7 and 49, respectively, is false, as is seen by considering the numerical example immediately following the proof of that theorem. Where would the analogous proof break down?
9. Show that if $k < 0$, Theorem 8.9 remains correct if the inequality (20) is replaced by

$$\sqrt{|k|} < u < \sqrt{\frac{\delta x_1 |k|}{2(\delta + 1)}}.$$

[*Hint:* Prove and use the fact that for $t > 0$, $\sqrt{1 + t} - 1 < t/2$.]
10. State and prove an analogue of Theorem 8.3, of the form, "if $x_1, \ldots, x_n \in \mathbf{R}$ and $t \in \mathbf{Z}^+$, then there are $q_1, \ldots, q_n, p \in \mathbf{Z}$, such that

$$|q_1 x_1 + \cdots + q_n x_n - p| < F(n, t), \qquad 1 \le \max_i(|q_i|) \le t."$$

Here F should be a strictly decreasing function of each of its arguments, with $\lim_{n \to \infty} F(n, t) = 0$ for each t. [*Hint:* Partition the unit interval into t subintervals, and use Dirichlet's principle.]
11. Find all solutions of $|\pi x - y| < 1/x$ in coprime integers x, y with $1 \le x \le 10$.
12. Show that if equation (17) has a solution in which either x or y is odd, then $d \equiv 5 \pmod 8$.
13. Suppose that δ, γ, ζ, and η are the minimal positive solutions described in Theorems 8.6, 8.7, and 8.8. Show that $\gamma = \eta/2$ if $\eta \equiv 0 \pmod 2$, and $\gamma = \eta^3/8$ if $\eta \equiv 1 + \sqrt{d} \pmod 2$.
14. In the notation of the text, suppose that $\delta = x_1 + y_1\sqrt{d}$. Show that γ exists if and only if both $\sqrt{(x_1 - 1)/2}$ and $y_1/2$ are integers, and the first divides the second.
15. If $x_n + y_n\sqrt{d} = \delta^n$ for $n \ge 0$, show that the sequences $\{x_n\}$ and $\{y_n\}$ can also be defined by the recursion relations

$$x_{n+1} = 2x_1 x_n - x_{n-1}$$
$$y_{n+1} = 2x_1 y_n - y_{n-1}$$

for $n \ge 1$. [*Hint:* $2x_n = \delta^n + \bar{\delta}^n = \delta(\delta^{n-1} + \delta^{-(n+1)})$.]
16. Show that if p is prime and $p \equiv 1 \pmod 4$, then the equation $x^2 - py^2 = 1-$ is solvable. [*Hint:* Let $x_1 + y_1\sqrt{d}$ be the fundamental solution of (15), so that

$$\frac{x_1 + 1}{2} \cdot \frac{x_1 - 1}{2} = p\left(\frac{y_1}{2}\right)^2.$$

Rule out the possibility that $(x_1 + 1)/2 = a^2$, $(x_1 - 1)/2 = pb^2$, for $a, b \in \mathbf{Z}$.]
17. For what solutions of (8) are x and y consecutive integers?

8.3 ALGEBRAIC NUMBER FIELDS AND ALGEBRAIC INTEGERS

The results of the preceding section, and perhaps of some others, will be better understood by looking more closely, and at the same time more broadly, at the domains $\mathbf{Z}[\sqrt{d}]$ and the fields associated with them.

The central fact from this new point of view is that $\sqrt{d}$ is an **algebraic number,** meaning that it is a root (in **C**) of a polynomial equation

$$p(x) = x^n + a_1x^{n-1} + \cdots + a_n = 0, \qquad a_1, \ldots, a_n \in \mathbf{Q}. \tag{21}$$

For $\sqrt{d}$ the equation is quadratic, namely $x^2 - d = 0$, and this case is especially simple in a number of ways, but for the moment let us consider the general case. (There is no loss in generality in supposing that $p(x)$ is **monic** (i.e., has leading coefficient 1), since the coefficients are rational numbers, not integers.) Suppose that α is a (real or) complex number satisfying equation (21); Gauss proved that there always is such a number, and the quadratic formula furnishes two of them when $n = 2$. By Theorem 2.8, $p(x)$ can be factored into polynomials irreducible over **Q**, and one of the factors must vanish when $x = \alpha$; for simplicity, let us then change notation and call this factor $p(x)$. If $f(x) \in \mathbf{Q}[x]$ is any other polynomial such that $f(\alpha) = 0$, the division theorem gives $f(x) = p(x)q(x) + r(x)$, where $r(x) = 0$ (the zero polynomial) or $\partial r < \partial p$; clearly $r(\alpha) = 0$. If $r(x) \neq 0$, then $(p(x), r(x)) = 1$, since $p(x)$ has no nonconstant divisors and $\partial r < \partial p$. But then there are $a(x), b(x) \in \mathbf{Q}[x]$ such that $a(x)p(x) + b(x)r(x) = 1$, contradicting $p(\alpha) = r(\alpha) = 0$. Hence $r(x) = 0$, and we have proved that α *is a zero of a unique monic irreducible polynomial* $p(x) \in \mathbf{Q}[x]$, *and* $p(x)$ *divides every polynomial* $f(x) \in \mathbf{Q}[x]$ *such that* $f(\alpha) = 0$. Then α is said to be an algebraic number of **degree** $\partial\alpha = n = \partial p$, and $p(x)$ is called its **defining polynomial**. (It is not at all obvious that there are any complex numbers which are not algebraic of some degree or other; this will be proved in the next chapter.)

The collection of all polynomials in α with coefficients in **Q** rather obviously forms a Ring, $\mathbf{Q}[\alpha]$. The elements of $\mathbf{Q}[\alpha]$ are complex numbers, and distinct polynomials $f(x)$ and $g(x)$ may give the same number, $f(\alpha) = g(\alpha)$; the result above shows that this happens exactly when $p(x) \mid (f(x) - g(x))$. The collection of all rational functions $f(\alpha)/g(\alpha)$ in α, where f, g are again polynomials over **Q** and $g(\alpha) \neq 0$, forms an **algebraic number field** $\mathbf{Q}(\alpha)$, the result of **adjoining** α to **Q**. Here is a pretty fact:

Theorem 8.10 $\mathbf{Q}[\alpha] = \mathbf{Q}(\alpha)$. *In fact, both of these objects coincide with the vector space over* **Q** *with basis* $1, \alpha, \ldots, \alpha^{n-1}$, *where* $n = \partial\alpha$.

Proof. To see that $\mathbf{Q}[\alpha] = \mathbf{Q}(\alpha)$, it suffices to know that $1/g(\alpha) \in \mathbf{Q}[\alpha]$ if $g(\alpha) \in \mathbf{Q}[\alpha]$ and $g(\alpha) \neq 0$. The last condition means that $p(x) \nmid g(x)$, where $p(x)$ is still the defining polynomial of α. Hence $(p(x), g(x)) = 1$, so $a(x)p(x) + b(x)g(x) = 1$ for some $a(x), b(x) \in \mathbf{Q}[x]$. Then $b(\alpha)g(\alpha) = 1$, and $1/g(\alpha) = b(\alpha)$.

So now we can restrict attention to $\mathbf{Q}[\alpha]$. By (21),

$$\alpha^n = -a_1\alpha^{n-1} - \cdots - a_{n-1}\alpha - a_n, \tag{22}$$

so α^n is a linear combination over **Q** of $1, \alpha, \ldots, \alpha^{n-1}$. Multiplying through in (22) by α and replacing α^n in the new first term on the right by its value from the present (22), we obtain α^{n+1} as a linear combination over **Q** of $1, \alpha, \ldots, \alpha^{n-1}$. Continuing, we see that every polynomial over $\mathbf{Q}$ in α can be reduced to (is equal to) a polynomial over $\mathbf{Q}$ in α of degree less than n. ∎

Another way to effect the reduction to a polynomial of degree less than n is to use the division theorem: if $f(x) = p(x)q(x) + r(x)$, then $f(\alpha) = r(\alpha)$, and either $r(x) = 0$ or $\partial r < \partial p$. In fact, this suggests another way of looking at the whole matter. For fixed irreducible $p(x) \in \mathbf{Q}[x]$ we have the notion of congruence $(\text{mod } p(x))$:

$$f(x) \equiv g(x) \ (\text{mod } p(x)) \quad \text{means} \quad p(x) \mid (f(x) - g(x)).$$

This relation partitions the ring $\mathbf{Q}[x]$ into residue classes $(\text{mod } p(x))$, and each residue class has, as just mentioned, a representative of degree less than n. The set of residue classes $(\text{mod } p(x))$ would form a Ring even if $p(x)$ were not irreducible; since it is, this Ring is a field, for exactly the same reasons that $\mathbf{Z}_p$ is a field. In fact, it is isomorphic to the field $\mathbf{Q}(\alpha)$, if α is a zero of $p(x)$, under the correspondence $\{f(x) \ (\text{mod } p(x))\} \leftrightarrow f(\alpha)$, where the symbol on the left indicates a residue class. If this is not already clear, it will become so by looking at a special case, say the polynomials $(\text{mod } x^2 + 1)$ (with a representative of the form $a + bx$ in each residue class, and $x^2 + 1 \equiv 0$) on the one hand, and the field $\mathbf{Q}(i)$ (with elements $a + bi$, and $i^2 + 1 = 0$) on the other. Using this approach, we could have defined a field isomorphic to $\mathbf{Q}(\alpha)$ without knowing or caring whether polynomials have zeros in $\mathbf{C}$.

A basic theorem, which we shall not pause to prove, is that *every finite extension of* $\mathbf{Q}$ (i.e., every field which is a finite-dimensional vector space over $\mathbf{Q}$) *is an algebraic number field* $\mathbf{Q}(\alpha)$, for suitable α. In particular, every "multiple" algebraic extension $K = \mathbf{Q}(\beta, \gamma, \ldots, \delta)$, resulting from the adjunction of finitely many algebraic numbers to $\mathbf{Q}$, has a "primitive element" α such that $K = \mathbf{Q}(\alpha)$.

The dimension of $F = \mathbf{Q}(\alpha)$ as a vector space over $\mathbf{Q}$ is equal to n, the degree of α. It is also called the **degree** of the field F over $\mathbf{Q}$, and we write $[F:\mathbf{Q}] = n$. The defining polynomial of α has n zeros (including α); they are called the **conjugates** of α, and we denote them by $\alpha_1 = \alpha, \alpha_2, \ldots, \alpha_n$. The product $\mathcal{N}\alpha = \alpha_1 \cdots \alpha_n$ called the **norm** of α, is a rational number, since $\mathcal{N}\alpha = (-1)^n a_n$, with a_n as in (21).

Since the ring $\mathbf{Q}[\alpha]$ has turned out to be the whole field $\mathbf{Q}(\alpha)$, we are at present without a suitable definition of the set of integers in the field. Presumably this should be a domain, and one which contains $\mathbf{Z}$. It is tempting to define the set of integers to be the ring $\mathbf{Z}^{(n)}[\alpha]$ of all polynomials of degree less than n in α, with coefficients in $\mathbf{Z}$, but unfortunately this is not in fact always a ring. For example, if $\alpha = 1/\sqrt{2}$, then $\alpha^3 = \frac{1}{2}\alpha \notin \mathbf{Z}^{(2)}[\alpha]$. It may be objected that $1/\sqrt{2}$ is obviously not an integer itself, but to prove this one needs to know first what the integers are, and we have come full circle. The key lies in defining *an integer*, rather than the set of all integers. An algebraic number α is said to be an **algebraic integer**, or simply an integer, if its defining polynomial (21) has coefficients $a_1, \ldots, a_n \in \mathbf{Z}$. (For clarity, we will refer to the elements of $\mathbf{Z}$ as rational integers.) We designate the **Ring of integers in** F by $\mathbf{Z}(F)$. When α is an integer, the relation (22) will obviously give every positive power of α as a linear combination of $1, \alpha, \ldots, \alpha^{n-1}$ with coefficients in $\mathbf{Z}$, so that then $\mathbf{Z}^{(n)}[\alpha]$ is a ring, and in fact is the ring $\mathbf{Z}[\alpha]$. But there is no reason that $\mathbf{Z}[\alpha]$ should coincide with the set of all integers in $\mathbf{Q}(\alpha)$, and it does not, in general. For a trivial example, note that $\sqrt{2} \notin \mathbf{Z}[2\sqrt{2}]$, although $\sqrt{2}$ is an integer in $\mathbf{Q}(2\sqrt{2})$. For a less obvious example, note that $\beta = (1 + \sqrt{5})/2$ is an integer, since it is a zero of $x^2 - x - 1$, but obviously $\beta \notin \mathbf{Z}[\sqrt{5}]$.

There is no way to prove that a definition is correct, of course, but the one just given of algebraic integers has two features in its favor. First, it is intrinsic: an algebraic number is never an integer in one algebraic number field and not in another. (In particular, no rational number $\notin \mathbf{Z}$ is an algebraic integer.) Second, it is as broadly encompassing as it can be, in the following sense: *the set $\mathbf{Z}(F)$ of all integers in a given algebraic number field F of degree n is a Ring (in particular it is an additive group) and is at the same time a finitely generated additive group, and every subset of F with these two properties is contained in $\mathbf{Z}(F)$.* Saying that G is a finitely generated additive group means that there are $\omega_1, \ldots, \omega_r \in G$ such that for every $\beta \in G$,

$$\beta = b_1\omega_1 + \cdots + b_r\omega_r \quad \text{for some } b_1, \ldots, b_r \in \mathbf{Z}; \tag{23}$$

a shorter name for the same concept is **module**. It is the case that *in an algebraic number field, every module has a basis*—a set of generators $\omega_1, \ldots, \omega_r$ which are linearly independent over $\mathbf{Z}$, so that if $\beta = 0$ in (23) then $b_1 = \cdots = b_r = 0$. The number r is then called the **rank** of the module. Thus a module of rank r is very similar to a vector space of dimension r, except that the coefficients (scalars) come from $\mathbf{Z}$ instead of from a field. As might be expected from its maximality property, *$\mathbf{Z}(F)$ is always of rank n, if $[F:\mathbf{Q}] = n$.* We shall not prove the italicized statements in this paragraph for $n > 2$, as they are somewhat beyond the scope of this book, but some of the proofs are sketched in the problems following.

With the integers in F defined, the stage is set for doing number theory in F. The **units** of $\mathbf{Z}(F)$ are the elements ε such that $\varepsilon^{-1} \in \mathbf{Z}(F)$. If the defining polynomial of ε is $x^m + e_1x^{m-1} + \cdots + e_m$, then that of ε^{-1} is

$$e_m^{-1}x^m f(x^{-1}) = x^m + \frac{e_{m-1}}{e_m}x^{m-1} + \cdots + \frac{1}{e_m},$$

so *ε is a unit if and only if $e_m = \pm 1$, or $\mathcal{N}\varepsilon = \pm 1$.* The units of $\mathbf{Z}(F)$ form a multiplicative group, $U(F)$; we shall have more to say about this group in the next section. Note that an algebraic integer is a unit or not, independently of which field it is encountered in, since units can be defined as integers whose reciprocals are also integers.

Two integers $\beta, \gamma \in \mathbf{Z}(F)$ are called **associates** if $\beta\gamma^{-1} \in U(F)$, and γ is said to be a **factor** of β if $\beta\gamma^{-1} \in \mathbf{Z}(F)$. β is **prime** if it has no factors other than units and associates. If $\gamma \mid \beta$, then either γ is an associate of β or $|\mathcal{N}\gamma| < |\mathcal{N}\beta|$, and from this it is easily deduced that every integer has a finite factorization as a product of

Ernst Kummer (1810–1893)

Kummer, along with the younger mathematicians Kronecker and Dedekind, initiated what might be called the algebraic theory of numbers, and their work, with that of Galois, forms the foundation of modern abstract algebra. Kummer spent much of his life studying the arithmetic of number fields generated by roots of unity, with applications to higher reciprocity laws and Fermat's equation. He obtained the first n-ic reciprocity law for general n, and he showed that Fermat's equation is unsolvable for a certain class of exponents which is probably infinite, although this has never been proved. (In particular, he settled the problem for all exponents less than 100.) One of the most abstract mathematicians of his day, he also did important work in geometry, atmospheric refraction, and ballistics.

primes. But now a new phenomenon arises, in that this factorization is not necessarily unique. (We give an example below, after we get to know the integers of certain fields more explicitly.) Not only is this aesthetically disappointing, at least at first, but it accounts for various insurmountable difficulties in rational arithmetic. For example, both Cauchy and Lamé gave fallacious proofs of Fermat's assertion concerning $x^n + y^n = z^n$, in which the only error lay in supposing that factorization is unique in the ring of integers of $\mathbf{Q}(\zeta)$, where ζ is a primitive nth root of unity. Kummer stayed with the problem and found a way to restore unique factorization, but new complications arose to stand in the way of a complete proof. Unfortunately, all this also lies beyond our present scope, and it is time to descend to the more elementary level of the algebraic numbers of degree 2. A final comment: uniqueness of factorization could not usually be restored in smaller rings than $\mathbf{Z}(F)$, and this further justifies our definition of integer.

PROBLEMS

Note. Some of these problems require elementary linear algebra, as indicated in the hints.

1. Prove that the ring of residue classes (mod $p(x)$) of polynomials in $\mathbf{Q}[x]$ is a field, if $p(x)$ is irreducible in $\mathbf{Q}[x]$.

2. Let $F = \mathbf{Q}(\alpha)$ be of degree n. Show that if $\mathbf{Z}(F)$ is indeed a module, it must be of rank n. [*Hint:* Show first that $m\alpha \in \mathbf{Z}(F)$ for suitable nonzero $m \in \mathbf{Z}$. Then show that the rank is $\geq n$ and also $\leq n$.]
3. Show that if F is a finite extension of $\mathbf{Q}$ and K is a finite extension of F, then K is a finite extension of $\mathbf{Q}$, and in fact $[K:F][F:\mathbf{Q}] = [K:\mathbf{Q}]$. [*Hint:* Suppose that F, as a vector space over $\mathbf{Q}$, has basis $\alpha_1, \ldots, \alpha_m$, and that K has a basis $\beta_1, \ldots, \beta_r$ over F. Consider the set of all products $\alpha_i\beta_j$.]
4. Let a and b be distinct square-free rational integers $\neq 1$. Show that the field $\mathbf{Q}(\sqrt{a} + \sqrt{b})$ contains $\sqrt{ab}$, and hence $\sqrt{a}$ and $\sqrt{b}$, and that in fact $\mathbf{Q}(\sqrt{a} + \sqrt{b}) = \mathbf{Q}(\sqrt{a}, \sqrt{b})$. Conclude, with the help of the assertion in Problem 3, that
$$[\mathbf{Q}(\sqrt{a} + \sqrt{b}):\mathbf{Q}] = 4.$$
Express $\sqrt{2}$ as a polynomial over $\mathbf{Q}$ in $\sqrt{2} + \sqrt{3}$.
5. a) Let F be an algebraic number field of degree n. Prove that every Ring R in F which is a module is contained in $\mathbf{Z}(F)$. [*Hint:* Suppose $\omega_1, \ldots, \omega_m$ is a basis for R, and suppose $\alpha \in R$. Then $\alpha\omega_i \in R$, $1 \leq i \leq m$, and these conditions give rise to a homogeneous $m \times m$ system of linear equations for the ω_i, so this system must have a singular coefficient matrix.] Compare Problem 2 of the next section.
   b) Show that $\mathbf{Z}(F)$ itself is a Ring. [*Hint:* Show that if $\alpha, \beta \in \mathbf{Z}(F)$, $\partial\alpha = m$, $\partial\beta = n$, then the set $\mathbf{Z}^{(m,n)}[\alpha, \beta]$ of polynomials over $\mathbf{Z}$, of degree $<m$ in α and of degree $<n$ in β, together with the zero polynomial, is a Ring, and apply (a).]

8.4 ARITHMETIC IN QUADRATIC FIELDS

By the quadratic formula from high school, every quadratic extension of $\mathbf{Q}$ is of the form $\mathbf{Q}(\sqrt{d})$, where initially $d \gtrless 0$ is rational, but can be taken to be a square-free integer different from 1 by observing that $\sqrt{ab^2/c} = \pm b\sqrt{ac}/c$, and hence that a square factor or denominator can be removed from the radical. Thus 1 and $\sqrt{d}$ always form a basis for $F = \mathbf{Q}(\sqrt{d})$, as a vector space over $\mathbf{Q}$; F is the set of numbers $a + b\sqrt{d}$, with $a, b \in \mathbf{Q}$, and each such representation is unique.

Theorem 8.11 *Let $d \neq 1$ be a square-free rational integer, and let $F = \mathbf{Q}(\sqrt{d})$. Then 1, $\sqrt{d}$ always form a basis for F, and they also form a basis for the module $\mathbf{Z}(F)$ if $d \equiv 2$ or $3 \pmod 4$, so that $\mathbf{Z}(F) = \mathbf{Z}[\sqrt{d}]$. If $d \equiv 1 \pmod 4$, then 1, $(1 + \sqrt{d})/2$ form a basis for $\mathbf{Z}(F)$, so that then $\mathbf{Z}(F)$ consists of the numbers $(a + b\sqrt{d})/2$, where $a, b \in \mathbf{Z}$ and $a \equiv b \pmod 2$.*

Proof. Suppose $\alpha = a + b\sqrt{d} \in \mathbf{Z}(F)$, where $a, b \in \mathbf{Q}$. If $b = 0$ then clearly $a \in \mathbf{Z}$. If $b \neq 0$, then the other zero of the defining polynomial of α is $\bar{\alpha} = a - b\sqrt{d}$, and evidently $\bar{\alpha} \in \mathbf{Z}(F)$. Since $\mathbf{Z}(F)$ is a ring, and since the only rational numbers in $\mathbf{Z}(F)$ are rational integers,

$$\alpha + \bar{\alpha} = 2a \in \mathbf{Z}(F), \text{ so } 2a \in \mathbf{Z}; \tag{24}$$

$$\alpha\bar{\alpha} = a^2 - db^2 \in \mathbf{Z}(F), \text{ so } a^2 - db^2 \in \mathbf{Z}; \tag{25}$$

$$(2a^2) - 4(a^2 - db^2) \in \mathbf{Z}(F), \text{ so } 4db^2 \in \mathbf{Z}. \tag{26}$$

If $a \in \mathbf{Z}$, then by (25), $b \in \mathbf{Z}$. In any case, since d is square-free, (26) implies that $2b \in \mathbf{Z}$. If $2a \equiv 1 \pmod 2$, then (25) implies that

$$0 \equiv 4(a^2 - db^2) \equiv 1 - d(2b)^2 \pmod 4,$$

so $2b \equiv 1 \pmod 2$ and $d \equiv 1 \pmod 4$. Conversely, if $2a$ and $2b$ are integers, and if $2a \equiv 2b \pmod 2$ and $d \equiv 1 \pmod 4$, then $4\alpha\bar{\alpha} \equiv 4a^2(1 - d) \equiv 0 \pmod 4$, so $\alpha\bar{\alpha} \in \mathbf{Z}$; hence

$$(x - \alpha)(x - \bar{\alpha}) = x^2 - (\alpha + \bar{\alpha})x + \alpha\bar{\alpha} \in \mathbf{Z}[x]$$

and $\alpha \in \mathbf{Z}(F)$. The theorem summarizes these facts in a slightly different way. ∎

For simplicity we put $\omega = \sqrt{d}$ when $d \equiv 2$ or $3 \pmod 4$, and $\omega = (1 + \sqrt{d})/2$ when $d \equiv 1 \pmod 4$, so that $1, \omega$ always form a basis for $\mathbf{Z}(F)$.

Thus we see that the quadratic domains $\mathbf{Z}[\sqrt{d}]$ with which we have heretofore been concerned are, from the present point of view, not always the "right" algebraic objects to look at. The domain $\mathbf{Z}[\sqrt{5}]$, for example, is a module of rank 2, but it is not the maximal such domain in $\mathbf{Q}(\sqrt{5})$. On the other hand, the Gaussian integers really do constitute the full ring of algebraic integers in $\mathbf{Q}(i)$.

If $d \equiv 2$ or $3 \pmod 4$, then the units of $\mathbf{Z}(F)$ are the elements $\alpha = a + b\omega$ such that

$$\alpha\bar{\alpha} = a^2 - db^2 = \pm 1.$$

If $d \equiv 1 \pmod 4$, they are the elements $\alpha = a + b\omega$ such that

$$\alpha\bar{\alpha} = \left(a + b\,\frac{1 + \sqrt{d}}{2}\right)\left(a + b\,\frac{1 - \sqrt{d}}{2}\right) = \frac{(2a - b)^2 - db^2}{4} = \pm 1,$$

so they are to be found among the solutions of the Pell equations

$$x^2 - dy^2 = \pm 4. \tag{27}$$

In fact, *every* solution x, y of (27) yields a unit $(x + y\sqrt{d})/2$, since $d \equiv 1 \pmod 4$ obviously implies $x \equiv y \pmod 2$ in (27). Invoking the information gained earlier about the structure of the set of solutions of these Pell equations, we obtain the first sentence in the following result; the second sentence is left to the reader.

Theorem 8.12 *The group $U(F)$ of units in a real quadratic domain $\mathbf{Z}(F)$ is a group having two generators, -1 and a fundamental solution of one of the Pell equations $x^2 - dy^2 = \pm 1$ or ± 4. The group of units of a nonreal quadratic domain is a finite cyclic group, generated by a root of unity.*

This is a special case of the following beautiful and important theorem due to Dirichlet: *The group of units in the ring of integers of any algebraic number field* $\mathbf{Q}(\alpha)$ *is finitely generated; there is one generator* ζ*—a root of unity—of finite order, and there are r generators* $\varepsilon_1, \ldots, \varepsilon_r$ *of infinite order, such that each unit* ε *has a unique representation in the form*

$$\varepsilon = \zeta^k \varepsilon_1^{k_1} \cdots \varepsilon_r^{k_r}, \qquad 1 \le k \le m, \quad k_1 \in \mathbf{Z}, \ldots, k_r \in \mathbf{Z}.$$

Here $r = r_1 + r_2 - 1$, where r_1 of the conjugates of α are real and $2r_2$ of them are nonreal complex numbers. (The latter automatically occur in complex-conjugate pairs, $\beta \pm \gamma i$.) Theorem 8.12 is Dirichlet's theorem for $n = 2$, in which case either $r_1 = 2$, $r_2 = 0$, and $r = 1$, or else $r_1 = 0$, $r_2 = 1$, $r = 0$. Once again we must omit the proof in the general case and return to the field $F = \mathbf{Q}(\sqrt{d})$.

G. Lejeune Dirichlet (1805–1859)

Born and raised near Cologne, Dirichlet attended the University of Paris at the time when Laplace, Legendre, Fourier, and Cauchy made it the world center of mathematics. During this period he mastered Gauss's *Disquisitiones*, the first person to do so even though 20 years had elapsed since it had appeared. His *Vorlesungen über Zahlentheorie* later made Gauss's work accessible to others. He was professor at Berlin for almost 30 years, succeeding Gauss at Göttingen only four years before his own death. His name is well known in complex analysis and the theory of Fourier series, and it has now appeared in several quite different contexts in the present book, all fundamental to the subject.

If β and γ are associates in $\mathbf{Z}(F)$, then $\beta = \gamma\varepsilon$, where $\varepsilon \in U(F)$, so $\mathcal{N}\varepsilon = \pm 1$ and $\mathcal{N}\beta = \pm\mathcal{N}\gamma$. Thus $\mathbf{Z}(F)$ is partitioned into equivalence classes of associates, and these classes coincide with the classes occurring in Theorem 8.9 when $d \equiv 2$ or 3 (mod 4), and the fundamental unit has norm $+1$. There is a corresponding theory when $d \equiv 1 \pmod 4$, of course.

As regards uniqueness of factorization into primes, the factorizations

$$6 = 2 \cdot 3 = (4 + \sqrt{10})(4 - \sqrt{10}) \tag{28}$$

will demonstrate that it fails in $\mathbf{Z}(F)$, where $F = \mathbf{Q}(\sqrt{10})$, if it is shown that 2, 3, $4 + \sqrt{10}$ and $4 - \sqrt{10}$ are prime in $\mathbf{Z}(F)$ and that neither of $4 \pm \sqrt{10}$ is an associate of 2.

First, note that for no $\alpha \in \mathbf{Z}(F)$ is $\mathcal{N}\alpha = \pm 2$ or ± 3, since the equations $x^2 - 10y^2 = \pm 2, \pm 3$ are unsolvable even when weakened to congruences (mod 10). Now in F, $2 = \alpha\beta$ implies $4 = \mathcal{N}\alpha\mathcal{N}\beta$, so that for $\alpha, \beta \in \mathbf{Z}(F)$, $|\mathcal{N}\alpha| = 1$ or $|\mathcal{N}\beta| = 1$, and the corresponding statement holds when the pair 2, 4 is replaced by 3, 9, or by $4 + \sqrt{10}$, 6, or by $4 - \sqrt{10}$, 6. Thus the four factors in (28) are prime. Neither of $4 \pm \sqrt{10}$ is an associate of 2 since $|\mathcal{N}(4 \pm \sqrt{10})| \neq |\mathcal{N}2|$.

Not even for quadratic fields (not to mention fields of higher degree!) is the entire story known as regards unique factorization. It was conjectured by Gauss that the only nonreal fields $\mathbf{Q}(\sqrt{-d})$ whose integers have unique factorization are those for $d = 1, 2, 3, 7, 11, 19, 43, 67, 163$; this was proved only in the latter half of the present century. By contrast, real quadratic fields with unique factorization continue to occur with considerable frequency, as far as computations have been carried, and Gauss conjectured that there are infinitely many of them. This has not been proved. It is known, however, that there are only finitely many quadratic fields F for which $\mathbf{Z}(F)$ is a Euclidean domain, at least when the function $s(\alpha)$ in Section 2.2 is taken to be $|\mathcal{N}\alpha|$; they are those for $d = -11, -7, -3, -2, -1, 2, 3, 5, 6, 7, 11, 13, 17, 19, 21, 29, 33, 37, 41, 57$, and 73. Of course, the existence of a Euclidean algorithm is in no way necessary for unique factorization.

PROBLEMS

1. Let $\omega = \sqrt{d}$ or $(1 + \sqrt{d})/2$, so that $1, \omega$ form a basis for the module $\mathbf{Z}(\mathbf{Q}(\sqrt{d}))$. Show that every submodule M which properly contains $\mathbf{Z}$ has a basis $1, m\omega$ for some nonzero $m \in \mathbf{Z}$. What is m when $M = \mathbf{Z}[\sqrt{d}]$?
2. For F quadratic, prove that every Ring R in F which is a module is contained in $\mathbf{Z}(F)$, perhaps in the following way:
   a) R is contained in a Ring R' of the same sort and with at most two generators, for the others are linearly dependent (over $\mathbf{Q}$) on one or two of them and can be eliminated;
   b) one generator (of any such Ring) is 1;
   c) if there is a second, its square must be in the Ring.

   Compare Problem 5(a) of the preceding section.
3. Prove the second sentence in Theorem 8.12.
4. Let $\zeta = (-1 + \sqrt{-3})/2$ and put $F = \mathbf{Q}(\zeta)$.
   a) Show that $\zeta^3 = 1$ and $\bar{\zeta} = \zeta^2 = -(\zeta + 1)$.
   b) Show that if $a, b \in \mathbf{Q}$, then $\mathcal{N}(a + b\zeta) = a^2 - ab + b^2$, so that $|\mathcal{N}(a + b\zeta)| < 1$ if $|a| \leq \frac{1}{2}$ and $|b| \leq \frac{1}{2}$.
   c) Show that $1, \zeta$ form a basis for $\mathbf{Z}(F)$.
   d) Describe explicitly the group of units in $\mathbf{Z}(F)$.

e) Show that $\mathbf{Z}(F)$ is a Euclidean domain.

f) Show that if $\alpha \in \mathbf{Z}(F)$ then $\mathcal{N}\alpha \not\equiv 2 \pmod 3$, and conclude that if $p \equiv 2 \pmod 3$ then p is prime in $\mathbf{Z}(F)$.

g) Show that the primes in $\mathbf{Z}(F)$ are (i) $1 - \zeta$, (ii) the rational primes $p \equiv 2 \pmod 3$, (iii) the two factors of each of the rational primes $p \equiv 1 \pmod 3$, and (iv) the associates of all these. [For (iii), show that if $p \mid (x^2 + 3)$, then p is divisible by some prime $\pi = a + b\zeta$, and $p = \mathcal{N}\pi$,]

5. Specialize the notion of Euclidean domain when $D = \mathbf{Z}(F)$, $F = \mathbf{Q}(\sqrt{d})$, by requiring that condition (ii) in Section 2.2 hold with $s(\alpha) = |\mathcal{N}\alpha|$. (Call D a norm-Euclidean domain then.) Show that (ii) holds if and only if, for every $\xi \in F$, there is an $\alpha \in \mathbf{Z}(F)$ such that $|\mathcal{N}(\xi - \alpha)| < 1$. Put $\xi = x + y\omega$, $\alpha = a + b\omega$, where $x, y \in \mathbf{Q}$ and $a, b \in \mathbf{Z}$, and put $u = x - a$, $v = y - b$. Show that what is required is that a and b exist for which

$$\left|\left(u + \frac{v}{2}\right)^2 - m\left(\frac{v}{2}\right)^2\right| < 1 \qquad \text{for } d \equiv 1 \pmod 4,$$

$$|u^2 - mv^2| < 1 \qquad \text{for } d \equiv 2 \text{ or } 3 \pmod 4.$$

Prove from this that F is Euclidean for $d = -11, -7, -3, -2, -1, 2, 3, 5, 13$, and furthermore that F is definitely not norm-Euclidean for any other negative values of d than those just listed.

NOTES AND REFERENCES

Section 8.1

Earlier proofs of Theorem 8.1, including Legendre's, proceeded by induction on some simple function of $|a|$, $|b|$, $|c|$. The present proof, which also provides a bound on the smallest nontrivial solution of (1), was given independently by Mordell [1951] and Skolem [1952].

For more on rational and integral points on conic sections and other curves of genus 0, see the books on Diophantine equations by Skolem [1938] and Mordell [1969].

Section 8.2

No simple criterion is known for the solvability of $x^2 - dy^2 = -1$, except that it can be related to the equally mysterious but easily answered question of whether the length of the period in the continued fraction expansion of $\sqrt{d}$ is even or odd. (See Section 9.5.) The 102,662 values of $d < 10^6$ for which it is solvable have been tabulated by Beach and Williams [1972].

The best upper bound known at present for a fundamental solution, valid for all $d \geq 5$, is $\log \zeta < (\frac{1}{2} \log d + 1)\sqrt{d}$, where $\mathcal{N}\zeta = 4$. This is an improvement by Hua [1942] of the beautiful inequality of I. Schur, $\zeta < d^{\sqrt{d}}$. See Problem 5 of Section 9.5 for a weaker theorem.

The recursion relations occurring in Problem 14 are not mere curiosities. They play an important role, for example, in the remarkable theorem of Matijasevič and Robinson [1975] to the effect that every Diophantine equation, in an arbitrary number of variables, is equivalent to (roughly, has the same solution set as) an equation in not more than 13 variables.

Section 8.3

Some books on algebraic numbers accessible to students with scanty algebraic background: Artin [1959], Borevich and Shafarevich [1966], LeVeque [1955], McCarthy [1966], Pollard and Diamond [1975]. Other books include Artin [1967], Lang [1970], Samuel [1967]. A beautiful classic, in German, is Hecke [1923].

For a (rather sophisticated) history of the theory of algebraic numbers, and in particular of the definition of algebraic integer, see the Historical Note in Bourbaki [1972].

There is a much-told tale, stemming from a 1910 memorial lecture on Kummer by Hensel, that Kummer had given a fallacious proof of the Fermat conjecture, assuming unique factorization of integers in fields generated by roots of unity. There are many reasons for doubting this story. The evidence for and against is engagingly laid out in Edwards [1975].

Section 8.4

For more on quadratic fields, see Cohn [1962], Hardy and Wright [1960], Niven and Zuckerman [1972].

The stories of who actually proved that the two lists given (of nonreal quadratic fields with unique factorization, and of all Euclidean quadratic fields) are complete and correct, are somewhat complicated. They can most easily be pieced together from Sections R14 and R12 of LeVeque [1974], but for sample references see Stark [1969] and Chatland and Davenport [1950].

In the nonreal quadratic case, it is known that a domain is Euclidean if and only if (ii) holds with $s(\alpha) = |\mathcal{N}\alpha|$, as in Problem 5; cf. Dubois and Steger [1958]. The proof fails in the real case because of the presence of infinitely many units; nothing seems to be known, except under unproved hypotheses.

9

Diophantine Approximation

9.1 FAREY SEQUENCES AND HURWITZ'S THEOREM

The subject encompassed by the chapter title is one of the major branches of number theory, and we can only touch on a few high spots. We shall develop only one small fragment of it in detail, namely the basic theory of continued fractions. This is a central tool in the subject, having applications outside as well, including the efficient calculation of fundamental units in real quadratic fields.

Briefly, Diophantine approximation is concerned with questions of which the following is typical: given a function (usually a polynomial) in several variables and with real coefficients, how close to zero can its value be made if the variables range over specified sets of integers, and how can this minimum absolute value be found? Generalizations leap to mind—allow several functions, complex coefficients, variables in a domain $\mathbf{Z}(F)$, p-adic absolute values, etc.—and some problems in the subject do not fit this paradigm at all, but it will get us started.

We have already proved one result of the kind described, in Theorem 8.3: if x is real and $t \in \mathbf{Z}^+$, there are integers p, q such that

$$|qx - p| \leq \frac{1}{t+1}, \qquad 1 \leq q \leq t.$$

(In this chapter, the letters p, q will not connote primes, merely integers.) Here the function is the simplest possible one, the linear function in two variables, $qx - p$, and we shall continue to be occupied with this case throughout most of this chapter. That this is really a theorem about approximation is brought out more clearly in its corollary, which asserts that the inequality

$$\left| x - \frac{p}{q} \right| < \frac{1}{q^2} \tag{1}$$

has infinitely many solutions $p/q \in \mathbf{Q}$, if x belongs to $\mathbf{R} \backslash \mathbf{Q}$ (the set of real-but-not-rational (= irrational) numbers.) Here the accuracy of the approximation of x by p/q is being compared with the "complexity" of the rational number p/q, which we choose to measure by the size of the denominator q. Since the rational numbers are dense on the real line (every interval contains some of them), there are rational numbers arbitrarily close to every irrational number. The arithmetically interesting question is, how close can they be, as a function of their complexity? Not only is it not obvious that the answer is the same for different irrational numbers, it is not true—and therein lies the interest! For example, suppose it turns out to be possible to find a set S of real numbers which contains all the algebraic numbers and is such that all members of S share a certain approximability property; then to show that π, say, is not algebraic, it would suffice to show that π allows an incompatible kind of rational approximation. The particular number π is too difficult, but we now find such a set S and show that there are real numbers not in it, so that **transcendental** (= nonalgebraic) numbers do exist.

Theorem 9.1 (Liouville) *If α is algebraic of degree $n > 1$, then for a suitable constant $c = c(\alpha) > 0$, the inequality*

$$\left|\alpha - \frac{p}{q}\right| > \frac{c}{q^n} \tag{2}$$

holds for every pair of integers p, q with $q > 0$.

Remark. The set S above can therefore be taken to be $\mathbf{Q}$ together with the set B of all real ξ with the property that for some constant $\omega = \omega(\xi)$, $|\xi - p/q| > q^{-\omega}$ for all $p, q \in \mathbf{Z}$, $q > 0$. At this stage it is conceivable that $S = \mathbf{R}$.

Proof. If the theorem is true for some c it is true for every smaller c, so we can suppose that $c < 1$. Then (2) holds for all p/q such that $|\alpha - p/q| > 1$, so we may suppose that $|\alpha - p/q| \leq 1$; thus p/q is bounded.

Let $f(x) = a_0x^n + \cdots + a_n$ be the irreducible polynomial with relatively prime coefficients in $\mathbf{Z}$, of which α is a zero. Then $f(p/q) \neq 0$ for all $p/q \in \mathbf{Q}$, since $\partial\alpha > 1$, so

$$\left|q^n f\left(\frac{p}{q}\right)\right| = |a_0p^n + a_1p^{n-1}q + \cdots + a_nq^n| \geq 1.$$

By the law of the mean,

$$f\left(\frac{p}{q}\right) = f\left(\frac{p}{q}\right) - f(\alpha) = \left(\frac{p}{q} - \alpha\right)f'(X),$$

where X lies between p/q and α and thus is bounded. Hence $|f'(X)|$ is bounded above, say by c_1, and we can choose $c_1 > 1$. Then

$$\left|\alpha - \frac{p}{q}\right| = \frac{|q^n f(p/q)|}{q^n|f'(X)|} > \frac{1}{c_1q^n},$$

so (2) holds with $c = 1/c_1$. ■

Theorem 9.2 *For every choice of signs in the various terms, the number*

$$\xi = 1 \pm \frac{1}{2^{1!}} \pm \frac{1}{2^{2!}} \pm \frac{1}{2^{3!}} \pm \cdots \pm \frac{1}{2^{n!}} \pm \cdots$$

is transcendental.

Proof. Make any choice of signs and keep it fixed. Put $q_n = 2^{(n-1)!}$ and

$$p_n = q_n(1 \pm 2^{-1!} \pm \cdots \pm 2^{-(n-1)!})$$

so that p_n/q_n is a partial sum of the series for ξ and

$$\left|\xi - \frac{p_n}{q_n}\right| \le 2^{-n!} + 2^{-(n+1)!} + \cdots,$$

where all terms are taken with positive sign in the series on the right. Now for $k \ge 1$, $2^{(n+k)!} \ge 2^{n!(n+k)} \ge (2^{n!})^k$, so the series on the right is dominated, term-for-term, by the series

$$2^{-n!} + 2^{-n!} + (2^{-n!})^2 + (2^{-n!})^3 + \cdots.$$

Summing this geometric series, we see that for each $n > 0$,

$$\left|\xi - \frac{p_n}{q_n}\right| \le 2^{-n!} + \frac{2^{-n!}}{1 - 2^{-n!}} < 2\frac{2^{-n!}}{1 - 2^{-n!}} \le 4 \cdot 2^{-n!} = 4q_n^{-n},$$

so $\xi \notin B$. Obviously $\xi \notin \mathbf{Q}$, since its base-2 expansion is not periodic, so $\xi \notin S$. Hence ξ is transcendental. ■

The complement—call it L—of S in $\mathbf{R}$ is called the set of **Liouville numbers.** They are the irrational numbers α for which (2) is false for each n, for suitable p/q. All elements of L are transcendental, and by Theorem 9.2, L is infinite. In fact, the theorem shows that L is uncountable. (See Section 9.6 if this is a new concept.) Still, the Liouville numbers are relatively rare; "almost all" real numbers, in a certain technical sense which need not concern us here (see Problem 2), belong to the set B defined above.

Before attempting to study the general question of how well rational numbers p/q with $q \le n$ can approximate a real number x, it is useful to dwell a moment on how these numbers p/q, for $q \le n$, are related to each other. For each n this is an infinite discrete set of rational numbers, which when arranged in increasing order of magnitude is called the **Farey sequence** of order n, and denoted by $\mathcal{F}_n$. For the first few values of n, fragments of the sequences look like this:

$$\mathscr{F}_1: \quad \ldots, \frac{-1}{1}, \frac{0}{1}, \frac{1}{1}, \frac{2}{1}, \ldots$$

$$\mathscr{F}_2: \quad \ldots, \frac{-1}{2}, \frac{0}{1}, \frac{1}{2}, \frac{1}{1}, \frac{3}{2}, \ldots$$

$$\mathscr{F}_3: \quad \ldots, \frac{-1}{3}, \frac{0}{1}, \frac{1}{3}, \frac{1}{2}, \frac{2}{3}, \frac{1}{1}, \frac{4}{3}, \ldots$$

$$\mathscr{F}_4: \quad \ldots, \frac{-1}{4}, \frac{0}{1}, \frac{1}{4}, \frac{1}{3}, \frac{1}{2}, \frac{2}{3}, \frac{3}{4}, \frac{1}{1}, \frac{5}{4}, \ldots$$

$$\mathscr{F}_5: \quad \ldots, \frac{-1}{5}, \frac{0}{1}, \frac{1}{5}, \frac{1}{4}, \frac{1}{3}, \frac{2}{5}, \frac{1}{2}, \frac{3}{5}, \frac{2}{3}, \frac{3}{4}, \frac{4}{5}, \frac{1}{1}, \frac{6}{5}, \ldots$$

$$\vdots$$

Clearly, the number of elements of $\mathscr{F}_n$ which lie between 0 and 1 inclusive is $1 + \varphi(1) + \varphi(2) + \cdots + \varphi(n)$.

The rational numbers p/q and r/s are said to be **adjacent in $\mathscr{F}_n$** if they are successive elements of $\mathscr{F}_n$.

Theorem 9.3

a) *If p/q and r/s are adjacent in $\mathscr{F}_n$ for some n, then $|ps - qr| = 1$.*

b) *If $|ps - qr| = 1$, then p/q and r/s are adjacent in $\mathscr{F}_n$ for*

$$\max(q, s) \le n < q + s,$$

and they are separated by the single element $(p + r)/(q + s)$ in $\mathscr{F}_{q+s}$.

Remark. This theorem on the one hand gives necessary and sufficient conditions that p/q and r/s be adjacent in $\mathscr{F}_n$, and on the other hand gives the law of formation of the new elements that appear in going from $\mathscr{F}_n$ to $\mathscr{F}_{n+1}$. The number $(p + r)/(q + s)$ is called the **mediant** of p/q and r/s.

Proof. We start with (b). Suppose that p/q and r/s are elements of $\mathscr{F}_n$ such that $qr - ps = \pm 1$; interchange their names if necessary, so that $r/s > p/q$ and $qr - ps = 1$. As t increases continuously from 0 to ∞, the number

$$f(t) = \frac{p + tr}{q + ts}$$

increases from p/q to r/s, and hence there is a one-to-one correspondence between the positive real numbers t and the points of the interval

$$\frac{p}{q} < x < \frac{r}{s}. \tag{3}$$

Moreover, it is clear that $f(t)$ is rational if and only if t is rational; since we are interested only in the rational numbers in the interval, we put $t = u/v$, where $(u, v) = 1$ and $u > 0$, $v > 0$. This gives

$$f\left(\frac{u}{v}\right) = \frac{vp + ur}{vq + us}.$$

Since

$$\begin{aligned} q(vp + ur) - p(vq + us) &= u(qr - ps) = u, \\ s(vp + ur) - r(vq + us) &= v(ps - qr) = -v, \end{aligned}$$

we have $(vp + ur, vq + us) = 1$. Thus we have shown that as u and v run over all pairs of relatively prime positive integers, the reduced fraction $(vp + ur)/(vq + us)$ runs over all rational numbers between p/q and r/s.

Among these fractions, the one with $u = v = 1$ is clearly the unique one of smallest denominator; it is the mediant of p/q and r/s, and

$$|(p + r)q - (q + s)p| = 1, \qquad |r(q + s) - s(p + r)| = 1.$$

Since $q + s > \max(q, s)$, part (b) of the theorem follows.

To prove (a), we proceed inductively. The sequence $\mathscr{F}_1$ consists of the integers $\ldots, -1/1, 0/1, 1/1, \ldots$, and $|a \cdot 1 - (a + 1) \cdot 1| = 1$, so that (a) is true for $n = 1$. If it is true for $n = m$, it is also true for $n = m + 1$, since the only elements of $\mathscr{F}_{m+1}$ not in $\mathscr{F}_m$ are certain mediants of adjacent elements of $\mathscr{F}_m$. The assertion follows by the induction principle. ■

Here is a simple application of the above ideas, yielding an improvement due to A. Hurwitz (1891) of the Corollary to Theorem 8.3.

Theorem 9.4 *Suppose that x is a real number, and that for arbitrary $n > 0$, x lies between adjacent elements r/s and u/v of $\mathscr{F}_n$. Then at least one of the three numbers r/s, u/v, and $l/m = (r + u)/(s + v)$ is a solution of the inequality*

$$\left| x - \frac{p}{q} \right| < \frac{1}{\sqrt{5}\, q^2}. \tag{4}$$

It follows that if x is irrational, (4) *always has infinitely many solutions.*

Proof. To be definite, suppose that

$$r/s < l/m < u/v.$$

If p/q is any of these three reduced fractions and $c \in \mathbf{R}^+$, designate by $I_c(p/q)$ the closed interval

$$\left[\frac{p}{q} - \frac{1}{cq^2}, \frac{p}{q} + \frac{1}{cq^2} \right]$$

on the real line. We wish to find the values of c such that the three intervals $I_c(r/s)$, $I_c(u/v)$, $I_c(l/m)$ together completely cover the interval $I = [r/s, u/v]$.

Now $I_c(r/s)$ and $I_c(u/v)$ intersect (or abut) if

$$\frac{r}{s} + \frac{1}{cs^2} \geq \frac{u}{v} - \frac{1}{cv^2},$$

$$\frac{1}{c}\left(\frac{1}{s^2} + \frac{1}{v^2}\right) \geq \frac{u}{v} - \frac{r}{s} = \frac{1}{vs},$$

$$c \leq \frac{v}{s} + \frac{s}{v}.$$

Putting $f(t) = t + 1/t$, this becomes

$$c \leq f\left(\frac{v}{s}\right).$$

Similarly, $I_c(r/s)$ and $I_c(l/m)$ intersect if

$$c \leq f\left(\frac{m}{s}\right) = f\left(1 + \frac{v}{s}\right).$$

Thus the left-hand portion $[r/s, l/m]$ of I is certainly completely covered by the three intervals I_c if

$$c \leq \max\left(f\left(\frac{v}{s}\right),\ f\left(1 + \frac{v}{s}\right)\right),$$

and *a fortiori* if

$$c \leq \min_{t>0} \{\max(f(t), f(1 + t))\} = c_0.$$

From the figure, the minimum occurs for that t_0 for which $f(t_0) = f(1 + t_0)$, namely $t_0 = (\sqrt{5} - 1)/2$, and $c_0 = \sqrt{5}$.

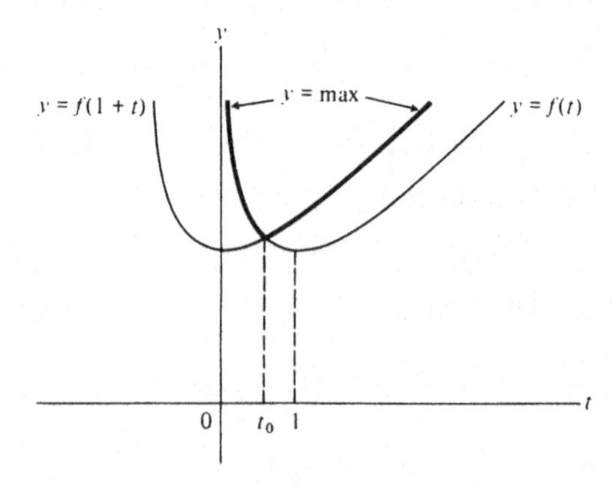

Figure 9.1

By the same reasoning, the right-hand portion $[l/m, u/v]$ is also covered by the three intervals I_c if $c \le \sqrt{5}$, so the entire interval $[r/s, u/v]$ is then covered. This proves the first part of the theorem, except that we must still exclude equality in (4). This is immediate; since $c = \sqrt{5}$ is irrational, two intersecting intervals I_c cannot merely abut, so every common point is an interior point of at least one of them.

The final sentence of the theorem results from the fact that if x is irrational, the interval defined by adjacent elements of $\mathscr{F}_n$ and in which x lies shrinks down on x from both sides as n increases, so any *fixed* rational approximation is neither r/s nor u/v in the theorem, for n sufficiently large. ■

Adolph Hurwitz (1859–1919)

Educated at Berlin and Leipzig, Hurwitz taught for eight years at Königsberg, where Hilbert and Minkowski were students, and spent the remainder of his life in Zürich. A student of Felix Klein, many of his papers were concerned with modular functions and with algebraic functions and their Riemann surfaces. He pioneered in integration on (special) topological groups, and proved a fundamental theorem about algebras over **R**. His criterion for polynomials all of whose zeros have nonpositive real parts is well known in stability theory. A friendly, unassuming man, he tended to work on individual problems rather than develop large theories.

PROBLEMS

1. According to Problem 2 of Section 6.5, $\varphi(t, n) = t\varphi(n) + O(\tau(n))$. Use this result to show that the Farey sequence $\mathscr{F}_n$ is rather evenly distributed on $[0, 1]$ for large n, in the sense that if the number of elements of $\mathscr{F}_n \cap [0, t]$ is $\Phi(t, n)$, then

$$\Phi(t, n) \sim t\Phi(1, n)$$

for each $t \in [0, 1]$.

2. Let ν be any real number larger than 2. For $c \in \mathbf{R}^+$, consider the interval

$$J_c(p/q) = \left[\frac{p}{q} - \frac{1}{cq^\nu}, \frac{p}{q} + \frac{1}{cq^\nu}\right]$$

of length $l(J_c(p/q))$. Show that if

$$\lambda(c) = \sum_{p/q \in [0,1]} l(J_c(p/q)) = \sum_{q=1}^{\infty} \sum_{\substack{0<p\leq q \\ (p,q)=1}} l(J_c(p/q)),$$

then $\lambda(c)$ is finite, and in fact $\lambda(c) \to 0$ as $c \to \infty$. Conclude that, given $\varepsilon > 0$, the set of all Liouville numbers in $[0, 1]$ is contained in a collection of intervals whose total "length" (an infinite series) is $< \varepsilon$. (This property is obviously not shared by the set of all irrational numbers in $[0, 1]$; it is in this sense that almost all numbers are not Liouville numbers.) The above argument proves a stronger theorem; state it.

3. Use the fact that $|p^2 - pq - q^2| \geq 1$ for all $p, q \in \mathbf{Z}$, not both zero, to show that Hurwitz's theorem is best possible, in the sense that there exists $x \in \mathbf{R}\backslash\mathbf{Q}$ for which the inequality (4) would not have infinitely many solutions if $\sqrt{5}$ were replaced by a larger constant.

9.2 BEST APPROXIMATIONS TO A REAL NUMBER

The reason that 22/7 is commonly used as an approximation to π is that it is remarkably close considering its simplicity (error about 0.00126); the simplest rational number closer to π is 179/57 (error about 0.00124). Two other unusually good approximations, already known in the sixteenth century in Europe, are 333/106 (error about $8.3 \cdot 10^{-5}$) and 355/113 (error about $2.6 \cdot 10^{-7}$; known in the fifth century in China). No fraction with denominator q, $106 < q < 113$, approximates π with an error even as small as that for 22/7, so obviously some denominators are much better to use than others. We now address the problem of exactly defining and then finding the unusually good approximations to an arbitrary real x.

We say that p/q is a **best approximation** to x if

$$(5) \qquad |qx - p| < |q'x - p'| \quad \text{for all } p', q' \text{ such that } 0 < q' \leq q \text{ and } p'/q' \neq p/q.$$

(Then also

$$\left|x - \frac{p}{q}\right| < \frac{q'}{q} \cdot \left|x - \frac{p'}{q'}\right| \leq \left|x - \frac{p'}{q'}\right|$$

for all such p', q', so that p/q is the element of $\mathscr{F}_q$ closest to x. This is not reversible: the element of $\mathscr{F}_n$ closest to x may not be a best approximation to x. For more on this matter, see Section 9.3, Problem 10.) The best approximations to x arrange themselves naturally by size of denominator into a sequence $\{P_k/Q_k\}$, and we wish to find a systematic method for determining all of them. Note that the definition (5) implies that $(p, q) = 1$, since (5) is false if $p = mp'$, $q = mq'$ with $m > 1$.

Since the Farey sequence of given order has the same appearance in any two intervals between consecutive integers, we may move x into $[0, 1)$ by subtracting off its greatest integer $[x] = \lambda_0$ and search for the best approximations to

$x' = x - \lambda_0$. Put $P_0 = \lambda_0$, $Q_0 = 1$; then P_0/Q_0 is a best approximation to x if and only if $0 \leqq x' < \frac{1}{2}$. (When x is rational there is a possibility of equality here ($x' = \frac{1}{2}$) and elsewhere. This will be discussed in detail at the end of this section; for the moment assume that if x is rational, its denominator is larger than that of other rational numbers under consideration.)

We argue as we did in proving Theorem 8.4, to show that there is an infinite sequence of best approximations when x is irrational. By Theorem 8.3, for every $\tau \in \mathbf{R}^+$ there are $p, q \in \mathbf{Z}$ such that

$$(6) \qquad 1 \le q \le [\tau] \le \tau \quad \text{and} \quad |qx - p| \le \frac{1}{[\tau] + 1} < \frac{1}{\tau}.$$

If $Q_0x - P_0 = 0$, then $x = P_0/Q_0$ and we are done. Otherwise, choosing $\tau = |Q_0x - P_0|^{-1}$, we see from (6) that there is a smallest q, $1 \le q \le |Q_0x - P_0|^{-1}$, such that $|qx - p| < |Q_0x - P_0|$ for some p. Designate this pair p, q as P_1, Q_1. (If P_0/Q_0 is not a best approximation to x, then $Q_1 = Q_0 = 1$, but hereafter the Q_k will strictly increase.) If $Q_1x - P_1 = 0$, we are done; otherwise, there is by (6) a smallest $q > Q_1$ such that $|qx - p| < |Q_1x - P_1|$ for some p, and $q \le |Q_1x - P_1|^{-1}$. Designate this new pair p, q by P_2, Q_2, and continue in this fashion. Thus for those k for which $Q_kx - P_k \ne 0$,

$$1 = Q_0 \le Q_1 < Q_2 < \cdots,$$

$$(7) \qquad |Q_{k+1}x - P_{k+1}| < |Q_kx - P_k| \qquad \text{for } k \ge 0,$$

$$(8) \qquad |qx - p| \ge |Q_kx - P_k| \qquad \text{for } 0 < q < Q_{k+1} \text{ and all } p, \text{ for } k \ge 1,$$

and

$$(9) \qquad Q_{k+1} \le |Q_kx - P_k|^{-1} \qquad \text{for } k \ge 0.$$

We now have a step-by-step characterization of the best approximations P_k/Q_k but, without an effective way to find the minimal solution of (6), it could hardly be described as an algorithm. For the latter we need the control provided by the following basic result:

$$(10) \qquad Q_{k+1}P_k - Q_kP_{k+1} = \pm 1 \qquad \text{for } k \ge 0.$$

This comes about by writing

$$(11) \qquad Q_{k+1}P_k - Q_kP_{k+1} = Q_k(Q_{k+1}x - P_{k+1}) - Q_{k+1}(Q_kx - P_k).$$

From this we see that if we put

$$\alpha_k = Q_kx - P_k \qquad \text{for } k \ge 0,$$

then

$$|Q_{k+1}P_k - Q_kP_{k+1}| \le Q_k|\alpha_{k+1}| + Q_{k+1}|\alpha_k| < 2Q_{k+1}|\alpha_k| < 2,$$

the final inequality being a consequence of (9). Since $Q_{k+1}P_k - Q_kP_{k+1}$ is a non-zero integer, (10) is proved.

To determine the sign in (10), note first that α_k and α_{k+1} cannot have the same sign. For if they did we would have, by (7),

$$|\alpha_k - \alpha_{k+1}| = |(Q_{k+1} - Q_k)x - (P_{k+1} - P_k)| < |\alpha_k|,$$

while $0 < Q_{k+1} - Q_k < Q_{k+1}$, contradicting (8). Hence we see from (11) and the behavior of* sgn α_k that $Q_{k+1}P_k - Q_kP_{k+1} = \operatorname{sgn} \alpha_{k+1}$, and since $\alpha_0 = 1 \cdot x - \lambda_0 > 0$, the assertion

$$Q_{k+1}P_k - Q_kP_{k+1} = (-1)^{k+1} \tag{12}$$

holds for $k \geq 0$. If we define $P_{-2} = 0$, $Q_{-2} = 1$, $P_{-1} = 1$, $Q_{-1} = 0$, then (12) continues to hold for $k \geq -2$; the advantage of this device is that while these P_{-k} and Q_{-k} have nothing to do with best approximations to x, they are also entirely independent of x and can conveniently be used as universal initial values in the recursive formulas that will now be obtained for the best approximations to x.

The importance of (12) lies in the fact that it constrains each new approximation P_{k+1}/Q_{k+1} to lie among the fractions u/v corresponding to solutions u, v of the linear Diophantine equation $P_kv - Q_ku = (-1)^{k+1}$, of which we already know one solution: $u = P_{k-1}$, $v = Q_{k-1}$. By Theorem 2.9, there is an integer—call it λ_{k+1}—such that

$$\begin{aligned} P_{k+1} &= P_{k-1} + \lambda_{k+1}P_k, \\ Q_{k+1} &= Q_{k-1} + \lambda_{k+1}Q_k, \end{aligned} \qquad \text{for } k \geq -1, \tag{13}$$

and $\lambda_{k+1} \geq 1$ since $Q_{k+1} > Q_k$. (13) gives

$$\alpha_{k+1} = \alpha_{k-1} + \lambda_{k+1}\alpha_k,$$

so, since $\operatorname{sgn} \alpha_k = (-1)^k$,

$$|\alpha_{k-1}| = |\alpha_{k+1}| + \lambda_{k+1}|\alpha_k|.$$

Thus

$$\left|\frac{\alpha_{k-1}}{\alpha_k}\right| = \lambda_{k+1} + \left|\frac{\alpha_{k+1}}{\alpha_k}\right|,$$

whence, since $-1 \leq \alpha_{k+1}/\alpha_k < 0$,

$$\lambda_{k+1} = \left[-\frac{\alpha_{k-1}}{\alpha_k}\right]. \tag{14}$$

Equations (13) and (14) give a recursive definition of the entire sequence of best approximations P_k/Q_k, with the aid of the auxiliary sequences λ_k and α_k. As we shall see, it is simpler to describe all these sequences in terms of the quantities

$$x_{k+1} = -\frac{\alpha_{k-1}}{\alpha_k}, \qquad k \geq 0, \tag{15}$$

---

* sgn $x = 1, 0$, or -1 according as $x > 0$, $x = 0$, or $x < 0$.

since it turns out that the latter can be expressed recursively in terms of x alone. In fact, we have

$$\alpha_{-1} = 0 \cdot x - 1 = -1 \quad \text{and} \quad \alpha_0 = 1 \cdot x - [x],$$

so

$$x_1 = \frac{1}{x - [x]},$$

and in general, for $k \geq 0$,

$$x_{k+1} = -\frac{\alpha_{k-1}}{Q_k x - P_k} = \frac{-\alpha_{k-1}}{(Q_{k-1}\lambda_k + Q_{k-2})x - (P_{k-1}\lambda_k + P_{k-2})}$$

$$= \frac{-\alpha_{k-1}}{\lambda_k \alpha_{k-1} + \alpha_{k-2}} = \frac{1}{\alpha_{k-2}/\alpha_{k-1} - \lambda_k},$$

so

$$x_{k+1} = \frac{1}{x_k - [x_k]} \qquad \text{for } k \geq 0.$$

The entire development has depended on the description in inequalities (7), (8), and (9) of the best approximations. These inequalities are valid up to the index κ, if there is one, for which $\alpha_\kappa = 0$, at which point $x = P_\kappa/Q_\kappa$ and the process stops. With this in mind, we summarize in the following theorem what has been proved.

Theorem 9.5 *Suppose that $x \in \mathbf{R}\backslash\mathbf{Z}$. Put $x_0 = x$, and define*

(16) $$x_k = \frac{1}{x_{k-1} - [x_{k-1}]} \qquad \text{for } 1 \leq k < \kappa + 1,$$

where κ is the smallest index, if such exists, for which $x_\kappa - [x_\kappa] = 0$, and $\kappa = \infty$ otherwise. Put

(14) $$\lambda_k = [x_k] \quad \text{for } 0 \leq k < \kappa + 1,$$

and define the rational numbers P_k/Q_k by the equations

(13) $$\begin{matrix} P_{-2} = 0, & P_{-1} = 1, & P_{k+1} = \lambda_{k+1}P_k + P_{k-1} \\ Q_{-2} = 1, & Q_{-1} = 0, & Q_{k+1} = \lambda_{k+1}Q_k + Q_{k-1} \end{matrix} \quad \text{for } -1 \leq k < \kappa.$$

Then the P_k/Q_k, for $1 \leq k < \kappa + 1$ (and for $k = 0$ if $Q_1 > 1$), constitute precisely the complete sequence of best approximations to x, and inequalities (7)–(9) are valid for these k. The relation

(15) $$x_k = -\frac{Q_{k-2}x - P_{k-2}}{Q_{k-1}x - P_{k-1}} \quad \text{for } 1 \leq k < \kappa + 1$$

holds, as well as

(12) $$Q_{k+1}P_k - Q_k P_{k+1} = (-1)^{k+1} \quad \text{for } -2 \leq k < \kappa.$$

To see how this theorem solves the problem of finding the best approximations to x, take $x = \pi \approx 3.14159265358$. Then $\lambda_0 = [\pi] = 3$, and

$$
\begin{aligned}
x_1 &= \frac{1}{x - [x]} \approx 7.06251335, & \lambda_1 &= [x_1] = 7, \\
x_2 &= \frac{1}{x_1 - [x_1]} \approx 15.996587, & \lambda_2 &= [x_2] = 15, \\
x_3 &= \frac{1}{x_2 - [x_2]} \approx 1.00342, & \lambda_3 &= [x_3] = 1, \\
x_4 &= \frac{1}{x_3 - [x_3]} \approx 292.624, & \lambda_4 &= [x_4] = 292, \\
&\vdots & &\vdots
\end{aligned}
$$

Using (13), we see that the first few best approximations to π, in the sense we have defined, are

$$\frac{3}{1}, \quad \frac{22}{7}, \quad \frac{333}{106}, \quad \frac{355}{113}, \quad \frac{103{,}993}{33{,}102}.$$

The algorithm defined by (16) is called the continued fraction algorithm, because it leads to expressions such as those for x and P_k/Q_k in (18) and (19) below, these being **finite continued fractions.**

Theorem 9.6 *If κ, $\{x_k\}$, $\{\lambda_k\}$, $\{P_k\}$, and $\{Q_k\}$ are determined as in Theorem 9.5, then the following relations hold:*

$$(17) \qquad x = \frac{P_{k-1}x_k + P_{k-2}}{Q_{k-1}x_k + Q_{k-2}}, \quad 1 \le k < \kappa + 1,$$

$$(18) \qquad x = \lambda_0 + \cfrac{1}{\lambda_1 + \cfrac{1}{\lambda_2 + \ddots + \cfrac{1}{\lambda_{k-1} + \cfrac{1}{x_k}}}}, \quad 1 \le k < \kappa + 1,$$

$$(19) \qquad \frac{P_k}{Q_k} = \lambda_0 + \cfrac{1}{\lambda_1 + \cfrac{1}{\lambda_2 + \ddots + \cfrac{1}{\lambda_{k-1} + \cfrac{1}{\lambda_k}}}}, \quad 0 \le k < \kappa + 1.$$

Proof. Relation (17) follows immediately from (15). The definitions of x_k and λ_k give

$$(20)\qquad \begin{aligned} x &= \lambda_0 + \frac{1}{x_1}, \\ x_1 &= \lambda_1 + \frac{1}{x_2}, \\ &\vdots \\ x_{k-1} &= \lambda_{k-1} + \frac{1}{x_k}, \end{aligned}$$

and if we successively eliminate $x_1, x_2, \ldots, x_{k-1}$, equation (18) results. To obtain (19), consider the equations (20) with x_k as an independent variable which assumes values greater than 1, with fixed $\lambda_0, \lambda_1, \ldots, \lambda_{k-1}$. Then x is a function of x_k, given more briefly by (18). Since P_{k-1}, P_{k-2}, Q_{k-1}, and Q_{k-2} depend only on $\lambda_0, \ldots, \lambda_{k-1}$, (17) and (18) are different expressions for the same functional relation. If in (17) x_k is given the value λ_k, then $x = P_k/Q_k$, and substituting these values in (18) gives (19). ■

The typographical complexity of the continued fractions in (18) and (19) is intolerable, and we abbreviate the right-hand side of equation (19), say, to

$$(21)\qquad \lambda_0 + \frac{1}{\lambda_1 +}\,\frac{1}{\lambda_2 +}\cdots\frac{1}{\lambda_k},$$

or even more simply to $\{\lambda_0; \lambda_1, \lambda_2, \ldots, \lambda_k\}$. (The semicolon is useful to distinguish, for example, between $\{0; 1, 2\} = 2/3$ and $\{1; 2\} = 3/2$.) The continued fraction (21) is **regular** if all the numerators are $+1$, all the λ's are integers, and $\lambda_1, \ldots, \lambda_k$ are positive. We shall mainly consider regular continued fractions in this book, and we sometimes omit the adjective when the meaning is clear. The quantities $\lambda_0, \ldots, \lambda_{k-1}$ in (18), being integral parts, are called the **partial quotients**, while x_k is called the **complete quotient**. The latter is always ≥ 1, but it may or may not be an integer.

We must now deal with the possibility of equality rather than the strict inequality in (5) and elsewhere. Suppose that we have reached a certain best approximation P_{k-1}/Q_{k-1}, and that there is an r/s with $s > Q_{k-1}$ such that $(r, s) = 1$ and

$$|sx - r| = |Q_{k-1}x - P_{k-1}|.$$

Then since $r/s \neq P_{k-1}/Q_{k-1}$, it must be that

$$sx - r = -(Q_{k-1}x - P_{k-1}).$$

Thus

$$(22)\qquad x = \frac{P_{k-1} + r}{Q_{k-1} + s},$$

so that x is the first rational number to appear between P_{k-1}/Q_{k-1} and r/s in the sequences $\mathscr{F}_s, \mathscr{F}_{s+1}, \ldots$. Hence $k = \kappa$, and if r/s is to be included among the best approximations, and if we put $r/s = P_\kappa/Q_\kappa$, then $P_{\kappa+1}/Q_{\kappa+1} = x$. Comparing equations (13) and (22), we see that $\lambda_{\kappa+1} = 1$, and by (19),

$$x = \lambda_0 + \frac{1}{\lambda_1 +} \cdots \frac{1}{\lambda_\kappa +} \frac{1}{1}.$$

If r/s is not included, then $x = P_\kappa/Q_\kappa$ and

$$x = \lambda_0 + \frac{1}{\lambda_1 +} \cdots \frac{1}{\lambda_\kappa + 1}.$$

(For example, $4/3 = \{1; 2, 1\} = \{1; 3\}$.) The two cases are distinguished by the fact that in the first, one complete quotient (the final one) is equal to 1, whereas in the second, $x_k > 1$ for all $k \leq \kappa$.

Now consider an *arbitrary* finite regular continued fraction

$$\{a_0; a_1, \ldots, a_n\}; \tag{23}$$

it clearly has a rational value, say x. The reduced fractions

$$\frac{p_0}{q_0} = a_0, \quad \frac{p_1}{q_1} = \{a_0; a_1\}, \quad \frac{p_2}{q_2} = \{a_0; a_1, a_2\}, \ldots, \tag{24}$$

which are analogous to the partial sums of a series, are called the **convergents** of the continued fraction (23). For $1 \leq k \leq n$ we have $x = \{a_0; a_1, \ldots, a_{k-1}, x_k'\}$, where

$$x_k' = \{a_k; a_{k+1}, \ldots, a_n\}, \tag{25}$$

and thus

$$x_k' = a_k + \frac{1}{x_{k+1}'}$$

for $k = 1, 2, \ldots, n-1$. Since $x_k' \geq 1$ for $k > n$, we have

$$x_1' = \frac{1}{x - a_0} = \frac{1}{x - [x]}, \qquad a_1 = [x_1'],$$

$$x_2' = \frac{1}{x_1' - a_1} = \frac{1}{x_1' - [x_1']}, \qquad a_2 = [x_2'],$$

$$\vdots$$

Thus the sequence $\{x_k'\}$ is identical with the sequence $\{x_k\}$ defined in Theorem 9.5, and $\{a_k\}$ is therefore identical with $\{\lambda_k\}$. Hence we have the following theorem.

Theorem 9.7 *The convergents (possibly excepting p_0/q_0) of any finite regular continued fraction are the best approximations to the value of the continued* *fraction. Every rational number can be expanded into a finite regular continued fraction, and this expansion is unique, except for the variation indicated by the identity*

$$x = \{a_0; a_1, \ldots, a_\kappa\} = \{a_0; a_1, \ldots, (a_\kappa - 1), 1\}, \quad a_\kappa > 1.$$

If the convergents and complete quotients are defined by (24) *and* (25), *and if* $\alpha_k = q_k x - p_k$, *then the following relations hold for* $k \le \kappa$:

$$x_0 = x, \quad x_k = \frac{1}{x_{k-1} - [x_{k-1}]} \quad \text{for } k > 0, \tag{26}$$

$$a_k = [x_k] \quad \text{for } k \ge 0, \tag{27}$$

$$\begin{cases} p_{-2} = 0, p_{-1} = 1, p_k = a_k p_{k-1} + p_{k-2} \\ q_{-2} = 1, q_{-1} = 0, q_k = a_k q_{k-1} + q_{k-2} \end{cases} \quad \text{for } k \ge 0, \tag{28}$$

$$q_k p_{k-1} - q_{k-1} p_k = (-1)^k \quad \text{for } k \ge -1, \tag{29}$$

$$q_k p_{k-2} - q_{k-2} p_k = (-1)^{k-1} a_k \quad \text{for } k \ge 0, \tag{30}$$

$$x = \frac{p_{k-1} x_k + p_{k-2}}{q_{k-1} x_k + q_{k-2}} \quad \text{for } k \ge 0, \tag{31}$$

$$x_k = -\frac{\alpha_{k-2}}{\alpha_{k-1}} \quad \text{for } k \ge 0. \tag{32}$$

This theorem simply recapitulates what we have proved up to now. The only new element is the relation (30), which is easily proved with the help of (28) and (29).

PROBLEMS

1. Carry through the procedure described in this section to find the first few best approximations to $\sqrt{3}$.
2. Find all the best approximations to 339/62.
3. Show that if $x = \frac{1}{2}(1 + \sqrt{5})$, then each λ_i is 1.
4. Find an $x \in \mathbf{R}$ and $n \in \mathbf{Z}^+$ such that the element of $\mathscr{F}_n$ closest to x is not a best approximation to x.
5. Draw the graphs of the functions

$$G_n(x) = \min_{p/q \in \mathscr{F}_n \cap [0,1]} \{|qx - p|\}, \quad x \in [0, 1]$$

for $n = 1, 2, 3$. For each of these values of n, and for each $p/q \in \mathscr{F}_n \cap [0, 1]$, describe the set of values of x for which p/q is a best approximation to x.
6. Prove that p_k and q_k, as defined by (28), are always relatively prime.

7. If $x = a/b$, where $(a, b) = 1$, and

$$x = a_0 + \frac{1}{a_1 +} \frac{1}{a_2 +} \cdots \frac{1}{a_\kappa},$$

then $p_\kappa = a$, $q_\kappa = b$. Use this and an identity of Theorem 9.7 to find a solution of the linear Diophantine equation $ax + by = c$. In particular, find a general solution of $247x + 77y = 31$.

8. Show that for $k \le \kappa$,

$$\frac{q_k}{q_{k-1}} = a_k + \frac{1}{a_{k-1} +} \cdots \frac{1}{a_2 +} \frac{1}{a_1}.$$

9. Show that elimination of the successive r_i's from the Euclidean algorithm,

$$a = bq_1 + r_1, \qquad b = r_1 q_2 + r_2, \qquad \text{etc.,}$$

leads to the finite expansion $a/b = \{q_1; q_2, \ldots\}$.

9.3 INFINITE CONTINUED FRACTIONS

Now consider the case that x is an irrational number ξ. The sequences $\{x_k\}$, $\{\lambda_k\}$, and $\{P_k/Q_k\}$ are now infinite, and we write

$$\xi = \lambda_0 + \frac{1}{\lambda_1 +} \frac{1}{\lambda_2 +} \cdots. \tag{33}$$

This equation must be understood as an abbreviation for the equation

$$\xi = \lim_{n\to\infty} \left(\lambda_0 + \frac{1}{\lambda_1 +} \cdots \frac{1}{\lambda_n}\right) = \lim_{n\to\infty} \frac{P_n}{Q_n}; \tag{34}$$

the convergents P_n/Q_n play a role here analogous to that of the partial sums of an infinite series. Equation (34) is true because the P_n/Q_n are still, of course, best approximations to x.

Conversely, if we start with an arbitrary infinite regular continued fraction

$$a_0 + \frac{1}{a_1 +} \frac{1}{a_2 +} \cdots, \tag{35}$$

we can show that the convergents

$$\frac{p_0}{q_0} = a_0, \quad \frac{p_1}{q_1} = a_0 + \frac{1}{a_1}, \quad \frac{p_2}{q_2} = a_0 + \frac{1}{a_1 +} \frac{1}{a_2}, \quad \ldots$$

always converge to an irrational number ξ. For take $n > 2$ and put

$$\frac{p}{q} = \frac{p_n}{q_n} = a_0 + \frac{1}{a_1 +} \cdots \frac{1}{a_n}.$$

Then by Theorem 9.7, the numbers $a_0, a_1, \ldots, a_{n-2}$ are uniquely determined by p/q, and the convergents in the expansion of p/q, which are also convergents of (35), satisfy the usual relations (28), (29), and (30) for $k \le n - 2$ and hence, since n is arbitrary, for all $k \ge 0$. From (30), we see that

$$\frac{p_{2k-2}}{q_{2k-2}} < \frac{p_{2k}}{q_{2k}} \quad \text{and} \quad \frac{p_{2k-1}}{q_{2k-1}} > \frac{p_{2k+1}}{q_{2k+1}},$$

so that

$$\frac{p_0}{q_0} < \frac{p_2}{q_2} < \frac{p_4}{q_4} < \cdots, \qquad \frac{p_1}{q_1} > \frac{p_3}{q_3} > \frac{p_5}{q_5} > \cdots.$$

By (29)

$$\frac{p_{2k}}{q_{2k}} < \frac{p_{2k+1}}{q_{2k+1}},$$

so that

$$\frac{p_{2k}}{q_{2k}} < \frac{p_{2l+1}}{q_{2l+1}}$$

for every $l \ge k$. Hence

$$\frac{p_0}{q_0} < \frac{p_2}{q_2} < \frac{p_4}{q_4} < \cdots < \frac{p_5}{q_5} < \frac{p_3}{q_3} < \frac{p_1}{q_1},$$

so that the sequences $\{p_{2k}/q_{2k}\}$ and $\{p_{2k+1}/q_{2k+1}\}$, being monotonic and bounded, are convergent. But (29) can be rewritten in the form

$$\frac{p_{k-1}}{q_{k-1}} - \frac{p_k}{q_k} = \frac{(-1)^k}{q_{k-1}q_k},$$

and since $q_k \to \infty$ as $k \to \infty$ we see that

$$\lim_{k \to \infty} \left(\frac{p_{2k}}{q_{2k}} - \frac{p_{2k+1}}{q_{2k+1}} \right) = 0,$$

and consequently $\lim p_k/q_k$ exists. Call this limit ξ, and put

$$\xi = a_0 + \frac{1}{a_1 +} \cdots \frac{1}{a_{n-1} +} \frac{1}{\xi_n}.$$

It follows, just as in the rational case, that

$$\xi_1 = \frac{1}{\xi - [\xi]}, \quad \xi_2 = \frac{1}{\xi_1 - [\xi_1]}, \quad \ldots,$$

and

$$a_0 = [\xi], \quad a_1 = [\xi_1], \quad \ldots,$$

so that the convergents p_k/q_k of (19) are the best approximations P_k/Q_k to ξ. From this we deduce the following assertions.

Theorem 9.8 *Every infinite regular continued fraction converges to an irrational number, the best approximations to which are afforded by the convergents of the continued fraction. Every irrational number can be expanded into an infinite regular continued fraction, and this expansion is unique. Moreover, if*

$$\xi = \{a_0; a_1, a_2, \ldots\} = \{a_0; a_1, \ldots, a_{k-1}, \xi_k\}$$

for $k > 0$, *and if the convergents* p_k/q_k *are defined by* (24), *then the relations* (26)–(32) *of Theorem* 9.7 *hold, with* x *and* x_k *replaced by* ξ *and* ξ_k, *respectively.*

As before, the numbers a_k are called the *partial quotients*, and the ξ_k the *complete quotients*, in the expansion.

Once the continued fraction expansion of ξ is known, the successive convergents can be computed very simply. For example, let $\xi = \sqrt{7}$. Then

$$\sqrt{7} = 2 + (\sqrt{7} - 2), \qquad a_0 = 2, \quad \xi_1 = (\sqrt{7} - 2)^{-1},$$

$$\frac{1}{\sqrt{7} - 2} = \frac{\sqrt{7} + 2}{3} = 1 + \frac{\sqrt{7} - 1}{3}, \qquad a_1 = 1, \quad \xi_2 = \left(\frac{\sqrt{7} - 1}{3}\right)^{-1},$$

$$\frac{3}{\sqrt{7} - 1} = \frac{\sqrt{7} + 1}{2} = 1 + \frac{\sqrt{7} - 1}{2}, \qquad a_2 = 1, \quad \xi_3 = \left(\frac{\sqrt{7} - 1}{2}\right)^{-1},$$

$$\frac{2}{\sqrt{7} - 1} = \frac{\sqrt{7} + 1}{3} = 1 + \frac{\sqrt{7} - 2}{3}, \qquad a_3 = 1, \quad \xi_4 = \left(\frac{\sqrt{7} - 2}{3}\right)^{-1},$$

$$\frac{3}{\sqrt{7} - 2} = \sqrt{7} + 2 = 4 + (\sqrt{7} - 2), \quad a_4 = 4, \quad \xi_5 = (\sqrt{7} - 2)^{-1}.$$

Since $\xi_5 = \xi_1$, also $\xi_6 = \xi_2$, $\xi_7 = \xi_3, \ldots$, so $\{\xi_k\}$ (and therefore also $\{a_k\}$) is periodic. Thus in this case we have a periodic expansion,

$$\sqrt{7} = 2 + \frac{1}{1+}\,\frac{1}{1+}\,\frac{1}{1+}\,\frac{1}{4+}\,\frac{1}{1+}\,\frac{1}{1+}\,\frac{1}{1+}\,\frac{1}{4+}\cdots.$$

Using the relations (28), we construct the following table:

k	-2	-1	0	1	2	3	4	5	6	$\cdots$
a_k			2	1	1	1	4	1	1	$\cdots$
p_k	0	1	2	3	5	8	37	45	82	$\cdots$
q_k	1	0	1	1	2	3	14	17	31	$\cdots$

Here the element $37 = p_4$, for example, is determined by multiplying $a_4 = 4$ by $p_3 = 8$ and adding $p_2 = 5$. Thus the best approximations to $\sqrt{7}$ are 3, 5/2, 8/3, 37/14, 45/17, ...

Relations (26)–(32) are the fundamental equations governing the theory of continued fractions. To them should be added the identity and the inequalities of the following theorem, which spell out the connection between the continued fraction expansion and the approximability properties of a real number.

Theorem 9.9 *If $\xi \in \mathbf{R}\backslash\mathbf{Q}$, then for $k \geq 0$,*

$$\xi - \frac{p_k}{q_k} = \frac{(-1)^k}{q_k(q_k\xi_{k+1} + q_{k-1})}. \tag{36}$$

Consequently,

$$\frac{1}{q_k(q_k + q_{k+1})} < \left|\xi - \frac{p_k}{q_k}\right| < \frac{1}{q_k q_{k+1}}, \tag{37}$$

and a fortiori,

$$\left|\xi - \frac{p_k}{q_k}\right| < \frac{1}{q_k^2}. \tag{38}$$

Proof. Equation (36) follows immediately from (31) and (29). Inequalities (37) follow from (36) with the help of the bounds

$$\begin{aligned} q_{k+1} &= q_k a_{k+1} + q_{k-1} < q_k\xi_{k+1} + q_{k-1} < q_k(a_{k+1} + 1) + q_{k-1} \\ &= q_{k+1} + q_k, \end{aligned}$$

and (38) results because $\{q_k\}$ is an increasing sequence. ∎

As a partial converse, we have

Theorem 9.10 *If $x \in R$ and*

$$\left|x - \frac{p}{q}\right| < \frac{1}{2q^2}, \tag{39}$$

then p/q is a convergent of the continued fraction expansion of x.

Proof. To be definite, suppose that $0 < x - p/q < 1/2q^2$; the case $0 < p/q - x < 1/2q^2$ is handled similarly. We show that p/q is a best approximation to x. Suppose the opposite, namely that there exists r/s such that $|sx - r| < |qx - p|$ and $s \leq q$. Clearly,

$$\left|\frac{p}{q} - \frac{r}{s}\right| \geq \frac{1}{qs}.$$

There are three cases. (a) If $r/s < p/q < x$, then

$$0 < \frac{p}{q} - \frac{r}{s} < x - \frac{r}{s} < \frac{q}{s} \cdot \frac{1}{2q^2} = \frac{1}{2qs};$$

this is impossible. (b) If $p/q < r/s < x$, then

$$0 < \frac{r}{s} - \frac{p}{q} < x - \frac{p}{q} < \frac{1}{2q^2} \leq \frac{1}{2qs};$$

also impossible. (c) If $x < r/s$, then since

$$\frac{r}{s} - x = \frac{r - sx}{s} < \frac{qx - p}{s} = \left(x - \frac{p}{q}\right) \cdot \frac{q}{s},$$

we obtain

$$0 \le \frac{r}{s} - \frac{p}{q} = \frac{r}{s} - x + x - \frac{p}{q} < \left(\frac{q}{s} + 1\right)\left(x - \frac{p}{q}\right) < \frac{q + s}{s \cdot 2q^2} \le \frac{1}{sq},$$

and this too is impossible. ■

These last two theorems show why continued fractions form a central tool in Diophantine approximation; in studying many questions involving the approximation of a single number, one can restrict attention entirely to continued fraction convergents.

PROBLEMS

1. Show by computation that $\sqrt{d}$ has a continued fraction expansion which is eventually periodic, for each nonsquare integer d with $2 \le d \le 15$. Examine the expansions you obtain, and make some conjectures.
2. Show that if both ξ and η have the initial partial quotients $a_0, a_1, \ldots, a_n$ and if $\xi < \theta < \eta$, then θ has these same initial partial quotients.
3. The recursion relation for q_k gives

$$q_1 = a_1, \quad q_2 = a_1 a_2 + 1, \quad q_3 = a_1 a_2 a_3 + a_1 + a_3, \quad \text{etc.}$$

   Show that always q_k results by summing all the terms that can be obtained from the product $a_1 a_2 \cdots a_k$ by omitting various numbers (≥ 0) of pairs of consecutive factors (With the understanding, as usual, that the empty product has value 1.) Deduce that the two rational numbers $\{a_0; a_1, \ldots, a_k\}$ and $\{b_0; a_k, \ldots, a_1\}$ always have the same denominator.
4. Find the exact interval on the real axis whose points have continued fraction expansions of the form $\{2; 1, 2, 1, 1, 4, 1, 1, \ldots\}$, in which only $a_0, \ldots, a_7$ are specified.
5. Show that $\xi \in \mathbf{R} \backslash \mathbf{Q}$ has bounded partial quotients a_k if and only if, for some $\delta > 0$, $|\xi - p/q| > \delta q^{-2}$ for every $p/q \in \mathbf{Q}$.
6. It is a consequence of Theorem 8.3 that if ξ is irrational, then to each positive integer t there corresponds at least one pair of integers x, y such that

$$\left|\xi - \frac{y}{x}\right| < \frac{1}{tx}, \qquad 1 \le x \le t.$$

   Show that this becomes false, for any irrational ξ and infinitely many t, if the second inequality above is replaced by $1 \le x \le t/2$. [*Hint:* Take $t = q_k + q_{k+1}$, and use Theorems 9.10 and 9.9.]

7. Suppose that $x \in \mathbf{R}$ has complete quotients x_k and convergents p_k/q_k. Show that x is a Liouville number if and only if for every $\omega > 0$ there is an index k such that $x_{k+1} > q_k^{\omega}$. Conclude that if e really does have the expansion mentioned at the beginning of the next section, then e is not a Liouville number.
8. Show that $\{0; 2^{1!}, 2^{2!}, \ldots, 2^{k!}, \ldots\}$ is transcendental. [*Hint:* Show first that $q_k \le 2^{k-1} a_k a_{k-1} \cdots a_1$ for $k \ge 1$.]
9. Show that at least one of every pair of successive convergents p_k/q_k ($k > 0$) satisfies (39), but that not every convergent need do so. [*Hint:* For the first part, assume the opposite and obtain the contradiction $(q_k q_{k+1})^{-1} \ge (q_k^{-2} + q_{k+1}^{-2})/2$.]
10. Given ξ with convergents p_k/q_k and partial quotients a_k, put

$$\frac{p_{k,a}}{q_{k,a}} = \frac{a p_{k-1} + p_{k-2}}{a q_{k-1} + q_{k-2}} \quad \text{for } a \ge 0,\ a \in \mathbf{Z},\ k \ge 1.$$

For $a = 0$ or $a = a_k$, these numbers are convergents; for $0 < a < a_k$, they are called **quasiconvergents**.

a) Prove that if k is even, then

$$\frac{p_{k-2}}{q_{k-2}} = \frac{p_{k,0}}{q_{k,0}} < \frac{p_{k,1}}{q_{k,1}} < \cdots < \frac{p_{k,a_k-1}}{q_{k,a_k-1}} < \frac{p_{k,a_k}}{q_{k,a_k}} = \frac{p_k}{q_k} < \xi < \frac{p_{k,a_k+1}}{q_{k,a_k+1}} < \frac{p_{k-1}}{q_{k-1}},$$

while the inequality signs are reversed when k is odd.

b) Show that always $q_{k-2} < q_{k,1} < \cdots < q_{k,a_k-1} < q_k$.

c) Show that if $0 < a < a_k$, then $p_{k,a-1}/q_{k,a-1}$ and $p_{k,a}/q_{k,a}$ are adjacent in the Farey sequence $\mathscr{F}_{q_{k,a}}$, as are $p_{k,a}/q_{k,a}$ and p_{k-1}/q_{k-1}.

d) Show that if for some k and a with $0 < a < a_k$,

$$\left|\xi - \frac{p}{q}\right| \le \left|\xi - \frac{p_{k,a}}{q_{k,a}}\right|$$

then either $q > q_{k,a}$ or $p/q = p_{k-1}/q_{k-1}$.

e) Show that if p/q lies between ξ and one of its convergents or quasiconvergents, then q is larger than the denominator of the latter.

f) Conversely, show that if p/q has the property that every rational number between ξ and p/q has denominator larger than q, then p/q is a convergent or quasi-convergent to ξ.

[*Hint:* The only tool needed for this entire investigation is Theorem 9.3, once the two inequalities involving ξ itself in (a) are understood.]

9.4 QUADRATIC IRRATIONALITIES

The problem of finding the best approximations to a real number x has thus been completely solved in terms of the regular continued fraction expansion of x. Of course, unless x is of a very special form it may be impossible to give the complete expansion of x, just as one cannot give the rule of formation for the digits occurring in the decimal expansion of π. But if a decimal approximation of x is known, a corresponding part of the continued fraction expansion of x can be determined. For example, from the series expansion for e, one can easily show that

$$2.7182 < e < 2.7183.$$

By a simple computation, one sees that

$$2.7182 = \{2; 1, 2, 1, 1, 4, 1, 1, 1, 3, 1, 9\},$$

and

$$2.7183 = \{2; 1, 2, 1, 1, 4, 1, 1, 19, 1, 1, 3\},$$

so that, by Problem 2 of the preceding section, $e = \{2; 1, 2, 1, 1, 4, 1, 1, \ldots\}$. (Actually, it is known that the sequence of partial quotients is

$$2, \quad 1, 2, 1, \quad 1, 4, 1, \quad 1, 6, 1, \quad \ldots, \quad 1, 2n, 1, \quad \ldots.)$$

There is, however, one simple case in which the complete expansion can be determined: that in which the partial quotients $a_0, a_1, a_2, \ldots$ constitute a sequence which is eventually periodic. Consider, for example, the continued fraction

$$\xi = 4 + \frac{1}{3+}\,\frac{1}{1+}\,\frac{1}{2+}\,\frac{1}{1+}\,\frac{1}{2+}\cdots,$$

where $a_{2n} = 1$ and $a_{2n+1} = 2$ for $n \geq 1$. To indicate the periodicity of the partial quotients, here and elsewhere, we place a vinculum above the period, and write $\xi = \{4; 3, \overline{1, 2}\}$. Then $\xi = \{4; 3, \xi_2\}$, where

$$\xi_2 = \{\overline{1; 2}\} = \{1; 2, \xi_2\},$$

so that

$$\xi_2 = 1 + \frac{1}{2 + 1/\xi_2} = 1 + \frac{\xi_2}{2\xi_2 + 1} = \frac{3\xi_2 + 1}{2\xi_2 + 1},$$

$$2\xi_2^2 - 2\xi_2 - 1 = 0,$$

$$\xi_2 = \frac{1 + \sqrt{3}}{2}.$$

(The plus sign is chosen since $\xi_2 > 0$.) Hence

$$\xi = 4 + \cfrac{1}{3 + \cfrac{1}{\cfrac{\sqrt{3} + 1}{2}}} = \frac{13\sqrt{3} + 21}{3\sqrt{3} + 5} = 6 - \sqrt{3}.$$

Conversely, we saw in the preceding section that the expansion of $\sqrt{7}$, for example, is eventually periodic: $\sqrt{7} = \{2; \overline{1, 1, 1, 4}\}$. We can now show that these are not isolated phenomena.

In the remainder of this chapter, it is convenient to use the term **quadratic irrationality** to mean a *real* algebraic number of degree 2 over **Q**.

Theorem 9.11 *Every eventually periodic regular continued fraction converges to a quadratic irrationality, and every quadratic irrationality has a regular continued fraction expansion which is eventually periodic.*

Proof. The first part is quite simple. Suppose that the first period begins with a_n, and let the length of the period be h, so that $a_{k+h} = a_k$ for $k \geq n$. Put

$$\xi = a_0 + \frac{1}{a_1 +} \cdots, \quad \text{and} \quad \xi_k = a_k + \frac{1}{a_{k+1} +} \cdots,$$

so that $\xi_{k+h} = \xi_k$ for $k \geq n$. By this and equation (31),

$$\xi = \frac{p_{n-1}\xi_n + p_{n-2}}{q_{n-1}\xi_n + q_{n-2}} = \frac{p_{n+h-1}\xi_n + p_{n+h-2}}{q_{n+h-1}\xi_n + q_{n+h-2}},$$

and therefore ξ_n satisfies a quadratic equation with integral coefficients. Since ξ_n is obviously not rational, it is a quadratic irrationality. By (31) again, the same is true of ξ itself, since if

$$A\xi_n^2 + B\xi_n + C = 0,$$

then

$$A(-q_{n-2}\xi + p_{n-2})^2 + B(-q_{n-2}\xi + p_{n-2})(q_{n-1}\xi - p_{n-1}) + C(q_{n-1}\xi - p_{n-1})^2 = 0,$$

and this is a quadratic equation in ξ.

The proof of the converse involves a little more computation. Suppose that

$$A\xi^2 + B\xi + C = 0,$$

where A, B, and C are integers and ξ is irrational. Then equation (31) gives

$$A(p_{k-1}\xi_k + p_{k-2})^2 + B(p_{k-1}\xi_k + p_{k-2})(q_{k-1}\xi_k + q_{k-2}) + C(q_{k-1}\xi_k + q_{k-2})^2 = 0,$$

or $A_k\xi_k^2 + B_k\xi_k + C_k = 0$, where the integers A_k, B_k, and C_k are given by the equations

$$\begin{aligned} A_k &= Ap_{k-1}^2 + Bp_{k-1}q_{k-1} + Cq_{k-1}^2, \\ B_k &= 2Ap_{k-1}p_{k-2} + B(p_{k-1}q_{k-2} + p_{k-2}q_{k-1}) + 2Cq_{k-1}q_{k-2}, \\ C_k &= Ap_{k-2}^2 + Bp_{k-2}q_{k-2} + Cq_{k-2}^2. \end{aligned}$$

If $f(x) = Ax^2 + Bx + C$, then

$$A_k = q_{k-1}^2 f\left(\frac{p_{k-1}}{q_{k-1}}\right), \qquad C_k = q_{k-2}^2 f\left(\frac{p_{k-2}}{q_{k-2}}\right).$$

Using Taylor's theorem, we have

$$A_k = q_{k-1}^2 \left\{ f(\xi) + f'(\xi)\left(\frac{p_{k-1}}{q_{k-1}} - \xi\right) + \tfrac{1}{2}f''(\xi)\left(\frac{p_{k-1}}{q_{k-1}} - \xi\right)^2 \right\},$$

since $f'''(x)$ is identically zero. Now $f(\xi) = 0$, and by (38),

$$\left|\xi - \frac{p_{k-1}}{q_{k-1}}\right| < \frac{1}{q_{k-1}^2}.$$

Hence

$$|A_k| < |f'(\xi)| + \frac{|f''(\xi)|}{2q_{k-1}^2},$$

and similarly

$$|C_k| < |f'(\xi)| + \frac{|f''(\xi)|}{2q_{k-2}^2},$$

so that $|A_k|$ and $|C_k|$ remain bounded as $k \to \infty$.

To see that $|B_k|$ is also bounded, we use the fact that all the quantities $B_k^2 - 4A_kC_k$ have the common value $B^2 - 4AC = D$. (This can be proved by a straightforward but tedious computation or, if one is acquainted with the theory of linear transformations, by noting that the quadratic form $A_kx'^2 + B_kx'y' + C_ky'^2$ is obtained from the form $Ax^2 + Bxy + Cy^2$ by the transformation

$$x = p_{k-1}x' + p_{k-2}y'$$
$$y = q_{k-1}x' + q_{k-2}y'$$

with determinant ± 1, and that two such forms have the same discriminant.) Since A_k and C_k are bounded and D is fixed, the quantity

$$B_k^2 = D + 4A_kC_k$$

must be bounded also.

Thus, there is a constant M such that

$$|A_k| < M, \qquad |B_k| < M, \qquad |C_k| < M$$

for all k. Since there are only finitely many triples of integers, each numerically smaller than M, there must be three indices, say n_1, n_2, and n_3, which give the same triple:

$$A_{n_1} = A_{n_2} = A_{n_3}, \qquad B_{n_1} = B_{n_2} = B_{n_3}, \qquad C_{n_1} = C_{n_2} = C_{n_3}.$$

Since the equation $A_{n_i}x^2 + B_{n_i}x + C_{n_i} = 0$ has only two roots, two of the numbers ξ_{n_1}, ξ_{n_2}, ξ_{n_3} must be equal; with proper naming, they can be taken to be ξ_{n_1} and ξ_{n_2}, where $n_1 < n_2$. If $n_2 - n_1 = h$, then $\xi_{n_1+h} = \xi_{n_1}$, and

$$\xi_{n_1+h+1} = \frac{1}{\xi_{n_1+h} - [\xi_{n_1+h}]} = \frac{1}{\xi_{n_1} - [\xi_{n_1}]} = \xi_{n_1+1},$$

$$\xi_{n_1+h+2} = \frac{1}{\xi_{n_1+h+1} - [\xi_{n_1+h+1}]} = \frac{1}{\xi_{n_1+1} - [\xi_{n_1+1}]} = \xi_{n_1+2},$$

and in general, $\xi_{k+h} = \xi_k$ for $k \geq n_1$. Thus the ξ_k's are eventually periodic. Since each a_k is determined exclusively by the corresponding ξ_k, the same is true of the a_k's. ■

A rather elaborate study has been made of the exact nature of the period, and we shall develop a little of what is known about it—when it commences, how long it is, and what shape it takes. The theory is especially simple for **reduced** quadratic irrationalities: ξ is reduced if it is larger than 1 while its conjugate $\bar{\xi}$ satisfies the inequality $-1 < \bar{\xi} < 0$.

Theorem 9.12 *Let ξ be a quadratic irrationality, with continued fraction expansion*

$$\xi = \{a_0; a_1, \ldots, a_{l-1}, \overline{a_l, \ldots, a_{l+h-1}}\}. \tag{40}$$

This expansion is purely periodic, so that $l = 0$, if and only if ξ is reduced.

Proof. As usual, let the complete quotients be ξ_k, so that

$$\xi = \{a_0; a_1, \ldots, a_{k-1}, \xi_k\} \quad \text{for } k \geq 1,$$

and suppose that $l \geq 0$ is the smallest index such that

$$\xi_{k+h} = \xi_k \quad \text{for } k \geq l. \tag{41}$$

We have

$$\xi_k = a_k + \frac{1}{\xi_{k+1}} \quad \text{and} \quad a_k = [\xi_k], \qquad k \geq 0,$$

and since the conjugate of a sum is the sum of the conjugates, and $a_k \in \mathbf{Z}$ for $k \geq 0$, we have

$$-\frac{1}{\overline{\xi_{k+1}}} = a_k + (-\overline{\xi_k}), \qquad k \geq 0.$$

(Here $\overline{\xi_k}$ denotes the conjugate of ξ_k, not the kth complete quotient of $\bar{\xi}$.) Thus if we put

$$\eta_k = (-\overline{\xi_k})^{-1} \qquad \text{for } k \geq 0,$$

then (since $\overline{\xi_0} = \bar{\xi}$) the conditions for ξ to be reduced become $\xi_0 > 1, \eta_0 > 1$, and the above equation becomes

$$\eta_{k+1} = a_k + \frac{1}{\eta_k}, \qquad k \geq 0. \tag{42}$$

Now suppose that ξ is reduced. Then $\eta_0 > 1$, and an easy induction using (42) shows that $\eta_k > 1$ for $k \geq 0$, so that

$$[\eta_{k+1}] = a_k, \qquad k \geq 0. \tag{43}$$

Finally, suppose $l > 0$. We obtain from (41) that $\eta_{l+h} = \eta_l$ and hence, by (43), $a_{l+h-1} = a_{l-1}$. But then by (42), $\eta_{l+h-1} = \eta_{l-1}$, so $\xi_{l+h-1} = \xi_{l-1}$. This is in contradiction with the minimality of l, so $l = 0$ if ξ is reduced.

To prove the converse, suppose that $l = 0$ in (40), so that $a_0 > 0$ since $a_h > 0$. Thus $\xi > 1$ and $\xi = \xi_h$, and hence

$$\xi = \{a_0; a_1, \ldots, a_{h-1}, \xi\} = \{\overline{a_0; a_1, \ldots, a_{h-1}}\}. \tag{44}$$

By (42)

$$\eta = \eta_h = \{a_{h-1}; \eta_{h-1}\} = \cdots = \{a_{h-1}; a_{h-2}, \ldots, a_0, \eta\},$$

so

$$\eta = \{\overline{a_{h-1}; a_{h-2}, \ldots, a_0}\} > 1, \tag{45}$$

so ξ is reduced. ■

Corollary *Suppose that $r > 1$ is a rational number but is not a rational square. Then the expansion of $\sqrt{r}$ has the form*

$$\sqrt{r} = \{a_0; \overline{a_1, \ldots, a_{h-1}, 2a_0}\},$$

where $a_1, \ldots, a_{h-1}$ has central symmetry: $a_1 = a_{h-1}, a_2 = a_{h-2}, \ldots$.

This is easily deduced from the facts that $\xi = \sqrt{r} + [\sqrt{r}]$ is reduced and that $\eta = (\xi - [\xi])^{-1} = \xi_1$, together with (44) and (45).

PROBLEMS

1. Give a complete proof of the Corollary to Theorem 9.12.

†2. Suppose that $\xi = (\sqrt{d} + A)/B$ with $A, B \in \mathbf{Z}$, $d \in \mathbf{Z}^+$, d not a square. (By the quadratic formula, every quadratic irrationality can be written in this form—but not in a unique way.)

   a) Show that if ξ as above is reduced, then

$$0 < A < \sqrt{d}, \qquad 0 < \sqrt{d} - A < B < \sqrt{d} + A < 2\sqrt{d}. \tag{46}$$

   [*Hint:* Use $\xi - \bar{\xi} > 0$, then $\xi + \bar{\xi} > 0$, then the individual inequalities for ξ and $\bar{\xi}$.]

   b) Show that if inequalities (46) hold, then ξ is reduced.

   c) Show that for fixed d, there are only finitely many reduced ξ's of the prescribed form, and that for each d there is at least one.

3. a) Show that if ξ is reduced, then so is every complete quotient ξ_k.

   b) Show that every reduced quadratic ξ can be written in the form $\xi = \xi_0 = (\sqrt{d} + A_0)/B_0$, where $A_0, B_0, d \in \mathbf{Z}^+$ and $(d - A_0^2)/B_0 \in \mathbf{Z}$. [*Hint:* Multiply by a/a if necessary.]

   c) Show that if ξ is as in (b), then for every $k \geq 0$, $\xi_k = (\sqrt{d} + A_k)/B_k$, where $A_k, B_k \in \mathbf{Z}^+$ and $(d - A_k^2)/B_k \in \mathbf{Z}$.

d) Show that $\ll \sqrt{d}\log d$ distinct pairs A_k, B_k can occur, in (c). [*Hint:* Show that for each $B \in (0, 2\sqrt{d})$, the inequalities (46) restrict A to an interval of length B or less, and conclude from Theorem 5.2 that $\ll \tau(B)$ integers in this interval satisfy the congruence implied by (c).] Deduce that the length h of the period of ξ is $\ll \sqrt{d}\log d$.

e) Infer that the length of the period in the expansion of $\sqrt{d}$, for $d \in \mathbf{Z}^+$ and d not a square, is $\ll \sqrt{d}\log d$.

f) Show that in the expansion of $\sqrt{d}$, each partial quotient a_k, for $0 \le k \le h$, satisfies $a_k < \sqrt{d}$.
(Continued in Section 9.5, Problem 5.)

9.5 APPLICATIONS TO PELL'S EQUATION AND TO FACTORIZATION

Theorem 9.13 *If N and d are integers with $d > 0$ and $|N| < \sqrt{d}$, and d is not a square, then all positive solutions of the Pell equation*

$$x^2 - dy^2 = N \tag{47}$$

are such that x/y is a convergent of the continued fraction expansion of $\sqrt{d}$.

Proof. Suppose that $x + y\sqrt{d}$ is a positive solution of (47). Then, if N is positive,

$$0 < x - y\sqrt{d} = \frac{N}{x + y\sqrt{d}} < \frac{\sqrt{d}}{x + y\sqrt{d}} = \frac{1}{(x/\sqrt{d}) + y} = \frac{1}{y((x/y\sqrt{d}) + 1)}.$$

Since $x/y > \sqrt{d}$, we have

$$\left|\sqrt{d} - \frac{x}{y}\right| < \frac{1}{2y^2}. \tag{48}$$

If N is negative, we deduce from the equation

$$y^2 - \frac{x^2}{d} = \frac{-N}{d}$$

the relations

$$0 < y - \frac{x}{\sqrt{d}} = \frac{-N/d}{y + (x/\sqrt{d})} < \frac{1}{y\sqrt{d} + x} = \frac{1}{x(1 + (y\sqrt{d}/x))}.$$

and

$$\left|\frac{1}{\sqrt{d}} - \frac{y}{x}\right| < \frac{1}{2x^2}. \tag{49}$$

Now if $\sqrt{d} = \{a_0; a_1, \ldots\}$, then $1/\sqrt{d} = \{0; a_0, a_1, \ldots\}$, so that the convergents of the continued fraction expansion of $1/\sqrt{d}$ are 0/1 and the reciprocals of the convergents of the continued fraction expansion of $\sqrt{d}$. Using this, the inequalities (48) and (49), and Theorem 9.10, we have the result. ∎

This theorem provides an effective method of determining all integers N, numerically smaller than $\sqrt{d}$, for which equation (47) is solvable, for it happens that the sequence $\{p_k^2 - dq_k^2\}$ is eventually periodic, and consequently only finitely many values of k need be examined. To see this, put $\xi = \sqrt{d}$ and

$$\sqrt{d} = \frac{p_{k-1}\xi_k + p_{k-2}}{q_{k-1}\xi_k + q_{k-2}}. \tag{50}$$

Solving for ξ_k and rationalizing the denominator, we can write

$$\xi_k = \frac{\sqrt{d} + r_k}{s_k}, \qquad k \geq 1,$$

where r_k and s_k are rational numbers. Substituting this back into (50), and replacing k by $k + 1$ throughout, we have

$$\sqrt{d} = \frac{p_k(\sqrt{d} + r_{k+1}) + p_{k-1}s_{k+1}}{q_k(\sqrt{d} + r_{k+1}) + q_{k-1}s_{k+1}},$$

or

$$(q_k r_{k+1} + q_{k-1}s_{k+1} - p_k)\sqrt{d} - (p_{k-1}s_{k+1} + p_k r_{k+1} - q_k d) = 0.$$

The rational and irrational parts must separately be zero, so

$$\begin{aligned} q_k r_{k+1} + q_{k-1}s_{k+1} &= p_k, \\ p_k r_{k+1} + p_{k-1}s_{k+1} &= q_k d. \end{aligned}$$

The determinant of this system is $q_k p_{k-1} - q_{k-1}p_k = (-1)^k$, so that

$$\begin{aligned} r_{k+1} &= (-1)^k(p_k p_{k-1} - q_k q_{k-1} d), \\ s_{k+1} &= (-1)^k(q_k^2 d - p_k^2). \end{aligned} \tag{51}$$

Now the numbers r_k and s_k are uniquely determined by ξ_k; since $\{\xi_k\}$ is eventually periodic, the same is true of $\{s_k\}$, and the eventual periodicity of $\{p_k^2 - dq_k^2\}$ follows from (51). (Incidentally, we see that r_k and s_k are always integers.)

The discussion of Pell's equation with $N = \pm 1$, in Chapter 8, had the serious drawback that no effective method was given for finding the fundamental solution, nor even of deciding when one exists for $N = -1$. The results obtained above entirely clarify these points: the first solution encountered, being the smallest, is the fundamental solution, and the equation $x^2 - dy^2 = -1$ is solvable if and only if a solution exists among the convergents of $\sqrt{d}$ up to the end of the second period. (For $\{s_k\}$ becomes periodic at the same point as $\{\xi_k\}$, and $\{(-1)^{k-1}s_k\}$ has period at most twice that of $\{s_k\}$.) It can be shown by the method sketched in Problem 4, below, that $s_k = 1$ for the first time at the end of the first period, so that the preceding convergent is the fundamental solution of one of the equations; if for this convergent $p_k^2 - dq_k^2 = 1$, then the equation $x^2 - dy^2 = -1$ is unsolvable, while if $p_k^2 - dq_k^2 = -1$, the convergents preceding ends of periods are alternately solutions for $N = -1$ and $N = 1$.

Legendre put equation (51) to quite a different use, namely to the factorization of large integers. Suppose N is to be factored, let m be an arbitrary but temporarily fixed positive integer, and put $d = mN$. Then (51) gives

$$q_k^2 mN = p_k^2 + (-1)^k s_{k+1}$$

and by Problem 2(a) of the preceding section, $0 < s_{k+1} < 2\sqrt{mN}$. Thus for a small multiplier m, s_{k+1} is much smaller than N, and perhaps it can be factored. If so, it can be written as $s_{k+1} = x_k y_k^2$, where x_k is square-free. Then if $p \mid N$, $p \mid (p_k^2 \pm x_k y_k^2)$ so $\pm x_k$ must be a quadratic residue of p, and in the usual way (Section 5.5) this sieves out the primes lying in half the residue classes (mod $4x_k$). If N is large, the length of the period in the expansion of $\sqrt{mN}$, which is the number of values of k available, is likely to be large, and of course various small values of m can be used.

This idea has been modified for use on modern computers, for example in the following way. Spend very little time on each s_k, discarding those which do not split completely into primes less than 1000, say. Keep track of the factorizations of the s_k's retained, and look for index sets I such that $\prod_{k \in I} (-1)^k s_k$ is a square, say S_I^2. Then

$$P_I^2 = \prod_{k \in I} p_{k-1}^2 \equiv \prod_{k \in I} (-1)^k s_k = S_I^2 \pmod{N}.$$

Compute P_I and S_I (mod N), and then $(P_I - S_I, N) = D_I$. Since $P_I^2 - S_I^2 = NM$, it is not unlikely, if N is not prime, that $D_I > 1$. If so, we have found a factor of N; if not, continue factoring the numbers s_k.

Using this approach, many very large numbers have recently been factored, including the seventh Fermat number $2^{2^7} + 1$, a number of 39 digits which had been known to be composite since 1905, but with no factors known.

PROBLEMS

1. For what N with $|N| < \sqrt{7}$ is the equation $x^2 - 7y^2 = N$ solvable?
2. Find the fundamental solution of $x^2 - 95y^2 = 1$; of $x^2 - 74y^2 = 1$.
3. Show, by induction or otherwise, that the numbers r_k and s_k defined in this section are positive integers.
4. a) Show that if the length of the period in the expansion of $\sqrt{d}$ is h, then $s_h = 1$, and hence that

$$p_{h-1}^2 - dq_{h-1}^2 = (-1)^h.$$

Thus $x^2 - dy^2 = -1$ is solvable if h is odd.

b) Using the fact that the numbers $\xi, \xi_1, \ldots, \xi_{h-1}$ are distinct, show that $s_k > 1$ if $1 \leq k \leq h - 1$. Deduce that the equation $x^2 - dy^2 = -1$ is solvable only if h is odd.

5. Using results from the problems of Section 9.4, conclude that there is a positive constant c such that the fundamental unit ε of every real quadratic field $Q(\sqrt{d})$ satisfies the inequality

$$\varepsilon < d^{c\sqrt{d}\log d}.$$

[*Hint:* Recall the inequality of Problem 8 of Section 9.3.]

9.6 EQUIVALENCE OF NUMBERS

Because each element of the sequence $\{\xi_k\}$ depends only on the preceding one, and because the defining rule

$$\xi_k = [\xi_k] + \frac{1}{\xi_{k+1}}$$

is the same for all $k \geq 1$, it is clear that if $\xi = \{a_0; a_1, a_2, \ldots\}$ and also

$$\xi = a_0 + \frac{1}{a_1 +} \cdots \frac{1}{a_{n-1} +} \frac{1}{\xi_n},$$

then

$$\xi_n = a_n + \frac{1}{a_{n+1} +} \cdots.$$

If we are interested in the possibility of finding infinitely many solutions of the inequality

$$\left|\xi - \frac{p}{q}\right| < \frac{1}{cq^2}, \qquad c \geq 2,$$

equation (36) and Theorem 9.10 show that we need only examine the numbers ξ_k with large k. For this reason, we shall term two irrational numbers ξ and ξ' **equivalent** if, for some j and k, $\xi'_j = \xi_k$. Then $\xi'_{j+m} = \xi_{k+m}$ for $m \geq 0$, and by the above remark, this means that if

$$\xi = \{a_0; a_1, \ldots, a_{k-1}, a_k, a_{k+1}, \ldots\},$$

then

$$\xi' = \{b_0; b_1, \ldots, b_{j-1}, a_k, a_{k+1}, \ldots\},$$

so that

$$\xi = \frac{p_{k-1}\xi_k + p_{k-2}}{q_{k-1}\xi_k + q_{k-2}}, \qquad \xi' = \frac{p'_{j-1}\xi_k + p'_{j-2}}{q'_{j-1}\xi_k + q'_{j-2}}. \tag{52}$$

Theorem 9.14 *Two irrational numbers ξ and ξ' are equivalent, in the sense that their continued fraction expansions are identical from some points on, if and only if there are integers A, B, C, and D such that*

$$\xi' = \frac{A\xi + B}{C\xi + D}, \quad \text{where } AD - BC = \pm 1. \tag{53}$$

Proof. Suppose ξ and ξ' are equivalent. Eliminating ξ_k from the equations (52) gives

$$\frac{-q_{k-2}\xi + p_{k-2}}{q_{k-1}\xi - p_{k-1}} = \frac{-q'_{j-2}\xi' + p'_{j-2}}{q'_{j-1}\xi' - p'_{j-1}}$$

or

$$\xi = \frac{A\xi' + B}{C\xi' + D},$$

where

$$A = p_{k-1}q'_{j-2} - p_{k-2}q'_{j-1}, \qquad B = p_{k-2}p'_{j-1} - p_{k-1}p'_{j-2},$$
$$C = q_{k-1}q'_{j-2} - q_{k-2}q'_{j-1}, \qquad D = q_{k-2}p'_{j-1} - q_{k-1}p'_{j-2}.$$

A simple calculation shows that

$$AD - BC = (p'_{j-1}q'_{j-2} - p'_{j-2}q'_{j-1})(p_{k-1}q_{k-2} - p_{k-2}q_{k-1}) = \pm 1,$$

so that (53) holds. (This can all be done with matrices, interpreting (53), for example, as

$$\begin{bmatrix} x' \\ y' \end{bmatrix} = \begin{bmatrix} A & B \\ C & D \end{bmatrix}\begin{bmatrix} x \\ y \end{bmatrix}, \qquad \xi' = x'/y', \qquad \xi = x/y.)$$

To complete the proof, suppose that equation (53) holds. By replacing A, B, C, and D by their negatives if necessary, we may suppose also that $C\xi + D > 0$. Substituting the value of ξ from (52) into (53) gives

$$\xi' = \frac{a\xi_k + b}{c\xi_k + d}, \tag{54}$$

where

$$a = Ap_{k-1} + Bq_{k-1}, \qquad b = Ap_{k-2} + Bq_{k-2},$$
$$c = Cp_{k-1} + Dq_{k-1}, \qquad d = Cp_{k-2} + Dq_{k-2},$$

and

$$ad - bc = (AD - BC)(p_{k-1}q_{k-2} - p_{k-2}q_{k-1}) = \pm 1.$$

By the inequality (38),

$$p_{k-1} = q_{k-1}\xi + \frac{\delta_{k-1}}{q_{k-1}}, \qquad p_{k-2} = q_{k-2}\xi + \frac{\delta_{k-2}}{q_{k-2}},$$

where

$$|\delta_{k-1}| < 1, \qquad |\delta_{k-2}| < 1.$$

Hence

$$c = (C\xi + D)q_{k-1} + \frac{C\delta_{k-1}}{q_{k-1}}, \qquad d = (C\xi + D)q_{k-2} + \frac{C\delta_{k-2}}{q_{k-2}}.$$

Since $C\xi + D$, q_{k-1}, and q_{k-2} are positive, and since $q_{k-1} > q_{k-2}$ and $q_k \to \infty$ with k, we have $c > d > 0$ for k sufficiently large. This is exactly the situation hypothesized in Theorem 9.15 below (with ξ_k in place of ξ), so it follows that b/d and a/c are successive convergents of the continued fraction expansion of ξ', for k sufficiently large:

$$a = p'_{j-1}, \quad b = p'_{j-2}, \quad c = q'_{j-1}, \quad d = q'_{j-2},$$

for suitable j. But then

$$\xi' = \frac{p'_{j-1}\xi_k + p'_{j-2}}{q'_{j-1}\xi_k + q'_{j-2}} = \frac{p'_{j-1}\xi'_j + p'_{j-2}}{q'_{j-1}\xi'_j + q'_{j-2}}$$

and it follows that $\xi_k = \xi'_j$, as was to be proved. ■

For the above proof we used the following theorem, which has some independent interest.

Theorem 9.15 *If a, b, c, and d are integers, and*

$$\xi' = \frac{a\xi + b}{c\xi + d}, \quad ad - bc = \pm 1, \quad \xi > 1, \quad c > d > 0,$$

then b/d and a/c are successive convergents of the continued fraction expansion of ξ'.

Proof. For each $p/q \in \mathscr{F}_n$, put

$$R_n(p/q) = \left\{x \in \mathbf{R}: \min_{r/s \in \mathscr{F}_n} (|sx - r|) = |qx - p|\right\}.$$

Clearly $p/q \in R_n(p/q)$, and it is shown below that $R_n(p/q)$ is contained in the interval lying between the two elements of $\mathscr{F}_n$ adjacent to p/q. If u/v is one of the two adjacent elements, then $|qx - p| = |vx - u|$ for that x for which $qx - p = -(vx - u)$, namely $x = (p + u)/(q + v)$, the mediant. So $R_n(p/q)$ is the interval connecting the two mediants straddling p/q.

Under the hypotheses of the theorem, a/c and b/d are adjacent in $\mathscr{F}_c$, and either

$$\frac{a}{c} < \xi' = \frac{a\xi + b}{c\xi + d} < \frac{a + b}{c + d} < \frac{b}{d}$$

or this entire inequality should be reversed. In either case, by what we have just proved, $\xi' \in R_c(a/c)$, so a/c is a best approximation to ξ'. On the other hand, $a/c \notin \mathscr{F}_{c-1}$, so a/c is the mediant at one end of the interval $R_{c-1}(b/d)$, so

$\xi' \in R_{c-1}(b/d)$, so b/d is also a best approximation to ξ', and in fact it is the one immediately preceding a/c. Hence b/d and a/c are successive convergents.

It remains to show that if p/q and r/s are adjacent in $\mathscr{F}_n$, then no point x between them belongs to any $R_n(t/u)$, if t/u is neither p/q nor r/s. Suppose that

$$\frac{t}{u} < \frac{p}{q} < x < \frac{r}{s};$$

the other possible order, in which $t/u > r/s$, is treated similarly. If $q \leq u$, then

$$0 < qx - p = q\left(x - \frac{p}{q}\right) < q\left(x - \frac{t}{u}\right) \leq u\left(x - \frac{t}{u}\right) = ux - t,$$

so if the assertion is false, it must be that $q > u$. But then if $qx - p \geq ux - t$, so that

$$x \geq \frac{p - t}{q - u},$$

we have

$$0 < r - sx \leq r - s\left(\frac{p - t}{q - u}\right) = \frac{(qr - sp) - (ur - st)}{q - u} \leq 0,$$

since $qr - sp = 1$ while $ur - st \geq 1$. This contradiction shows that the set $R_n(p/q)$ does not extend past the two elements to which p/q is adjacent, for every $p/q \in \mathscr{F}_n$, so each such set is an interval. ■

We shall use the symbol "$\cong$" to designate equivalence in the regular continued fraction sense. The notion of equivalence, together with equation (36), can be used to gain new insight concerning the so-called **Markov constant** $M(\xi)$, which is the upper limit of those numbers λ such that the inequality

$$\left|\xi - \frac{p}{q}\right| < \frac{1}{\lambda q^2}$$

has infinitely many solutions p, q. From (36), it is clear that $|q_k^2(\xi - p_k/q_k)|$ is approximately inversely proportional to ξ_k, so that $M(\xi)$ will probably have its smallest value for those ξ for which $a_k = 1$ for all large k. Now if

$$\xi = 1 + \frac{1}{1 +}\,\frac{1}{1 +}\cdots,$$

then

$$\xi = 1 + \frac{1}{\xi}, \quad \xi = \frac{1 + \sqrt{5}}{2}.$$

These remarks lead one to expect that the first part of the following theorem might be true.

Theorem 9.16 *If $\xi \cong \xi'$, then $M(\xi) = M(\xi')$. If $\xi \cong (1 + \sqrt{5})/2$, then $M(\xi) = \sqrt{5}$. If ξ is irrational and not equivalent to $(1 + \sqrt{5})/2$, then $M(\xi) \geq \sqrt{8}$. If $\xi \cong \sqrt{2}$, then $M(\xi) = \sqrt{8}$. If ξ is not equivalent to either $(1 + \sqrt{5})/2$ or $\sqrt{2}$, and is irrational, then $M(\xi) \geq 17/6$.*

Proof. By (36), $M(\xi) = \limsup_{k\to\infty} (\xi_{k+1} + q_{k-1}/q_k)$. Now

$$\frac{q_{k-1}}{q_k} = \cfrac{1}{a_k + \cfrac{q_{k-2}}{q_{k-1}}},$$

and iterating this reduction formula gives $q_{k-1}/q_k = \{0; a_k, a_{k-1}, \ldots, a_1\}$. Thus

$$M(\xi) = \limsup_{k\to\infty} (\{a_{k+1}; a_{k+2}, \ldots\} + \{0; a_k, a_{k-1}, \ldots, a_1\}). \tag{55}$$

If $\xi' \cong \xi$, then $\xi'_j = \xi_k$ and $a'_j = a_k$ for all sufficiently large j and k for which $j - k$ has a suitable fixed value h. If the convergents of ξ' are p'_j/q'_j, then for such j and k the continued fraction expansions of q_{k-1}/q_k and q'_{j-1}/q'_j have the same partial quotients at the beginning, and the interval of agreement can be made arbitrarily long by choosing j and k sufficiently large. Hence, although q_{k-1}/q_k and q'_{k-1}/q'_k may not even have limits individually,

$$\lim_{\substack{k\to\infty\\ j-k=h}} \left(\frac{q_{k-1}}{q_k} - \frac{q'_{j-1}}{q'_j}\right) = 0,$$

so that

$$\lim_{\substack{k\to\infty\\ j-k=h}} \left\{\left(\xi_k + \frac{q_{k-1}}{q_k}\right) - \left(\xi'_j + \frac{q'_{j-1}}{q'_j}\right)\right\} = 0,$$

and so $M(\xi) = M(\xi')$.

To prove the second assertion of the theorem, we need only notice that

$$M\left(\frac{1 + \sqrt{5}}{2}\right) = \lim_{k\to\infty} \left\{\left(1 + \frac{1}{1 +} \cdots\right) + \underbrace{\left(\frac{1}{1 +} \frac{1}{1 +} \cdots \frac{1}{1}\right)}_{k \text{ terms}}\right\}$$

$$= \frac{1 + \sqrt{5}}{2} + \frac{1}{(1 + \sqrt{5})/2} = \sqrt{5},$$

by (55).

To prove the third part, we may suppose that $a_{k+1} \geq 2$ for infinitely many indices k. If $a_{k+1} \geq 3$ for infinitely many k, it is clear from (55) that $M(\xi) \geq 3$. Since $\sqrt{8} < 3$, we need only consider those ξ's for which a_k is either 1 or 2 for all large k. If there are infinitely many 1's and 2's, there are infinitely many values of

k such that $a_k = 1$, $a_{k+1} = 2$. But then, since the value of a continued fraction is always at least equal to its convergent with index 2,

$$a_{k+1} + \frac{1}{a_{k+2} +} \cdots \geq 2 + \frac{1}{a_{k+2} + 1/a_{k+3}} \geq 2 + \frac{1}{2 + 1/1} = \frac{7}{3}$$

and

$$\frac{1}{a_k +} \frac{1}{a_{k-1} +} \cdots \frac{1}{a_1} \geq \frac{1}{1 + 1/a_{k-1}} \geq \frac{1}{1 + 1/1} = \frac{1}{2},$$

so that

$$M(\xi) \geq \frac{7}{3} + \frac{1}{2} = \frac{17}{6} = 2.833\ldots > \sqrt{8}.$$

On the other hand, if $a_k = 2$ for all large k, then

$$\xi \cong 1 + \frac{1}{2 +} \frac{1}{2 +} \cdots = \sqrt{2},$$

and

$$M(\xi) = \lim_{k \to \infty} \left\{ \left(2 + \frac{1}{2 +} \cdots \right) + \left(\underbrace{\frac{1}{2 +} \frac{1}{2 +} \cdots \frac{1}{2}}_{k \text{ terms}} \right) \right\}$$

$$= (\sqrt{2} + 1) + (\sqrt{2} - 1) = \sqrt{8}. \quad \blacksquare$$

Note that Hurwitz's theorem is a consequence of Theorem 9.16. The earlier proof of that theorem was a good one in the sense that it required very little apparatus and no prior knowledge that $\sqrt{5}$ was the right constant. But once we have available the whole machinery of continued fractions, we gain much greater insight into the question of what numbers can be approximated only very poorly by rational numbers.

For the remainder of this section we suppose that the reader is familiar with the notion of a countably infinite set, and with the fact that the real numbers are not countable.

Theorem 9.16 contains the first two of an infinite sequence of assertions about the values less than 3 assumed by $M(\xi)$. A. A. Markov showed that there are only countably many such values, that their sole limit point is 3, and that each such value corresponds precisely to the set of numbers equivalent to a certain quadratic irrationality. There is no really simple proof known. We can show, however, that the set of ξ for which $M(\xi) = 3$ is of quite a different type.

Theorem 9.17 *There are uncountably many numbers ξ such that $M(\xi) = 3$, and such that no two of the ξ's are equivalent.*

Proof. Let $r_1, r_2, \ldots$ be a strictly increasing sequence of positive integers, and let

$$(56) \qquad \xi = \{1; \underbrace{1, \ldots, 1}_{r_1}, 2, 2, \underbrace{1, \ldots, 1}_{r_2}, 2, 2, 1, \ldots\},$$

where there are r_1 partial quotients 1, then two 2's, then r_2 1's, then two 2's, then r_3 1's, etc. Thus two blocks consisting entirely of 1's are always separated by two 2's, and the blocks of 1's become steadily longer. Let

$$\beta_k = \xi_{k+1} + \frac{q_{k-1}}{q_k} = \left(a_{k+1} + \frac{1}{a_{k+2} +} \cdots\right) + \left(\frac{1}{a_k +} \frac{1}{a_{k-1} +} \cdots \frac{1}{a_1}\right).$$

If we choose k so that $a_{k+1} = 1$, then clearly $\xi_{k+1} < 2$, $q_{k-1}/q_k < 1$, and $\beta_k < 3$. If k runs through a sequence of indices such that $a_{k+1} = a_{k+2} = 2$, then since the $r_i \to \infty$,

$$\beta_k = \left(2 + \frac{1}{2 +} \frac{1}{1 +} \frac{1}{1 +} \cdots\right) + \left(\frac{1}{1 +} \frac{1}{1 +} \cdots \frac{1}{1}\right)$$

$$\to \left(2 + \frac{1}{2 + (\sqrt{5} - 1)/2}\right) + \frac{\sqrt{5} - 1}{2} = 3.$$

On the other hand, if k runs through a sequence of indices for which $a_k = a_{k+1} = 2$, then

$$\beta_k = \left(2 + \frac{1}{1 +} \frac{1}{1 +} \cdots\right) + \left(\frac{1}{2 +} \frac{1}{1 +} \frac{1}{1 +} \cdots \frac{1}{1}\right)$$

$$\to 2 + \frac{1}{(1 + \sqrt{5})/2} + \frac{1}{2 + (\sqrt{5} - 1)/2} = 3.$$

Hence $M(\xi) = \limsup \beta_k = 3$.

To complete the proof, it suffices to show that the set of inequivalent ξ's defined as in (56) is not countable. Now ξ and ξ' are equivalent if and only if the sequences $r_1, r_2, \ldots$ and $r_1', r_2', \ldots$ associated with them are identical from some points on, so that we can transfer the notion of equivalence from the numbers ξ and ξ' to the sequences $\{r_k\}$ and $\{r_k'\}$. Suppose that the inequivalent sequences among all the increasing sequences of positive integers can themselves be arranged in a sequence, say $R_1, R_2, \ldots$, where R_i stands for the sequence $r_{i1}, r_{i2}, \ldots$, with $r_{i1} < r_{i2} < \cdots$. With proper naming we can suppose that R_1 is the sequence $1, 2, 3, \ldots$ of all positive integers in order. If $i > 1$, R_i is not equivalent to R_1, and there are therefore infinitely many positive integers not included in it.

For $i > 1$ let $S_i = \{s_{ik}\}$ be the sequence complementary to R_i; that is, the positive integers, ordered by size, which do not occur in R_i. Each S_i is an infinite sequence. Now define a sequence T as follows. Pick t_1 in S_2, and then successively choose $t_2, t_3, \ldots$ so that

$$\begin{array}{llll} t_1 \in S_2, & t_1 < t_2 \in S_3, & & \\ 1 + t_2 < t_3 \in S_2, & t_3 < t_4 \in S_3, & t_4 < t_5 \in S_4, & \\ 1 + t_5 < t_6 \in S_2, & t_6 < t_7 \in S_3, & t_7 < t_8 \in S_4, & t_8 < t_9 \in S_5, \\ \vdots & & & \end{array}$$

From this scheme it is apparent that T is an increasing sequence of integers, infinitely many of which are contained in S_k, and therefore not contained in R_k, for arbitrary $k \geq 2$. Hence T is certainly not equivalent to any of $R_2, R_3, \ldots$. Since each element $t_3, t_6, t_{10}, \ldots$ of T which lies in S_2 exceeds its predecessor in T by more than one, T is also not equivalent to R_1. Hence T is not equivalent to any R_k, contrary to the hypothesis that the sequence $\{R_k\}$ contains an element equivalent to any increasing sequence of positive integers. ■

Corollary *There are real transcendental numbers ξ with $M(\xi) = 3$.*

For the real algebraic numbers are certainly countable: order the polynomials $p(x) = a_0 + a_1x + \cdots + a_nx^n \in \mathbf{Z}[x]$ according to the size of $H(p) = n + |a_0| + \cdots + |a_n|$, so that $H(p)$ is nondecreasing; there are only finitely many $p(x)$ with given value of $H(p)$, and each $p(x)$ has only finitely many zeros, which can be listed in arbitrary order. Delete the duplications, and the nonreal elements. ■

Thus one must take care not to be misled by the Liouville numbers. It is true that every number which can be approximated sufficiently well by rational numbers is transcendental, but there are also transcendental numbers all of whose rational approximations are almost as bad as is possible in view of Hurwitz's theorem.

PROBLEMS

1. Are the numbers $\sqrt{5}$ and $(1 + \sqrt{5})/2$ equivalent? What about $\sqrt{3}$ and $(1 + \sqrt{3})/2$?
2. Show that every quadratic irrationality is equivalent to a reduced one. Is the latter unique?
3. Show that two conjugate quadratic irrationalities have reverse periods; for example, $(4 + \sqrt{10})/3 = \{2; \overline{2, 1, 1}\}$ and $(4 - \sqrt{10})/3 = \{0; 3, \overline{1, 1, 2}\}$. [*Hint:* Follow the path $\xi \to \xi_k$ (reduced) $\to -1/\bar{\xi}_k \to \bar{\xi}$.]

9.7 THE TRANSCENDENCE OF e

Nearly 30 years elapsed after Liouville's proof of the existence of transcendental numbers before a proof was found of the transcendence of a number given "naturally"; this was C. Hermite's proof for e. Then in 1882, F. Lindemann proved that π is transcendental, thus settling in the negative the Greek question of whether, given a circle, a square can be constructed with ruler and compass such that the two figures have equal area. Actually, Lindemann proved a quite general theorem about values of the exponential function, which implies that $z \neq 0$ and e^z cannot simultaneously be algebraic numbers; for $z = 1$ this gives Hermite's theorem again, and for $z = \pi i$ yields the transcendence of π, since $e^{\pi i} = -1$ is algebraic.

A sizable portion of Hermite's long career was devoted to those two nineteenth-century sirens, invariant theory and the theory of elliptic and abelian functions, but he also made many other beautiful advances in algebra, analysis, and number theory. He brought analysis into the arithmetic theory of forms, providing a reduction theory for "n-ary" quadratic form; he solved the general quintic equation in terms of elliptic functions, and he obtained a weaker version of the structure theorem on units (Section 8.4) independently of Dirichlet. He gave of himself generously to his own students and, through voluminous correspondence, to young mathematicians throughout the world.

Charles Hermite (1822–1901)

Hermite's proof was long and difficult, and a number of simpler proofs were found later. Perhaps it became something of a contest; at any rate, one issue of *Mathematische Annalen* in 1893 contained three simple proofs, by Hilbert, Hurwitz, and P. Gordan. We present Hilbert's proof, which is as elegant as it is mysterious.

Theorem 9.18 *e is transcendental.*

Proof. We see by integration by parts that for $k \in \mathbf{Z}^+$,

$$\int_0^\infty x^k e^{-x}\,dx = -x^k e^{-x}\Big]_0^\infty + k\int_0^\infty x^{k-1}e^{-x}\,dx$$

$$= k\int_0^\infty x^{k-1}e^{-x}\,dx,$$

so that if I_k is the value of the integral on the left, then

$$I_k = kI_{k-1} = k(k-1)I_{k-2} = \cdots = k!\,I_0 = k!.$$

As a student and young faculty member at Königsberg (his birthplace, now Kaliningrad), Hilbert was a student of Lindemann's and a friend of Hurwitz and Minkowski; not unpredictably, most of his work until 1897 was in algebra (invariants and ideal theory) and number theory. What was supposed to have been a report on the state of algebraic number theory (the *Zahlbericht* of 1897) became instead a creative reworking of the entire subject, the touchstone of twentieth-century research. Hilbert was also a powerful analyst, as his solution of Waring's problem demonstrated. He held a chair at Göttingen from 1895 until his retirement in 1930.

David Hilbert (1862–1943)

Since $k! \mid (k+1)!$, it follows that if $p(x) \in \mathbf{Z}[x]$ has constant term $p(0)$, then for $m \geq 0$,

$$\int_0^\infty x^m p(x) e^{-x}\, dx \equiv p(0) m! \pmod{(m+1)!}, \tag{57}$$

and consequently this last integral does not vanish if $(m+1) \nmid p(0)$.

With this preliminary fact available, we can proceed to the proof. Suppose that e is algebraic, satisfying the equation

$$a_0 + a_1 e + \cdots + a_n e^n = 0, \qquad a_0, \ldots, a_n \in \mathbf{Z},\ a_0 \neq 0. \tag{58}$$

Let r be a positive integer, to be specified more exactly later, and with these values of n and r define the quantity

$$\int_b^c = \int_b^c x^r \{(x-1)\cdots(x-n)\}^{r+1} e^{-x}\, dx, \quad \text{for } 0 \leq b < c \leq \infty.$$

Then upon multiplying through in (58) by $\int_0^\infty$, we obtain

$$P_1 + P_2 = 0,$$

where

$$\begin{aligned} P_1 &= a_0 \int_0^\infty + a_1 e \int_1^\infty + \cdots + a_n e^n \int_n^\infty, \\ P_2 &= \qquad\quad a_1 e \int_0^1 + \cdots + a_n e^n \int_0^n. \end{aligned} \tag{59}$$

We shall reach a contradiction by showing that $P_1/r!$ is a nonzero integer while $|P_2|/r! < 1$, for suitable r.

By making the change of variable $x - k = y$, we obtain

$$a_k e^k \int_k^\infty = a_k \int_k^\infty x^r\{(x-1)\cdots(x-n)\}^{r+1} e^{-(x-k)}\,dx$$

$$= a_k \int_0^\infty (y+k)^r\{(y+k-1)\cdots(y+k-n)\}^{r+1} e^{-y}\,dy$$

$$= \begin{cases} a_0 \int_0^\infty y^r p_0(y) e^{-y}\,dy & \text{for } k = 0, \\ a_k \int_0^\infty y^{r+1} p_k(y) e^{-y}\,dy & \text{for } 0 < k \le n, \end{cases}$$

where $p_0(y), \ldots, p_n(y)$ are certain polynomials in $\mathbf{Z}[y]$. Thus by (57), all terms in the expression (59) for P_1 are integers, and all terms after the first are multiples of $(r+1)!$, so

$$P_1 \equiv a_0 p_0(0) r! \equiv a_0(-1)^{n(r+1)}(n!)^{r+1} r! \pmod{(r+1)!}.$$

Hence if

$$(r + 1, a_0 n!) = 1, \tag{60}$$

then P_1 is a nonzero integer, and is a multiple of $r!$.

To bound $|P_2|$, define

$$M = \max_{0 \le x \le n} |x(x-1)\cdots(x-n)|,$$

$$N = \max_{0 \le x \le n} |(x-1)\cdots(x-n)e^{-x}|.$$

Then for $1 \le k \le n$,

$$\left|a_k \int_0^k\right| \le |a_k| \int_0^k M^r N\,dx = k|a_k|M^r N,$$

so

$$|P_2| \le (|a_1|e + 2|a_2|e^2 + \cdots + n|a_n|e^n)M^r N.$$

Since $M^r = o(r!)$ for fixed M, it follows that $|P_2| < r!$ for all sufficiently large r. For some such values of r, (60) holds, and we have a contradiction. Hence (58) is false. ■

NOTES AND REFERENCES

Section 9.1

Theorem 9.1, published by J. Liouville in 1844, provided the first proof of the existence of transcendental numbers. It was also the first indication that transcendency and approximation questions are connected. Theorem 9.2 reflects the spirit of the other early existence proof, due to G. Cantor (1874), who pointed out that the set of algebraic numbers is countable while **R** is not.

Joseph Liouville (1809–1882) wrote some 400 papers in analysis, mathematical physics, algebra, and number theory. Two of his most important contributions to mathematics were not research articles; he initiated and edited for 39 years the *Journal de mathématiques pures et appliquées,* and he devoted three years to assimilating, organizing, and publishing Galois's works, which were left in a chaotic state when the author was killed in a duel at the age of 19.

Section 9.2

The "derivation" here of the continued fraction algorithm as the procedure for finding best approximations was originally inspired by Cassels, Ledermann, and Mahler [1951], who generalized the Farey sequence to study the approximation of complex numbers by Gaussian rationals. But most of what is proved in this chapter about continued fractions was known to Euler and Lagrange.

Section 9.3

The most complete account of continued fractions is that of Perron [1929]. For an extensive account in English, see Chrystal [1889].

Section 9.4

Almost nothing is known about the continued fraction expansions of algebraic numbers of degrees larger than 2. The partial quotients cannot increase too rapidly, because of the result in Problem 7 of Section 9.3, for example, but it is not even known whether they can remain bounded, nor whether they can be unbounded, for any nonquadratic number.

Theorem 9.12 is, slightly surprisingly, due to Galois.

Section 9.5

The return to continued fraction identities as an aid to the factorization of large integers can be credited to M. A. Morrison and J. Brillhart, who factored F_7 in 1971. Their expository paper [1975] gives many details and applications.

Section 9.6

Markov came to the quantity $M(\xi)$ in his study (1880) of the minima of indefinite binary quadratic forms. For an exposition of this theory, along with many other topics in Diophantine approximation, see Cassels [1957]. An encyclopaedic account of the entire subject up to its publication date is to be found in Koksma [1936].

Section 9.7

In another two pages, Hilbert gave a similar proof for the transcendence of π. It involves an integral in the complex plane rather than on the real line, and the basic theorem on symmetric polynomials, and so lies outside our scope.

For the older literature on transcendental numbers, see Koksma [1936]; for more recent work see LeVeque [1974], vol. 3. For an explanation of what is behind Hilbert's proof, and an analysis of other proofs of the transcendence of e and π, see Mahler [1976].

Bibliography

Ankeny, N. C., and C. A. Rogers [1951]. "A conjecture of Chowla." *Annals of Math.* (2) **53:** 541–550.

Artin, E. [1959]. *Theory of Algebraic Numbers.* Göttingen.

———. [1967]. *Algebraic Numbers and Algebraic Functions.* Gordon & Breach Science Publishers, New York-London-Paris.

Bachman, G. [1964]. *Introduction to p-adic Numbers and Valuation Theory.* Academic Press, New York-London.

Bachmann, P. [1902]. *Niedere Zahlentheorie*, 2 vols. Teubner, Leipzig. Reprinted in 1 volume by Chelsea Publ. Co., New York, 1968.

Beach, B. D., and H. C. Williams [1972]. "A numerical investigation of the Diophantine equation $x^2 - dy^2 = -1$." Proc. Third Southeastern Conference on Combinatorics, Graph Theory, and Computing, 37–68. Florida Atlantic University, Boca Raton, Florida.

Bell, E. T. [1937]. *Men of Mathematics.* Simon & Schuster, New York.

———. [1945]. *The Development of Mathematics*, 2nd edition. McGraw-Hill, New York.

Birkhoff, G., and S. MacLane [1965]. *A Survey of Modern Algebra.* Macmillan, New York.

Borevich, Z. I., and I. R. Shafarevich [1966]. *Number Theory.* (Translated from the 1964 Russian edition.) Pure and Applied Mathematics, vol. 20. Academic Press, New York-London.

Bourbaki, N. [1972]. *Elements of Mathematics. Commutative Algebra.* Hermann, Paris; Addison-Wesley, Reading, Mass.

Boyer, C. B. [1968]. *A History of Mathematics.* J. Wiley & Sons, New York.

Brauer, A., and R. L. Reynolds [1951]. "On a theorem of Aubry-Thue." *Canad. Journ. Math.* **3:** 367–374.

Brillhart, J., D. H. Lehmer, and J. L. Selfridge [1975]. "New primality criteria and factorizations of $2^m \pm 1$." *Math. of Comp.* **29:** 620–647.

Brent, R. P. [1975]. "Irregularities in the distribution of primes and twin primes." *Math. of Comp.* **29:** 43–56.

Burgess, D. A. [1957]. "The distribution of quadratic residues and nonresidues." *Mathematika* **4**: 106–112.

Buxton, M., and S. Elmore [1976]. "An extension of lower bounds for odd perfect numbers." *Notices Amer. Math. Soc.* **23**: A–55.

Carmichael, R. D. [1907]. "On Euler's φ-function." *Bull. Amer. Math. Soc.* **13**: 241–243.

Cassels, J. W. S. [1957]. *An Introduction to Diophantine Approximation.* Cambridge Tracts in Math. and Math. Physics. No. 45. Cambridge Univ. Press. Reprinted by Hafner, 1972.

Cassels, J. W. S., W. Ledermann, and K. Mahler [1951]. "Farey sections in $k(i)$ and $k(\rho)$." *Philos. Trans. Royal Soc. London*, Ser. A **243**: 585–626.

Chatland, H., and H. Davenport [1950]. "Euclid's algorithm in real quadratic fields." *Canad. Math. J.* **2**: 289–296.

Chebyshev, P. L. [1851]. "Sur la fonction qui determine la totalité des nombres premiers inférieurs à une limite donné. *J. math. pures et appl.* (I) **17** (1852): 341–365. Oeuvres I, 29–48. (Orig. publ. in Russian.)

———. [1852]. "Memoire sur les nombres premiers." *Ibid.*, 366–390. Oeuvres I, 51–70.

Chrystal, G. [1889]. *Algebra*, vol. 2. Adam & Black, Edinburgh. Reprinted, Chelsea Publ. Co. 1959.

Cohn, H. [1962]. *A Second Course in Number Theory.* John Wiley & Sons, New York.

Davenport, H. [1962]. *Analytic Methods for Diophantine Equations and Diophantine Inequalities.* Ann Arbor Publishers, Ann Arbor, Mich.

Dickson, L. E. [1909]. "On the congruence $x^n + y^n + z^n \equiv 0 \pmod{p}$." *Journ. für reine u. angew. Math.* **135**: 134–141, 181–188.

———. [1919]. *History of the Theory of Numbers.* Carnegie Institution, Washington; reprinted by Chelsea Publ. Co., New York, 1966.

Dubois, D. W., and A. Steger [1958]. "A note on division algorithms in imaginary quadratic number fields." *Canad. J. Math.* **10**: 285–286.

Edwards, H. M. [1975]. "The background of Kummer's proof of Fermat's last theorem for regular primes." *Arch. for History of Exact Sciences*, **14**: 219–236.

Erdös, P. [1945]. "On the least primitive root of a prime p." *Bull. Amer. Math. Soc.* **51**: 131–132.

Erdös, P., and R. L. Graham [1972]. "On a linear diophantine problem of Frobenius." *Acta Arith.* **21**: 399–408.

Estermann, T. [1962]. "Note on a paper of Rotkiewicz." *Acta Arith.* **8**: 465–467.

Frobenius, G. [1878]. "Theorie der linearen Formen mit ganzen Coefficienten." *Journ. für reine u. angew. Math.* **86**: 146–208. Gesamm. Abh. I, 482–544.

———. [1914]. "Über das quadratische Reziprozitätsgesetz." *Sitzungsber. Königl. Preuss. Akad. Wiss. Berlin*, 335–349; Gesamm. Abh. III, 628–647.

Gauss, C. F. [1801]. *Disquisitiones Arithmeticae.* German transl. by H. Maser (with appendices giving Gauss's later number-theoretic work); reprinted by Chelsea Publ. Co., 1965. English transl. Yale Univ. Press, 1966.

Gelfond, A. O., and Yu. V. Linnik [1962]. *Elementary Methods in Analytic Number Theory.* (Russian) Moscow. Translated into English (Rand McNally, 1965; Pergamon Press, 1966) and French (Gauthier-Villars, 1965).

Goldstein, L. J. [1971]. "Density questions in algebraic number theory." *Amer. Math. Monthly* **78**: 342–351.

Guy, R. K. [1975]. "How to factor a number." *Proc. Fifth Manitoba Conference on Numer. Math. and Comput.*, Univ. of Manitoba, Winnipeg. 36 pp.

Halberstam, H., and H.-E. Richert [1974]. *Sieve Methods.* Academic Press, New York-London.

Hardy, G. H., and E. M. Wright [1960]. *An Introduction to the Theory of Numbers.* 4th edition. Oxford University Press.

Hasse, H. [1923]. "Über die Darstellbarkeit von Zahlen durch quadratischen Formen im Körper der rationalen Zahlen." *Journ. für reine u. angew. Math.* **152**: 129–148.

———. [1963]. *Zahlentheorie.* 2nd edition. Akademie-Verlag, Berlin.

Hauptman, H., E. Vegh, and J. Fisher [1970]. *Table of all primitive roots for primes less than 5000.* Naval Research Lab., Washington, D.C.

Hecke, E. [1923]. *Vorlesungen über die Theorie der algebraischen Zahlen.* Leipzig. Reprinted by Chelsea Publ. Co., New York, 1948.

Hensel, K. [1913]. *Zahlentheorie.* Göschen, Berlin-Leipzig.

Herstein, I. [1964]. *Topics in Algebra.* Xerox College Publishing, Lexington, Mass.

Hooley, C. [1967]. "On Artin's conjecture." *Journ. für reine u. angew. Math.* **225**: 209–220.

Hua Lo-Keng [1942]. "On the least solution of Pell's equation." *Bull. Amer. Math. Soc.* **48**: 731–735.

Ingham, A. E. [1932]. *The distribution of prime numbers.* Cambridge Tract No. 30, Cambridge. Reprinted by Hafner, 1971.

Inkeri, K. [1946]. "Untersuchungen über die Fermatsche Vermutung." *Ann. Acad. Sci. Fennicae. Ser. A. I. Math. Phys.* **33**: 60 pp.

Ireland, K., and M. I. Rosen [1972]. *Elements of Number Theory, Including an Introduction to Equations over Finite Fields.* Bogden and Quigley, Tarrytown-on-Hudson.

Jacobi, C. G. J. [1829]. *Fundamenta Nova Theoriae Functionum Ellipticarum.* Werke I, 49–239.

Killgrove, R., and K. Ralston [1959]. "On a conjecture concerning the primes." *Math. Tables and Aids to Comput.* **13**: 121–122.

Knuth, D. E. [1969]. *The Art of Computer Programming, vol. 2: Seminumerical Algorithms.* Addision-Wesley Publishing Co., Reading, Mass.

Koksma, J. F. [1936]. *Diophantische Approximationen.* Ergebn. d. Math. IV 4, J. Springer, Berlin.

Kronecker, L. [1875]. "Zur Geschichte des Reciprocitätsgestezes." Werke **2**, 3–10. Reprinted by Chelsea Publ. Co., New York, 1968.

Landau, E. [1909]. *Handbuch der Lehre von der Verteilung der Primzahlen*, 2 vols. Teubner. Reprinted by Chelsea Publishing Co., 1953.

———. [1958]. *Elementary Number Theory*. Chelsea Publishing Co., New York. (Translated from *Vorlesungen über Zahlentheorie*, vol. I; S. Hirzel, 1927.)

Lang, S. [1970]. *Algebraic Number Theory*. Addison-Wesley Publ. Co., Reading, Mass.

Legendre, A. M. [1798]. *Essai sur la théorie des nombres*. Paris.

Lehmer, D. H. [1927]. "Tests for primality." *Bull. Amer. Math. Soc.* **33:** 327–340.

———. [1932]. "On Euler's totient function." *Bull. Amer. Math. Soc.* **38:** 745–757.

———. [1959]. "On the exact number of primes less than a given limit." *Ill. J. Math.* **3:** 381–388.

———. [1969]. "Computer technology applied to the theory of numbers." *Studies in Number Theory* (MAA Studies in Mathematics, vol. 6). Mathematical Association of America, Washington, D.C., pp. 117–151.

Lehmer, D. H., and E. Lehmer [1941]. "On the first case of Fermat's last theorem." *Bull. Amer. Math. Soc.* **47:** 139–142.

LeVeque, W. J. [1955]. *Topics in Number Theory*, vol. 2. Addison-Wesley Publ. Co., Reading, Mass.

———. [1974]. *Reviews in Number Theory*, vols. 1–6. Amer. Math. Soc., Providence, R.I.

Lewis, D. J. [1969]. "Diophantine equations: p-adic methods." *Studies in Number Theory*, Mathematical Association of America, Washington, D.C., pp. 25–75.

Littlewood, J. E. [1914]. "Sur la distribution des nombres premiers." *Comptes Rendus Acad. Sci. Paris.* **158:** 263–266.

Mahler, K. [1961]. *Lectures on Diophantine Approximations. Part I: g-adic Numbers and Roth's Theorem.* University of Notre Dame.

———. [1973]. *Introduction to p-adic Numbers and their Functions.* Cambridge Tracts in Mathematics, No. 64. Cambridge University Press.

———. [1976]. *Lectures on Transcendental Numbers.* Lecture Series in Mathematics, vol. 546. Springer-Verlag, New York.

Matijasevič, Yu., and J. Robinson [1975]. "Reduction of an arbitrary diophantine equation to one in 13 unknowns." *Acta Arithm.* **27:** 521–553.

McCarthy, P. J. [1966]. *Algebraic Extensions of Fields.* Blaisdell Publ. Co., Waltham, Mass.

Minkowski, H. [1890]. "Unter die Bedingungen, unter welcher zwei quadratische Formen mit rationalen Koeffizienten ineinander rational transformiert werden können." *Journ. für reine u. angew. Math.* **106:** 5–26. *Gesamm. Abh.* I: 219–239.

Mordell, L. J. [1951]. "On the equation $ax^2 + by^2 - cz^2 = 0$. *Monatshefte für Math.* **55:** 323–327.

———. [1969]. *Diophantine Equations.* Pure and Applied Math. vol. 30. Academic Press. New York.

Morrison, M. A., and J. Brillhart [1975]. "A method of factoring and the factorization of F_7." *Math. of Comp.* **29:** 183–205.

Niven, I., and H. S. Zuckerman [1972]. *An Introduction to the Theory of Numbers.* (3rd edition.) John Wiley & Sons, New York.

Niven, I., and B. Powell [1976]. "Primes in certain arithmetic progressions." *Amer. Math. Monthly* **83:** 467–469.

Perron, O. [1929]. *Die Lehre von den Kettenbrüchen.* Teubner, Leipzig.

Pollard, H., and H. G. Diamond [1975]. *The Theory of Algebraic Numbers.* Carus Math. Monographs, vol. 9. Math. Assn. Amer.

Robinson, R. M. [1957]. "The converse of Fermat's theorem." *Amer. Math. Monthly* **64:** 703–710.

Rota, G.-C. [1964]. "On the foundations of combinatorial theory. I. Theory of Möbius functions." *Z. Wahrscheinlichkeitstheorie* **2:** 340–368.

Samuel, P. [1967]. *Algebraic Theory of Numbers.* Houghton Mifflin Co., Boston, Mass.

Schur, I. [1917]. "Über die Kongruenz $x^m + y^m \equiv z^m \pmod{p}$." *Jahresber. deutsch. math. Verein.* **25:** 114–117.

Serre, J.-P. [1973]. *A Course in Arithmetic.* Graduate Texts in Math. vol. 7. Springer-Verlag, New York.

Shanks, D. [1972]. "Five number-theoretic algorithms." *Proc., 2nd Manitoba Confer. on Numer. Math.*, 51–70. Univ. of Manitoba, Winnipeg.

Siegel, C. L. [1929]. *Über einige Anwendungen diophantischer Approximationen.* Abhandl. Preuss. Akad. Wissensch. Physik.-Math. Klasse, Nr. 1.

Skewes, S. [1955]. "On the difference $\pi(x) - \mathrm{li}(x)$." (II) *Proc. London Math. Soc.* (3) **5:** 48–70.

Skolem, T. [1938]. *Diophantische Gleichungen.* Ergebnisse der Math. V-4, Springer, Berlin. Reprinted by Chelsea Publishing Co.

———. [1952]. "On the diophantine equation $ax^2 + by^2 + cz^2 = 0$." *Rendiconti Mat.* (5) **11:** 88–102.

Stark, H. M. [1969]. "On the 'gap' in a theorem of Heegner." *J. Number Theory* **1:** 16–27.

Taussky, O. [1970]. "Sums of squares." *Amer. Math. Monthly* **77:** 805–830.

Tijdeman, R. [1976]. "On the equation of Catalan." *Acta Arithmetica* **29:** 197–209.

Trost, E. [1934]. "Zur Theorie der Potenzreste." *Nieuw Archief Wisk.* (Amsterdam) **18:** 58–61.

Uhler, H. S. [1948]. "On all of Mersenne's numbers, particularly M_{193}." *Proc. Nat. Acad. Sci. USA* **34:** 102–103.

Uspensky, J. V., and M. A. Heaslet [1939]. *Elementary Number Theory.* McGraw-Hill, New York.

Val'fish, A. Z. [1956]. "On the representations of numbers by sums of squares. Asymptotic formulas." *American Mathematical Society. Translations* (2) **3:** 163–248.

Wagstaff, S. [1976]. "Fermat's last theorem is true for any exponent less than 100000." *Notices, Amer. Math. Soc.* **23:** A–53

Western, A. E., and J. C. P. Miller [1968]. *Tables of Indices and Primitive Roots.* Royal Society Math. Tables, vol. 9. Cambridge Univ. Press.

Appendix

FACTOR TABLE

	00	03	06	09	12	15	18	21	24	27	30	33	36	39	42	45	48	51	54	57	60	63	66	69	72	75	78	81	84	87
01	0	7	0	17	0	19	0	11	7	37	0	0	13	47	0	7	0	0	11	0	17	0	7	67	19	13	29	0	31	7
07	0	0	0	0	17	11	13	7	29	0	31	0	0	0	7	0	11	0	0	13	0	7	0	0	0	0	37	11	7	0
11	0	0	13	0	7	0	0	0	0	0	0	7	23	0	0	13	17	19	7	0	0	0	11	0	0	7	73	0	13	31
13	0	0	0	11	0	17	7	0	19	0	23	0	0	7	11	0	0	0	0	29	7	59	17	31	0	11	13	7	47	0
17	0	0	0	7	0	37	23	29	0	11	7	31	0	0	0	0	0	7	0	0	11	0	13	0	7	0	0	0	19	23
19	0	11	0	0	23	7	17	13	41	0	0	0	7	0	0	0	61	0	0	7	13	71	0	11	0	73	7	23	0	0
23	0	17	7	13	0	0	0	11	0	7	0	0	0	0	41	0	7	47	11	59	19	0	37	7	31	0	0	0	0	11
29	0	7	17	0	0	11	31	0	7	0	13	0	19	0	0	7	11	23	61	17	0	0	7	13	0	0	0	11	0	7
31	0	0	0	7	0	0	0	0	11	0	7	0	0	0	0	23	0	7	0	11	37	13	19	29	7	17	41	47	0	0
37	0	0	7	0	0	29	11	0	0	7	0	47	0	31	19	13	7	11	0	0	0	0	0	7	0	0	17	79	11	0
41	0	11	0	0	17	23	7	0	0	0	0	13	11	7	0	19	47	53	0	0	7	17	29	11	13	0	0	7	23	0
43	0	7	0	23	11	0	19	0	7	13	17	0	0	0	0	7	29	37	0	0	0	0	7	53	0	19	11	17	0	7
47	0	0	0	0	29	7	0	19	0	41	11	0	7	0	31	0	37	0	13	7	0	11	17	0	0	0	7	0	0	0
49	7	0	11	13	0	0	43	7	31	0	0	17	41	11	7	0	13	19	0	0	23	7	61	0	11	0	47	29	7	13
53	0	0	0	0	7	0	17	0	11	0	43	7	13	59	0	29	23	0	7	11	0	0	0	17	0	7	0	31	79	0
59	0	0	0	7	0	0	11	17	0	31	7	0	0	37	0	47	43	7	53	13	73	0	0	0	7	0	29	41	11	19
61	0	19	0	31	13	7	0	0	23	11	0	0	7	17	0	0	0	13	43	7	11	0	0	0	53	0	7	0	0	0
67	0	0	23	0	7	0	0	11	0	0	0	7	19	0	17	0	31	0	7	73	0	0	59	0	13	7	0	0	0	11
71	0	7	11	0	31	0	0	13	7	17	37	0	0	11	0	7	0	0	0	29	13	23	7	0	11	67	17	0	43	7
73	0	0	0	7	19	11	0	41	0	47	7	0	0	29	0	17	11	7	13	23	0	0	0	19	7	0	0	11	37	31
77	7	13	0	0	0	19	0	7	0	0	17	11	0	41	7	23	0	31	0	53	59	7	11	0	19	0	0	13	7	67
79	0	0	7	11	0	0	0	0	37	7	0	31	13	23	11	19	7	0	0	0	0	0	0	7	29	11	0	0	61	0
83	0	0	0	0	0	0	7	37	13	11	0	17	29	7	0	0	19	71	0	0	7	13	41	0	0	0	0	7	17	0
89	0	0	13	23	0	7	0	11	19	0	0	0	7	0	0	13	0	0	11	7	0	0	0	29	37	0	7	19	13	11
91	7	17	0	0	0	37	31	7	47	0	11	0	0	13	7	0	67	29	17	0	0	7	0	0	23	0	13	0	7	59
97	0	0	17	0	0	0	7	13	11	0	19	43	0	7	0	0	59	0	23	11	7	0	37	0	0	71	53	7	29	19

	01	04	07	10	13	16	19	22	25	28	31	34	37	40	43	46	49	52	55	58	61	64	67	70	73	76	79	82	85	88
01	0	0	0	7	0	0	0	31	41	0	7	19	0	0	11	43	13	7	0	0	0	37	0	0	7	11	0	59	0	13
03	0	13	19	17	0	7	11	0	0	0	29	41	7	0	13	0	0	11	0	7	17	19	0	47	67	0	7	13	11	0
07	0	11	7	19	0	0	0	0	23	7	13	0	11	0	59	17	7	41	0	0	31	43	19	7	0	0	0	29	47	0
09	0	0	0	0	7	0	23	47	13	53	0	7	0	19	31	11	0	0	7	37	41	13	0	43	0	7	11	0	67	23
13	0	7	23	0	13	0	0	0	7	29	11	0	47	0	19	7	17	13	37	0	0	11	7	0	71	23	41	43	0	7
19	7	0	0	0	0	0	19	7	11	0	0	13	0	0	7	31	0	17	0	11	29	7	0	0	13	19	0	0	7	0
21	11	0	7	0	0	0	17	0	0	7	0	11	61	0	29	0	7	23	0	0	0	0	11	7	0	0	89	0	0	0
27	0	7	0	13	0	0	41	17	7	11	53	23	0	0	0	7	13	0	0	0	11	0	7	0	17	29	0	19	0	7
31	0	0	17	0	11	7	0	23	0	19	31	47	7	29	61	11	0	0	0	7	0	59	53	79	0	13	7	0	19	0
33	7	0	0	0	31	23	0	7	17	0	13	0	0	37	7	41	0	0	11	19	0	7	0	13	0	17	0	0	7	11
37	0	19	11	17	7	0	13	0	43	0	0	7	37	11	0	0	0	0	7	13	17	41	0	31	11	7	0	0	0	0
39	0	0	0	0	13	11	7	0	0	17	43	19	0	7	0	0	11	13	29	0	7	47	23	0	41	0	17	7	0	0
43	11	0	0	7	17	31	29	0	0	0	7	11	19	13	43	0	0	7	23	0	0	17	11	0	7	0	13	0	0	37
49	0	0	7	0	19	17	0	13	0	7	47	0	23	0	0	0	7	29	31	0	11	0	17	7	0	0	0	73	83	0

51	0	11	0	0	7	13	0	0	0	0	23	7	11	0	19	0	0	59	7	0	0	0	43	11	0	7	0	37	17	53
57	0	0	0	7	23	0	19	37	0	0	7	0	13	0	0	0	0	7	0	0	47	11	29	0	7	13	73	23	43	17
61	7	0	0	0	0	11	37	7	13	0	29	0	0	31	7	59	11	0	67	0	61	7	0	23	17	47	19	11	7	0
63	0	0	7	0	29	0	13	31	11	7	0	0	53	17	0	0	7	19	0	11	0	23	0	7	37	79	0	0	0	0
67	0	0	13	11	0	0	7	0	17	47	0	0	0	7	11	13	0	23	19	0	7	29	67	37	53	11	31	7	13	0
69	13	7	0	0	37	0	11	0	7	19	0	0	0	13	17	7	0	11	0	0	31	0	7	0	0	0	13	0	11	7
73	0	11	0	29	0	7	0	0	31	13	19	23	7	0	0	0	0	0	0	7	0	0	13	11	73	0	7	0	0	19
79	0	0	19	13	7	23	0	43	0	0	11	7	0	0	29	0	13	0	7	0	37	11	0	0	47	7	79	17	23	13
81	0	13	11	23	0	41	7	0	29	43	0	59	19	7	13	31	17	0	0	0	7	0	0	73	11	0	23	7	0	83
87	11	0	0	0	19	7	0	0	13	0	0	11	7	61	41	43	0	17	37	7	23	13	11	19	83	0	7	0	31	0
91	0	0	7	0	13	19	11	29	0	7	0	0	17	0	0	0	7	11	0	43	41	0	0	7	19	0	61	0	11	17
93	0	17	13	0	7	0	0	0	0	11	31	7	0	0	23	13	0	67	7	71	11	43	0	41	0	7	0	0	13	0
97	0	7	0	0	11	0	0	0	7	0	23	13	0	17	0	7	19	0	29	0	0	73	7	47	13	43	11	0	0	7
99	0	0	17	7	0	0	0	11	23	13	7	0	29	0	53	37	0	7	11	17	0	67	13	31	7	0	19	43	0	11

	02	05	08	11	14	17	20	23	26	29	32	35	38	41	44	47	50	53	56	59	62	65	68	71	74	77	80	83	86	89
03	7	0	11	0	23	13	0	7	19	0	0	31	0	11	7	0	0	0	13	0	0	7	0	0	11	0	53	19	7	29
09	11	0	0	0	0	0	7	0	0	0	0	11	13	7	0	17	0	0	71	19	7	23	11	0	31	13	0	7	0	59
11	0	7	0	11	17	29	0	0	7	41	13	0	37	0	11	7	0	47	31	23	0	17	7	13	0	11	0	0	79	7
17	7	11	19	0	13	17	0	7	0	0	0	0	11	23	7	53	29	13	41	61	0	7	17	11	0	0	0	0	7	37
21	13	0	0	19	7	0	43	11	0	23	0	7	0	13	0	0	0	17	7	31	0	0	19	0	41	7	13	53	37	11
23	0	0	0	0	0	0	7	23	43	37	11	13	0	7	0	0	0	0	0	0	7	11	0	17	13	0	71	7	0	0
27	0	17	0	7	0	11	0	13	37	0	7	0	43	0	19	29	11	7	17	0	13	61	0	0	7	0	23	11	0	79
29	0	23	0	0	0	7	0	17	11	29	0	0	7	0	43	0	47	73	13	7	0	0	0	0	17	59	7	0	0	0
33	0	13	7	11	0	0	19	0	0	7	53	0	0	0	11	0	7	0	43	17	23	47	0	7	0	11	29	13	89	0
39	0	7	0	17	0	37	0	0	7	0	41	0	11	0	23	7	0	19	0	0	17	13	7	11	43	71	0	31	53	7
41	0	0	29	7	11	0	13	0	19	17	7	0	23	41	0	11	71	7	0	13	79	31	0	37	7	0	11	19	0	0
47	13	0	7	31	0	0	23	0	0	7	17	0	0	11	0	47	7	0	0	19	0	0	41	7	11	61	13	17	0	23
51	0	19	23	0	0	17	7	0	11	13	0	53	0	7	0	0	0	0	0	11	7	0	13	0	0	23	83	7	41	0
53	11	7	0	0	0	0	0	13	7	0	0	11	0	0	61	7	31	53	0	0	13	0	7	23	29	0	0	0	17	7
57	0	0	0	13	31	7	11	0	0	0	0	0	7	0	0	67	13	11	0	7	0	79	0	17	0	0	7	61	11	13
59	7	13	0	19	0	0	29	7	0	11	0	0	17	0	7	0	0	23	0	59	11	7	19	0	0	0	0	13	7	17
63	0	0	0	0	7	41	0	17	0	0	13	7	0	23	0	11	61	31	7	67	0	0	0	13	17	7	11	0	0	0
69	0	0	11	7	13	29	0	23	17	0	7	43	53	11	41	19	37	7	0	47	0	0	0	67	7	17	0	0	0	0
71	0	0	13	0	0	7	19	0	0	0	0	0	7	43	17	13	11	41	53	7	0	0	0	71	31	19	7	11	13	0
77	0	0	0	11	7	0	31	0	0	13	29	7	0	0	11	17	0	19	7	43	0	0	13	0	0	7	41	0	0	47
81	0	7	0	0	0	13	0	0	7	11	17	0	0	37	0	7	0	0	13	0	11	0	7	43	0	31	0	17	0	7
83	0	11	0	7	0	0	0	0	0	19	7	0	11	47	0	0	13	7	0	31	61	29	0	11	7	43	59	83	19	13
87	7	0	0	0	0	0	0	7	0	29	19	17	13	53	7	0	0	0	11	0	0	7	71	0	0	13	0	0	7	11
89	17	19	7	29	0	0	0	0	0	7	11	37	0	59	67	0	7	17	0	53	19	11	83	7	0	0	0	0	0	89
93	0	0	19	0	0	11	7	0	0	41	37	0	17	7	0	0	11	0	0	13	7	19	61	0	59	0	0	7	0	17
99	13	0	29	11	0	7	0	0	0	0	0	59	7	13	11	0	0	0	41	7	0	0	0	23	0	11	7	37	0	0

An excerpt from J. C. Burkhardt's *Tables des diviseurs* . . . of 1817; reprinted with permission from G. S. Carr, *Formulas and Theorems in Mathematics* (New York: Chelsea Publishing Company, 1970), where the table extends to 99,000.

The table gives the smallest divisor of every integer from 1 to 9000 which is not a multiple of 2, 3, or 5, except that the entry for a prime p is 0 rather than p. The first two digits of the integer are in the large-type rows, and the last two digits in the left column. For example, $30{,}012 = 4 \cdot 7503 = 4 \cdot 3 \cdot 41 \cdot 61$.

COMPUTER-PLOTTED GRAPHS

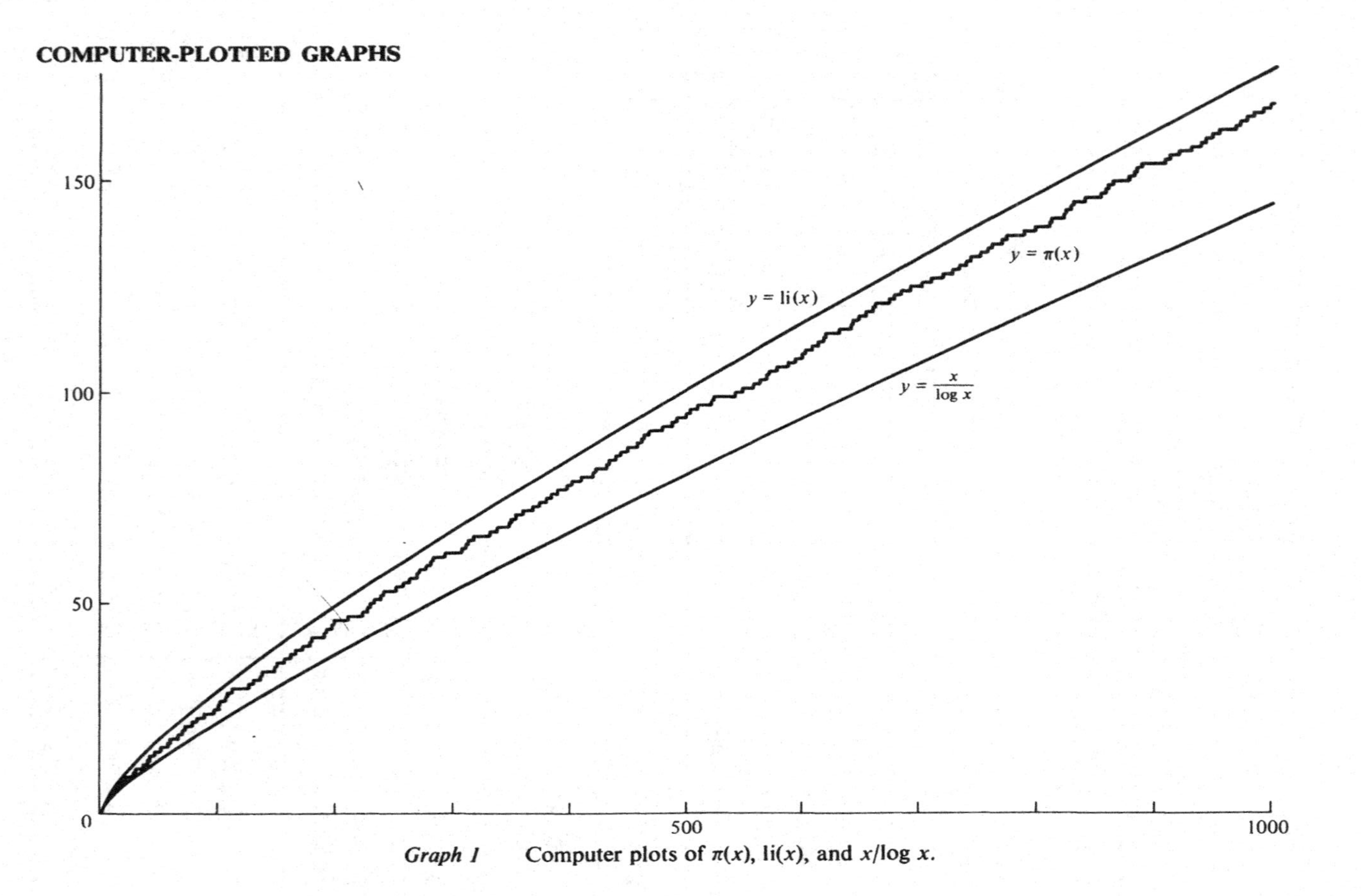

Graph 1 Computer plots of $\pi(x)$, $\mathrm{li}(x)$, and $x/\log x$.

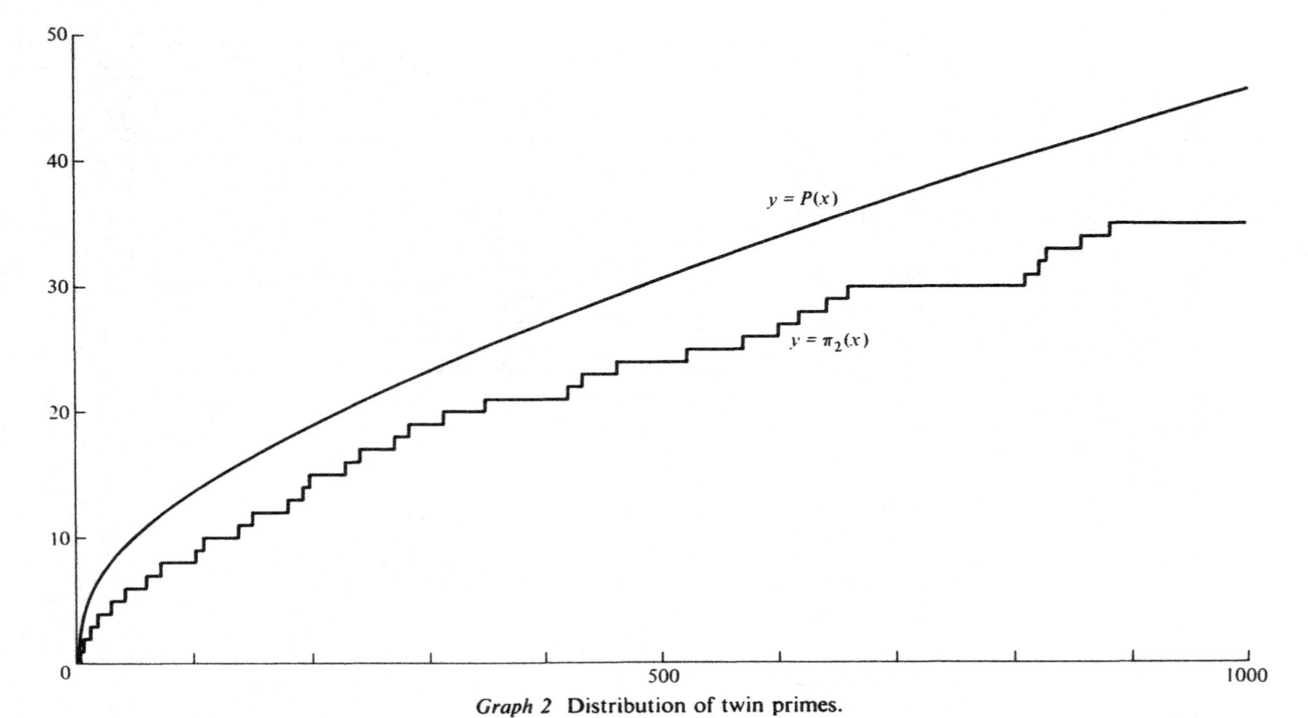

Graph 2 Distribution of twin primes.

The step function gives the number $\pi_2(x)$ of integers n for which $n - 1$ and $n + 1$ are both prime. The curve is the graph of the function

$$P(x) = c \int_2^x \frac{dt}{\log^2 t}, \qquad \text{where } c = 2 \prod_{p>2} (1 - (p-1)^{-2}) = 1.3203 \ldots.$$

It is conjectured that $\pi_2(x) \sim P(x)$.

TWO-WAY TABLE OF INDICES

$p = 3$

0	1	2
N	0	1	2
0		0	1

$p = 5$

0	1	2	4	3
N	0	1	2	3	4
0		0	1	3	2

$p = 7$

0	1	3	2	6	4	5
N	0	1	2	3	4	5	6
0		0	2	1	4	5	3

$p = 11$

0	1	2	4	8	5	10	9	7	3	6
N	0	1	2	3	4	5	6	7	8	9
0		0	1	8	2	4	9	7	3	6
1	5									

$p = 13$

0	1	2	4	8	3	6	12	11	9	5
1	10	7								
N	0	1	2	3	4	5	6	7	8	9
0		0	1	4	2	9	5	11	3	8
1	10	7	6							

$p = 17$

0	1	3	9	10	13	5	15	11	16	14
1	8	7	4	12	2	6				
N	0	1	2	3	4	5	6	7	8	9
0		0	14	1	12	5	15	11	10	2
1	3	7	13	4	9	6	8			

$p = 19$

0	1	2	4	8	16	13	7	14	9	18
1	17	15	11	3	6	12	5	10		
N	0	1	2	3	4	5	6	7	8	9
0		0	1	13	2	16	14	6	3	8
1	17	12	15	5	7	11	4	10	9	

$p = 23$

0	1	5	2	10	4	20	8	17	16	11
1	9	22	18	21	13	19	3	15	6	7
2	12	14								
N	0	1	2	3	4	5	6	7	8	9
0		0	2	16	4	1	18	19	6	10
1	3	9	20	14	21	17	8	7	12	15
2	5	13	11							

$p = 29$

0	1	2	4	8	16	3	6	12	24	19
1	9	18	7	14	28	27	25	21	13	26
2	23	17	5	10	20	11	22	15		
N	0	1	2	3	4	5	6	7	8	9
0		0	1	5	2	22	6	12	3	10
1	23	25	7	18	13	27	4	21	11	9
2	24	17	26	20	8	16	19	15	14	

$p = 31$

0	1	3	9	27	19	26	16	17	20	29
1	25	13	8	24	10	30	28	22	4	12
2	5	15	14	11	2	6	18	23	7	21
N	0	1	2	3	4	5	6	7	8	9
0		0	24	1	18	20	25	28	12	2
1	14	23	19	11	22	21	6	7	26	4
2	8	29	17	27	13	10	5	3	16	9
3	15									

For each p the table is in two parts, separated by the horizontal midline, and in each part the first digit of N is in the left column and the second digit is on the midline. The two parts are inverse to one another; the top part gives ind N, while the bottom part gives the number whose index is N. The smallest positive primitive root is always used; it is in the 01 position at the top. *Example:* For $p = 29$, the smallest primitive root is 2; every number whose index is 13 is $\equiv 18 \pmod{29}$, and ind $25 \equiv 11 \pmod{28}$.

$p = 37$

0	1	2	4	8	16	32	27	17	34	31
1	25	13	26	15	30	23	9	18	36	35
2	33	29	21	5	10	20	3	6	12	24
3	11	22	7	14	28	19				
N	0	1	2	3	4	5	6	7	8	9
0		0	1	26	2	23	27	32	3	16
1	24	30	28	11	33	13	4	7	17	35
2	25	22	31	15	29	10	12	6	34	21
3	14	9	5	20	8	19	18			

$p = 41$

0	1	6	36	11	25	27	39	29	10	19
1	32	28	4	24	21	3	18	26	33	34
2	40	35	5	30	16	14	2	12	31	22
3	9	13	37	17	20	38	23	15	8	7
N	0	1	2	3	4	5	6	7	8	9
0		0	26	15	12	22	1	39	38	30
1	8	3	27	31	25	37	24	33	16	9
2	34	14	29	36	13	4	17	5	11	7
3	23	28	10	18	19	21	2	32	35	6
4	20									

$p = 43$

0	1	3	9	27	38	28	41	37	25	32
1	10	30	4	12	36	22	23	26	35	19
2	14	42	40	34	16	5	15	2	6	18
3	11	33	13	39	31	7	21	20	17	8
4	24	29								
N	0	1	2	3	4	5	6	7	8	9
0		0	27	1	12	25	28	35	39	2
1	10	30	13	32	20	26	24	38	29	19
2	37	36	15	16	40	8	17	3	5	41
3	11	34	9	31	23	18	14	7	4	33
4	22	6	21							

$p = 47$

0	1	5	25	31	14	23	21	11	8	40
1	12	13	18	43	27	41	17	38	2	10
2	3	15	28	46	42	22	16	33	24	26
3	36	39	7	35	34	29	4	20	6	30
4	9	45	37	44	32	19				
N	0	1	2	3	4	5	6	7	8	9
0		0	18	20	36	1	38	32	8	40
1	19	7	10	11	4	21	26	16	12	45
2	37	6	25	5	28	2	29	14	22	35
3	39	3	44	27	34	33	30	42	17	31
4	9	15	24	13	43	41	23			

$p = 53$

0	1	2	4	8	16	32	11	22	44	35
1	17	34	15	30	7	14	28	3	6	12
2	24	48	43	33	13	26	52	51	49	45
3	37	21	42	31	9	18	36	19	38	23
4	46	39	25	50	47	41	29	5	10	20
5	40	27								
N	0	1	2	3	4	5	6	7	8	9
0		0	1	17	2	47	18	14	3	34
1	48	6	19	24	15	12	4	10	35	37
2	49	31	7	39	20	42	25	51	16	46
3	13	33	5	23	11	9	36	30	38	41
4	50	45	32	22	8	29	40	44	21	28
5	43	27	26							

$p = 59$

0	1	2	4	8	16	32	5	10	20	40
1	21	42	25	50	41	23	46	33	7	14
2	28	56	53	47	35	11	22	44	29	58
3	57	55	51	43	27	54	49	39	19	38
4	17	34	9	18	36	13	26	52	45	31
5	3	6	12	24	48	37	15	30		
N	0	1	2	3	4	5	6	7	8	9
0		0	1	50	2	6	51	18	3	42
1	7	25	52	45	19	56	4	40	43	38
2	8	10	26	15	53	12	46	34	20	28
3	57	49	5	17	41	24	44	55	39	37
4	9	14	11	33	27	48	16	23	54	36
5	13	32	47	22	35	31	21	30	29	

$p = 61$

0	1	2	4	8	16	32	3	6	12	24
1	48	35	9	18	36	11	22	44	27	54
2	47	33	5	10	20	40	19	38	15	30
3	60	59	57	53	45	29	58	55	49	37
4	13	26	52	43	25	50	39	17	34	7
5	14	28	56	51	41	21	42	23	46	31
N	0	1	2	3	4	5	6	7	8	9
0		0	1	6	2	22	7	49	3	12
1	23	15	8	40	50	28	4	47	13	26
2	24	55	16	57	9	44	41	18	51	35
3	29	59	5	21	48	11	14	39	27	46
4	25	54	56	43	17	34	58	20	10	38
5	45	53	42	33	19	37	52	32	36	31
6	30									

$p = 67$

0	1	2	4	8	16	32	64	61	55	43
1	19	38	9	18	36	5	10	20	40	13
2	26	52	37	7	14	28	56	45	23	46
3	25	50	33	66	65	63	59	51	35	3
4	6	12	24	48	29	58	49	31	62	57
5	47	27	54	41	15	30	60	53	39	11
6	22	44	21	42	17	34				
N	0	1	2	3	4	5	6	7	8	9
0		0	1	39	2	15	40	23	3	12
1	16	59	41	19	24	54	4	64	13	10
2	17	62	60	28	42	30	20	51	25	44
3	55	47	5	32	65	38	14	22	11	58
4	18	53	63	9	61	27	29	50	43	46
5	31	37	21	57	52	8	26	49	45	36
6	56	7	48	35	6	34	33			

$p = 71$

0	1	7	49	59	58	51	2	14	27	47
1	45	31	4	28	54	23	19	62	8	56
2	37	46	38	53	16	41	3	21	5	35
3	32	11	6	42	10	70	64	22	12	13
4	20	69	57	44	24	26	40	67	43	17
5	48	52	9	63	15	34	25	33	18	55
6	30	68	50	66	36	39	60	65	29	61
N	0	1	2	3	4	5	6	7	8	9
0		0	6	26	12	28	32	1	18	52
1	34	31	38	39	7	54	24	49	58	16
2	40	27	37	15	44	56	45	8	13	68
3	60	11	30	57	55	29	64	20	22	65
4	46	25	33	48	43	10	21	9	50	2
5	62	5	51	23	14	59	19	43	4	3
6	66	69	17	53	36	67	63	47	61	41
7	35									

$p = 73$

0	1	5	25	52	41	59	3	15	2	10
1	50	31	9	45	6	30	4	20	27	62
2	18	17	12	60	8	40	54	51	36	34
3	24	47	16	7	35	29	72	68	48	21
4	32	14	70	58	71	63	23	42	64	28
5	67	43	69	53	46	11	55	56	61	13
6	65	33	19	22	37	39	49	26	57	66
7	38	44								
N	0	1	2	3	4	5	6	7	8	9
0		0	8	6	16	1	14	33	24	12
1	9	55	22	59	41	7	32	21	20	62
2	17	39	63	46	30	2	67	18	49	35
3	15	11	40	61	29	34	28	64	70	65
4	25	4	47	51	71	13	54	31	38	66
5	10	27	3	53	26	56	57	68	43	5
6	23	58	19	45	48	60	69	50	37	52
7	42	44	36							

$p = 79$

0	1	3	9	27	2	6	18	54	4	12
1	36	29	8	24	72	58	16	48	65	37
2	32	17	51	74	64	34	23	69	49	68
3	46	59	19	57	13	39	38	35	26	78
4	76	70	52	77	73	61	25	75	67	43
5	50	71	55	7	21	63	31	14	42	47
6	62	28	5	15	45	56	10	30	11	33
7	20	60	22	66	40	41	44	53		
N	0	1	2	3	4	5	6	7	8	9
0		0	4	1	8	62	5	53	12	2
1	66	68	9	34	57	63	16	21	6	32
2	70	54	72	26	13	46	38	3	61	11
3	67	56	20	69	25	37	10	19	36	35
4	74	75	58	49	76	64	30	59	17	28
5	50	22	42	77	7	52	65	33	15	31
6	71	45	60	55	24	18	73	48	29	27
7	41	51	14	44	23	47	40	43	39	

$p = 83$

0	1	2	4	8	16	32	64	45	7	14
1	28	56	29	58	33	66	49	15	30	60
2	37	74	65	47	11	22	44	5	10	20
3	40	80	77	71	59	35	70	57	31	62
4	41	82	81	79	75	67	51	19	38	76
5	69	55	27	54	25	50	17	34	68	53
6	23	46	9	18	36	72	61	39	78	73
7	63	43	3	6	12	24	48	13	26	52
8	21	42								
N	0	1	2	3	4	5	6	7	8	9
0		0	1	72	2	27	73	8	3	62
1	28	24	74	77	9	17	4	56	63	47
2	29	80	25	60	75	54	78	52	10	12
3	18	38	5	14	57	35	64	20	48	67
4	30	40	81	71	26	7	61	23	76	16
5	55	46	79	59	53	51	11	37	13	34
6	19	66	39	70	6	22	15	45	58	50
7	36	33	65	69	21	44	49	32	68	43
8	31	42	41							

$p = 89$

0	1	3	9	27	81	65	17	51	64	14
1	42	37	22	66	20	60	2	6	18	54
2	73	41	34	13	39	28	84	74	44	43
3	40	31	4	12	36	19	57	82	68	26
4	78	56	79	59	88	86	80	62	8	24
5	72	38	25	75	47	52	67	23	69	29
6	87	83	71	35	16	48	55	76	50	61
7	5	15	45	46	49	58	85	77	53	70
8	32	7	21	63	11	33	10	30		
N	0	1	2	3	4	5	6	7	8	9
0		0	16	1	32	70	17	81	48	2
1	86	84	33	23	9	71	64	6	18	35
2	14	82	12	57	49	52	39	3	25	59
3	87	31	80	85	22	63	34	11	51	24
4	30	21	10	29	28	72	73	54	65	74
5	68	7	55	78	19	66	41	36	75	43
6	15	69	47	83	8	5	13	56	38	58
7	79	62	50	20	27	53	67	77	40	42
8	46	4	37	61	26	76	45	60	44	

$p = 97$

0	1	5	25	28	43	21	8	40	6	30
1	53	71	64	29	48	46	36	83	27	38
2	93	77	94	82	22	13	65	34	73	74
3	79	7	35	78	2	10	50	56	86	42
4	16	80	12	60	9	45	31	58	96	92
5	72	69	54	76	89	57	91	67	44	26
6	33	68	49	51	61	14	70	59	4	20
7	3	15	75	84	32	63	24	23	18	90
8	62	19	95	87	47	41	11	55	81	17
9	85	37	88	52	66	39				
N	0	1	2	3	4	5	6	7	8	9
0		0	34	70	68	1	8	31	6	44
1	35	6	42	25	65	71	40	89	78	81
2	69	5	24	77	76	2	59	18	3	13
3	9	46	74	60	27	32	16	91	19	95
4	7	85	39	4	58	45	15	84	14	62
5	36	63	93	10	52	87	37	55	47	67
6	43	64	80	75	12	26	94	57	61	51
7	66	11	50	28	29	72	53	21	33	30
8	41	88	23	17	73	90	38	83	92	54
9	79	56	49	20	22	82	48			

GREEK ALPHABET

Α	α	Alpha	Ν	ν	Nu
Β	β	Beta	Ξ	ξ	Xi
Γ	γ	Gamma	Ο	o	Omicron
Δ	δ	Delta	Π	π	Pi
Ε	ε, ϵ	Epsilon	Ρ	ρ	Rho
Ζ	ζ	Zeta	Σ	σ	Sigma
Η	η	Eta	Τ	τ	Tau
Θ	θ, ϑ	Theta	Υ	υ	Upsilon
Ι	ι	Iota	Φ	ϕ, φ	Phi
Κ	κ	Kappa	Χ	χ	Chi
Λ	λ	Lambda	Ψ	ψ	Psi
Μ	μ	Mu	Ω	ω	Omega

LIST OF SYMBOLS

Index